Soft Wheat:

Production, Breeding, Milling, and Uses

Edited by

W. T. Yamazaki
U.S. Department of Agriculture
Science and Education Administration
Soft Wheat Quality Laboratory
Wooster, Ohio, U.S.A.

C. T. Greenwood
Department of Chemistry
School of Natural Resources
The University of the South Pacific
Suva, Fiji

Published by the

American Association of Cereal Chemists, Inc.
St. Paul, Minnesota

Library of Congress Catalog Card Number: 80-65826

ISBN 0-913250-17-1

Printed in the United States of America

American Association of Cereal Chemists, Inc.

Volume VI
Monograph Series

CONTRIBUTORS

R. E. Allan, United States Department of Agriculture, Science and Education Administration, Wheat Breeding and Production Research, Pullman, Washington, U.S.A.

L. J. Brenneis (deceased), Quality Control, Mennel Milling Co., Fostoria, Ohio, U.S.A.

N. F. Derera, Plant Breeding Institute, University of Sydney, Narrabri, New South Wales, Australia

M. Ford, Cereal Processing Department, RHM Research Ltd., The Lord Rank Research Centre, High Wycombe, England

C. T. Greenwood, Department of Chemistry, School of Natural Resources, University of the South Pacific, Suva, Fiji

R. Guinet, Ecole de Boulangerie et de Pâtisserie de Paris, Paris, France

K. W. Kingswood, Cereal Processing Department, RHM Research Ltd., The Lord Rank Research Centre, High Wycombe, England

H. J. Loving (retired), Quality Control, Mennel Milling Co., Fostoria, Ohio, U.S.A.

F. G. H. Lupton, Plant Breeding Institute, Cambridge University, Cambridge, England

S. Nagao, Quality Control and Assurance, Flour Milling Department and Cereal Science Research Division, Central Research Laboratory, Nisshin Flour Milling Co., Ltd., Tokyo, Japan

F. L. Patterson, Plant Genetics, Department of Agronomy, Purdue University, West Lafayette, Indiana, U.S.A.

W. Seibel, Institute of Baking Technology, Federal Research Center of Grain and Potato Processing, Detmold, Federal Republic of Germany.

W. T. Yamazaki, United States Department of Agriculture, Science and Education Administration, Soft Wheat Quality Laboratory, Wooster, Ohio, U.S.A.

CREDITS: Cover photographs courtesy of Nabisco Inc., Fair Lawn, New Jersey. Printed by permission of Nabisco Inc.

PREFACE

Wheat is as old as farming and has been grown throughout the world—except in climatic extremes—and this cereal forms an essential and staple foodstuff. Cultivars have been developed to suit particular environments, especially in attempts to improve their bread-baking properties at the same time that plant adaptability is being enhanced. In response to demands for bread-type wheats, the world export market has been dominated by the "hard-milling" type of wheat produced in the United States, Canada, and Argentina. Many previous publications—including the classic AACC Monograph—have emphasized its use for bread making. But in other parts of the world, softer milling cultivars exist, and these have important uses for various baked and other products.

It is the purpose of this Monograph to attempt to consolidate information on this latter important crop. We appreciate that it is only possible to make a broad definition of the term "soft wheat," and here we use it in the general milling context but exclude durum wheat.

Production, development (breeding), milling, and end-product uses for soft wheat are discussed in this volume. In keeping with the worldwide distribution of soft wheat and its multifarious products, some of which are universal and others characteristic of a particular region, contributing authors represent these major producing and utilizing regions. They have given of their time and talent, and we are grateful for their interest and generosity.

We also wish to thank the Directors of the American Association of Cereal Chemists, sponsors of this Monograph, for providing the impetus and momentum for the initiation of the project, and the editorial staff of the Association, especially Steven C. Nelson, Manager of Publications and Communications, for their technical assistance. A number of secretaries and typists participated in the preparation of manuscripts, and we are indeed very grateful to them. One who made a major contribution in this area is Miss Carrie H. Voiers, Soft Wheat Quality Laboratory, and we wish publicly to thank her for her efforts in typing and checking drafts of the chapters.

The Editors

CONTENTS

CHAPTER PAGE

1. Soft Wheat Production. By W. T. YAMAZAKI, M. FORD, K. W. KINGSWOOD, and C. T. GREENWOOD 1

I. Introduction 1

II. Soft Wheat and Hard Wheat 3

III. World Soft Wheat Production 4
A. Africa, B. Asia, C. Europe, D. North America, E. Oceania, F. South America, G. The Union of Soviet Socialist Republics

IV. International Trade in Soft Wheat 27

2. Soft Wheat Breeding in the United States. By F. L. PATTERSON and R. E. ALLAN 33

Part One: Culture and Breeding in the Eastern United States

I. Introduction 33
A. Adaptation, B. Culture, C. Harvest and Marketing

II. Soft Wheat Improvement Before 1925 36
A. Introduction and Evaluation, B. Cultivar Distribution, C. Breeding

III. Soft Wheat Improvement, 1925–1949 40
A. Breeding Objectives, B. Breeding for Quality, C. Cultivar Distribution, D. Breeding for Disease Resistance, E. Breeding for Insect Resistance

IV. Soft Wheat Improvement, 1950 to Present 44
A. Breeding for Quality, B. Cultivar Distribution, C. Hybrid Wheat, D. Breeding for Disease Resistance, E. Breeding for Insect Resistance, F. Progress in Breeding

Part Two: Culture and Breeding in the Western United States

I. Introduction 57

II. Soft Wheat Breeding 59
A. Early Wheat Improvement in the West, B. Early Quality Studies and Observations

III. Soft Wheat Improvement, 1919–1944 61
A. Cultivar Distribution, B. Cultivar Quality Traits, C. Identification of Adaptive Traits for the Region, D. Breeding to Control Common Bunt, E. California Backcross Wheat Cultivars, F. Relative Changes in Yield

IV. Soft Wheat Improvement, 1945–1959 71
A. Cultivar Distribution, B. Establishment of the Western Wheat Quality Laboratory, C. Cultivar Quality Traits, D. Significance of Facultative Wheats, E. Neutralization of Common Bunt, F. Changes in Wheat Management, G. Relative Changes in Yield

V. Soft Wheat Improvement, 1960 to Present 77
A. Cultivar Distribution, B. Stripe Rust Epiphytotics and Breeding for Resistance, C. Identification of the Value of Semidwarf Growth Habit, D. Advantages and Disadvantages of Semidwarfs, E. Semidwarf Wheats and Their Predominance, F. Management of Semidwarf Wheat, G. Diseases Associated with Semidwarf Culture, H. Breeding for Dwarf Bunt Resistance and Loss of Export Markets, I. Foot Root Becomes the Major Disease, J. New Wheats for Specific Production Situations, K. Impact of Increased Irrigation on Wheat Breeding, L. Breeding Wheats for Early Seeding and Erosion Control, M. Hybrid Wheat Research, N. Quality Research Directly Aiding Plant Breeding, O. Export Quality Requirements of Western Soft Wheat, P. The High-Alpha-Amylase Problem, Q. Relative Changes in Yield

Appendix: Common and Scientific Names of Diseases, Insects, and Plants 91

3. **Soft Wheat Breeding in Europe and Australia.** By F. G. H. LUPTON and N. F. DERERA 99

Part One: Breeding in Europe

I. History of Soft Wheat Cultivation 99
A. Early Efforts at Wheat Breeding, B. Development of Modern Plant Breeding

II. Genetic Relationships and Breeding Systems 104
A. Construction of the Ideotype, B. Choice of Parental Material, C. Mutation Breeding, D. Selection Methods in Segregating Populations, E. Breeding for Disease Resistance, F. Breeding for Processing Quality

III. Multiplication and Maintenance of New Varieties 117

IV. Future Prospects 118

Part Two: Breeding in Australia

I. History of Cultivation 119

II. Use of Genetic Variability 120

III. Quality Aspects of Soft Wheats 121

IV. Current Status and Outlook 123

4. **Milling in the European Economic Community.** By M. FORD and K. W. KINGSWOOD 129

I. Introduction 129

II. Soft Wheat—Milling Implications 130
A. The Wheat Kernel, B. Kernel Texture, C. Water Absorption and Starch Damage, D. Extraction Rate

III. EC Wheat Marketing 136
A. France, B. West Germany, C. United Kingdom

IV. EC Grain Market Regulations 139
A. Grain Prices, B. Testing of Wheat

V. EC Milling 144
A. Milling Principles, B. Milling Developments, C. Milling in France, D. Milling in West Germany, E. Milling in the United Kingdom, F. Milling in Other EC Countries

VI. Outlook 165

5. Soft Wheat Uses in the United States. By H. J. LOVING and L. J. BRENNEIS 169

I. Introduction and History 169

II. Baked Foods from Soft Wheat Flours 171
A. Cakes, B. Pies, C. Cookies and Crackers, D. Sweet Goods, E. Ice Cream Cones and Sugar Wafers, F. Pretzels, G. Prepared Mixes, H. Toaster Pastries

III. Ingredients Used in Baked Foods 192
A. Sweeteners, B. Shortening, C. Eggs, D. Milk, E. Leavening, F. Salt, Flavors, and Colors

IV. Nonbaked Foods from Soft Wheats 196
A. Cake Donuts, B. Breakfast Cereals, C. Soup Thickeners, D. Batters and Breaders, E. Fried Noodles, F. Bulgur

V. Nonfood Uses of Soft Wheat Flours 203
A. Glue Extenders, B. Feed Ingredients, C. Miscellaneous

VI. Comparison of Soft Red and Soft White, Eastern and Western Soft White Wheats 203

6. Soft Wheat Uses in Europe. By C. T. GREENWOOD, R. GUINET, and W. SEIBEL 209

Part One: Soft Wheat Uses in the United Kingdom

I. Baked Products in the United Kingdom 209

II. Bread Making 210
A. Introduction, B. Essentials of Bread Making, C. Properties of Bread, D. Bread Varieties

III. Flour Confectionery 221
A. Introduction, B. Ingredients, C. Aeration, D. Flour Confectionery Manufacture

IV. Biscuits 227
A. Introduction, B. Types of Biscuits, C. The Biscuit-Making Process, D. Wafers

Part Two: Soft Wheat Uses in France

I. Wheat 232
A. Production and Use, B. Usage and Grading

II. Flour 234
A. Definition, B. Milling, C. Flour Classification

III. Bread Products 236
A. Traditional French Bread, B. Regional Breads, C. Fine Bread Products, D. The French and Their Bread

IV. Confectionery Products 245

V. Biscuit Products 248

VI. The Future of Biscuit and Pastry Products in France 248

Part Three: Soft Wheat Uses in the Federal Republic of Germany

I. Wheat Production and Wheat Quality 249
A. Production and Yield, B. Quality Evaluation of Wheat Cultivars, C. Production of Quality Wheat, D. Grading of Wheat

II. Soft Wheat Milling 252
A. Milling Grist, B. Principles of Milling, C. Flour and Whole-Meal Regulations, D. Flour Treatment, E. Flour and Whole-Meal Quality, F. Production of Ready Mixes

III. Baked Products 255
A. Bread, B. Other Baked Goods (*Feine Backwaren*), C. Consumption

7. **Soft Wheat Uses in the Orient.** By S. NAGAO 267

I. Introduction 267

II. Soft Wheat Consumption in the Orient 267
A. Japan, B. South Korea, C. Hong Kong, D. Taiwan, E. The People's Republic of China, F. The Philippines, G. Indonesia, H. Singapore, I. Malaysia, J. Thailand, K. India, L. Sri Lanka, M. Pakistan, N. Bangladesh, O. Afghanistan

III. Baked Products from Soft Wheat Flour 277
A. *Castilla*, B. Sponge Cakes, C. Baked Japanese Buns, D. Baked Chinese Buns, E. Biscuits and Cookies, F. Chinese Cookies, G. Crackers, H. Japanese Crackers, I. *Chapati* and *Nan*, J. Other Baked Products

IV. Nonbaked Uses of Soft Wheat Flour 291
A. *Tempura*, B. Japanese Noodles, C. Soft Wheat Noodles in Other Countries, D. Steamed Japanese Buns, E. Steamed Chinese Buns, F. Donuts, G. Other Nonbaked Products

V. Nonfood Uses of Soft Wheat 302
A. Industrial Uses, B. Feed Uses

Soft Wheat:

Production, Breeding, Milling, and Uses

CHAPTER 1

SOFT WHEAT PRODUCTION

W. T. YAMAZAKI
United States Department of Agriculture
Science and Education Administration
Soft Wheat Quality Laboratory, Wooster, Ohio, U.S.A.

M. FORD
K. W. KINGSWOOD
RHM Research Ltd.
The Lord Rank Research Centre, High Wycombe, England

C. T. GREENWOOD[1]
University of the South Pacific
Suva, Fiji

I. INTRODUCTION

Wheat and rice are the most important sources of caloric intake by peoples of the world. For millennia, bread, leavened or unleavened, has been the usual form in which wheat is consumed. In addition, it is eaten as noodles, alimentary pastes, confectionery products, and numerous baked and otherwise processed foods. Wheat is grown on every continent.

The three most important species from the standpoint of production and for food use are *Triticum aestivum* (common wheat), *T. durum* (durum wheat), and *T. compactum* (club wheat).

Common wheat comprises by far the largest proportion of the types and classes of wheat grown. It has been of great historic importance as a food, having accompanied people in their migrations. In the course of domestication, wheat has been extensively modified, either naturally or purposefully, to improve its adaptability. Cultivars of the species are adapted to a wide range of climatic conditions, from the subtropic to the cool, temperate and from the continental, with its hot summer, cold winter, and low precipitation, to the mild winter, cool spring, and high rainfall of the maritime.

Cultivars are distinguished by certain kernel characteristics. Some have red or brownish, others white or pale yellow seed coats. More to the point of this book,

[1]Formerly, Flour Milling and Baking Research Association, Chorleywood, England

some have hard (or vitreous), others soft (or mealy) kernel texture. The wheats with soft texture (including those with semihard kernels grown in certain parts of the world) are the subject of this discussion.

Durum wheat, used principally for making of macaroni, spaghetti, and similar pasta products in certain countries and for unleavened bread in others, is grown in much smaller quantity than common wheat. Probably because many of its cultivars are adapted to and grown in regions with continental climate, the crop normally consists of kernels that are vitreous and high in protein content. Although kernel vitreosity and high protein content are not necessarily correlated with inherently hard kernel texture (52), the properties are often found together in durum wheat. In the breeding of durum wheat, hard kernel texture is an attribute that is often actively promoted, because that quality is considered advantageous for good crop utilization. Durum is not considered a soft wheat and will not be discussed further.

Club wheat is grown commercially in the Pacific Northwest of the United States and in certain areas of the Near East, especially in Turkey. Whereas it is not an important crop from the standpoint of world production, it is a significant part of the wheat economy where it is grown. Almost all club wheat cultivars of the United States are soft-textured and low to medium in protein content. This species is discussed in this volume from both the breeding and use standpoints.

Wheat improvement is directed toward increasing and stabilizing field yield and disease and insect resistance, and improving utilization quality. Because of major efforts in cultivar development, significant gains have been made in productivity, yield stability, and product quality. In the matter of adaptation, it appears that in areas with continental climate, the common spring, the cold-adapted common winter, and the spring durum wheats predominate. Crops of these classes are usually characterized by medium to high protein content and hard-textured grain. Certain spring wheats are also grown where winters are mild and the crop does not undergo vernalization. Cultivars grown under these conditions may have soft or hard textured kernels.

Fall-sown wheats that survive the winter with good stands generally outyield spring wheats grown in the same area. Most hard winter wheat cultivars appear to possess greater tolerance to environmental stress than those of soft wheat, especially in terms of extreme temperature variations and in drought tolerance, and will survive where the latter may not. This does not apply universally, however.

Soft wheat cultivars, of either winter or spring habit, are generally adapted to the mild temperate zones where rainfall is adequate to plentiful. Under such conditions, field yield is high, grain protein content is low to medium, and kernels are mealy or soft-textured. This type of wheat is unique in a number of ways.

In contrast to rice, wheat is not usually consumed in the form in which it is grown. It is processed to various extents in different parts of the world and thus made more amenable to further treatment, usually baking. As a food source, wheat is used in bread, confectionery baked goods, and other products in most of the wheat-eating world, in *chapati* in the Indian subcontinent, and in unleavened bread elsewhere.

Products are usually baked from indigenous wheats, and quality considerations for the goods have been established by the consumer. As new products are introduced and tastes change, however, wheat requirements may also change. For example, people in parts of western Europe were formerly satisfied with bread from domestic soft wheat. They have turned, however, to high-protein hard wheat as a superior source of raw material to satisfy their increasingly sophisticated demand for quality bread.

II. SOFT WHEAT AND HARD WHEAT

The terms "soft wheat" and "hard wheat" are applied to grain having soft and hard kernel texture, respectively, although the terms may refer in certain parts of the world to common and durum wheats, respectively. Kernel texture more often than not is a cultivar trait (9,56), although environmental factors may modify texture. When ground, soft-textured grain yields relatively large quantities of finely granulated flour, whereas hard wheat grain produces a coarser product under similar grinding conditions (30). This difference in fracturing behavior of the respective endosperms has been attributed (52) to the presence of proteinaceous material on the surface of starch granules of hard wheat cultivars. This material promotes adhesion of endosperm protein to the starch surface within the cells. The adhesion is relatively strong; hence, application of stress to hard wheat endosperm results in fracture taking place along cell walls rather than through cell contents. Soft wheat endosperm under similar stress fractures across cells, releasing cell contents. The starch granules can be separated easily from the protein matrix; therefore, the flour contains relatively large quantities of free starch granules.

The distinction between "true softness" and "yellow berry," or grain from a hard-kernel cultivar grown under humid conditions or in soil deficient in nitrogen was noted by Freeman in 1918 (24) but remains to this day a source of some misunderstanding and confusion. Briggle and Reitz (15) pointed out the hazard of assigning a textural characteristic to a cultivar on the basis of its "yellow berry" kernels.

The practicality of measuring kernel texture by a grinding and sieving test was determined by Cutler and Brinson (18), who also indicated the heritable nature of endosperm granularity. The genetic nature of texture was confirmed by a number of workers, including Berg (9), Symes (56), Moss et al (39), and Simmonds (52). It is thus quite apparent that texture is a heritable trait, relatively unaffected by protein content, but modifiable by season, location, and field conditions during grain development, though not to the extent that cultivar identity is lost.

Simmonds (52) pointed out the distinction between hardness and vitreosity. Hardness is a cultivar trait. Vitreosity, on the other hand, is influenced by environmental factors such as light, temperature, and grain drying rate, as well as protein content. However, not all grain exhibits the same degree of vitreousness at the same protein content.

In some parts of the world such as the United States, Canada, Japan, and Australia, the term "soft wheat" implies grain utility for certain products as much as it refers to texture itself and, to some extent, protein content. Flour from wheat of this class is not used for bread but rather for cookies (biscuits), cakes, saltines, sugar wafers, noodles, and similar items. The properties of flour from soft wheat are considered uniquely advantageous for the production of such goods for which product tenderness is an important quality factor.

Although the association between fine flour granularity and tenderness in confectionary products is recognized by millers and bakers in Western Europe, all domestic grain of the species *T. aestivum* is called "soft wheat," and the term "hard wheat" is reserved for wheat of *T. durum*. Thus, the soft wheats of the region are used for the production of flour for the entire spectrum of products made from wheat, with the possible exception of alimentary pastes.

In Eastern Europe, a distinction is made between hard and soft types among the common wheats, but the basis for the differentiation appears to be the suitability of the grain for baking bread. Three categories are used for wheat classification: 1) wheats that are eminently suitable for bread production by themselves; 2) those that can be used for bread, provided they are blended with wheat of the first category; and 3) those that are not suited for bread. Wheat of the last category is often referred to as soft wheat.

In this monograph the term "soft wheat" includes the finely granulating cultivars produced in North and South America, Australia, and Asia and all cultivars of *T. aestivum* of Western Europe. Although no direct mention will be made of properties and uses of soft-textured wheats grown in other regions (principally because of lack of information), it is presumed that such wheats can be used to advantage for products discussed in this monograph. Specifically excluded from extended discussion are the hard wheats of North and South America, the durum wheats, the bread wheats of Eastern Europe and the U.S.S.R., the wheats of Africa, the Near East, the Indian subcontinent, and those of Queensland in Australia.

III. WORLD SOFT WHEAT PRODUCTION

The distribution of soft wheat throughout the world coincides with the climate, vegetation, and soil patterns referred to as "humid," "broadleaf and mixed broadleaf-coniferous forests," and "gray-brown podzolic," respectively, by Blumenstock and Thornthwaite (10). Historically then, soft wheat culture is associated with a climate characterized by moderate rainfall but additionally with temperatures or wind patterns that do not result in rapid evaporation of surface moisture. Schlehuber and Tucker (47) stated that, compared to hard wheats, soft wheat cultivars are lower in protein content because they are adapted to and are grown in areas with mild climate and adequate rainfall. The lower protein content is attributed to lower nitrogen availability, resulting from rainfall percolating down through the soil and carrying with it mobile nitrates that may otherwise be utilized.

It is difficult to obtain a reliable estimate of the worldwide production of soft wheat. Wheat production and classification statistics in most parts of the world do not distinguish whether the crop is hard or soft textured. The presence of intermediate textured cultivars, and the possibility that a cultivar can appear to be of one class one year and be sufficiently influenced by environment to be placed in the other class another year, add to the difficulty in classification.

Table I presents total wheat and estimated soft wheat production in major wheat-growing nations. An attempt has been made to determine the approximate percentage of soft wheat from data provided by various sources.

TABLE I

Total Wheat Production[a] and Estimated Soft Wheat Production, 1977 Crop

Nation	Total Wheat Production (1,000 te)	Estimated Soft Wheat Production (1,000 te)	(%)	Reference for Soft Wheat
Africa				
Algeria	1,500	0	0	(54)
Egypt	1,690		n.a.	
Morocco	1,530	0	0	(54)
South Africa	1,830	100	4	Personal communication
Other	1,440		n.a.	
Total	7,990	100		
Asia				
Afghanistan	2,640	1,210	46	Personal communication
People's Rep. China	40,000	16,800	42	See text
India	29,080	300	1	Personal communication
Iran	5,000	3,500	70	Personal communication
Pakistan	8,940	900	10	Personal communication
Syria	1,000	0	0	
Turkey	13,500	2,030	15	See text
Other	2,760	720	26	
Total	102,920	25,460		
Europe				
Austria	1,100	···	n.a.	
Bulgaria	3,100	430	14	Personal communication
Czechoslovakia	5,000	1,750	35	See text
France	17,050	16,540	97	Personal communication
Germany, East	3,100	1,550	50	Personal communication
Germany, West	7,130	3,000	42	(11)
Greece	1,700	1,360	80	Personal communication
Hungary	5,310	···	n.a.	
Italy	6,650	4,660	70	Personal communication
Poland	5,600	2,800	50	(58)
Romania	6,500	···	n.a.	
Spain	3,990	···	n.a.	
Sweden	1,560	1,360	87	Personal communication
United Kingdom	5,270	5,270	100	Personal communication
Yugoslavia	5,620	4,220	75	Personal communication
Other	3,680	2,940	80	
Total	82,360	45,880		
North America				
Canada	19,650	980	5	Personal communication
Mexico	2,350	350	15	Personal communication
United States	55,130	15,440	28	(43)
Other	50	···	0	
Total	77,180	16,770		
Oceania				
Australia	9,200	4,780	52	Personal communication
Other	370	110	29	
Total	9,570	4,890		

(continued)

TABLE I–*Continued*

Nation	Total Wheat Production (1,000 te)	Estimated Soft Wheat Production (1,000 te)	(%)	Reference for Soft Wheat
U.S.S.R.	90,000	0	0	Personal communication
South America				
Argentina	6,000	0	0	Personal communication
Brazil	2,000	1,900	95	(45)
Other	1,570	860	55	Personal communication
Total	9,570	2,760		
World Total	379,600	95,860	25.3	

[a]Source: Agricultural Statistics, U.S. Department of Agriculture, U.S. Government Printing Office, Washington, DC, 1978.

A. Africa

Little soft wheat is grown in Africa. In Algeria, Morocco, and Tunisia, durum wheat comprises 70 to 75% of the crop; the remainder is bread wheat (40,54). Similarly, the wheat grown in Egypt is preponderantly of the hard type. The Republic of South Africa grows enough wheat to have become self-sufficient in recent years, but it is estimated that only about 4% of the crop is soft wheat (P. C. J. Mares, personal communication).

B. Asia

About 25.8% of the world wheat crop was grown in Asia (excluding the U.S.S.R.) in 1976 (57), with the People's Republic of China, India, and Turkey supplying a considerable portion. Wheat is a staple food for the people in western and central Asia and in northern China.

AFGHANISTAN

Wheat is the major food and principal caloric source of the people. Cultivated since antiquity, many species of wheat are grown, but *T. aestivum* is by far the most widely distributed. Wheat accounts for approximately 60% of all grains produced and is grown on almost every farm, largely for home consumption in the form of *chapati* or bread. Only about one fourth of the national crop is marketed, mostly in the cities, where it is an important economic factor. Wheat production during the period 1973-1975 averaged more than 2,800,000 metric tons (te) on about 2,250,000 ha (23).

Both winter and cold-tolerant spring wheats are grown in the valleys where they normally overwinter under a blanket of snow. More than one half of the area devoted to wheat culture is irrigated. Yields under irrigation are substantially greater than from the rainfed fields. The planting of spring-type Mexican cultivars has contributed to yield increases (M. S. Jalalyar, personal communication).

Soft Wheat: Production, Milling, and Uses
AACC Monograph VI

March, 1981
Editors' Note: page 6, section **A. Africa**

More recent information indicates that most wheat in Egypt, grown on irrigated land in the Delta region, is soft in kernel texture.

Like its neighbors, Afghanistan makes no official distinction between soft and hard textured wheats. However, a review of the national germ plasm collection showed that about 27.4% of the lines are identifiable as soft wheat. In addition, a recent survey of the wheat market at Kabul revealed that about 46% of the wheat production was classifiable as soft wheat and ranged in color from white to light red (M. S. Jalalyar, personal communication).

INDIA

Although wheat constitutes a major crop in India, reaching a production of 28,336,000 te in 1976 (57), very little of it is soft wheat. Soft wheat culture is restricted to small pockets in the state of Madhya Pradesh. However, the quantity of available grain is sufficient to fulfill the requirements of the domestic biscuit and cake industry (M. V. Rao, personal communication).

IRAN

Wheat has been grown in Iran for centuries, and remains the main staple of the nation, providing an estimated 50% of the caloric intake. It also accounts for almost a third of the total farm income (33). In 1975, the crop was estimated at 4.7 million te. This was a record crop, owing to favorable weather conditions, grown on 4.3 million ha of largely nonirrigated land (32).

About 45% of the wheat area is occupied by winter or semiwinter varieties, much of it in the highlands (Azerbaijan, Khorasan, Kermanshah). High-yielding varieties such as Bezostaya, a hard red winter selection introduced from the U.S.S.R. several years ago, have outyielded local varieties such as Roshan and Omid (27). Spring wheats are found mostly in the coastal regions of the north.

Practically all of the harvest is used as food or seed for the following crop. Very little goes into animal feedstuffs (49).

Little information is available about specific varieties; however, Rashid, a spring wheat (26), has soft kernel texture.[2] Seeborg (49) stated that Australia has been a major supplier of wheat for Iranian bread. This importation is probably FAQ wheat, and consequently soft to semihard in kernel texture, a type to which the Iranians are presumably accustomed. An implication is that a significant portion of wheat harvested in Iran may be classified as soft, and indeed, it is estimated that about 70% of the crop is considered soft (T. Mahlooji, personal communication). The principal form in which wheat is consumed is unleavened bread such as *sangak, barberi, taftoon,* and *lavash* (49). Operators of stone mills prefer white wheats in preparing meal for bread, but the modern roller mills have the flexibility to adapt their milling technique to accommodate red wheats as well (49).

PAKISTAN

Wheat production in Pakistan has increased significantly in the past several years, partly because of the introduction of Mexican cultivars. These new cultivars, typified by Mexi-Pak, and modified forms from further breeding to adapt to conditions in Pakistan were grown on 35% of the wheat area in 1967-1968 (25), on 57% in 1973-1974 (37), and on almost 80% in 1975 (M. H.

[2]Soft Wheat Quality Laboratory, unpublished data.

Chaudhary, personal communication). Some soft wheat cultivars being grown include Penjamo 62, and Lerma Rojo 64. However, soft wheat accounts for only 5 to 10% of the total wheat production, and most of it is used by the biscuit and confectionery industry (M. H. Chaudhary, personal communication).

PEOPLE'S REPUBLIC OF CHINA

With an estimated 43 million tons of production in 1976, the People's Republic of China is the largest wheat-producing nation in Asia. The northern winter wheat region, which includes the North China Plains (Hopeh, Shantung, Kiangsu, Honan, Shensi, Shansi, and portions of Anhwei and Hupeh provinces), constitutes 60% of the total wheat production area. Varieties grown in this region are hard in kernel texture. South of this region, principally in the Ch'ang Chiang (Yangtze) basin, and separated from the northern hard wheat region by the Huai Ho (river), is the soft wheat region. The basin comprises about 23% of the Chinese wheat area. The kernel texture of the crop there is regarded as soft to semihard. To the southwest in the fertile Szechwan basin lies 10% of the total wheat area; all of the wheat there is soft-textured. The north Hunan plains provide 1% of China's wheat of the early maturing, soft-textured varieties. Thus, about 34% of the wheat acreage is devoted to soft wheat culture. To complete the picture, spring wheat accounts for approximately 10% of the wheat area. The Manchurian plains (Heilungkiang, Kirin, and Liaoning provinces), account for 4%, and the remaining 6% are southwest on the southern fringe of the Gobi desert in northern Hopeh, Shansi, and Kansu provinces and in Inner Monogolia (17).

Soft wheat is often grown as a part of a double-cropping system. In the Ch'ang Chiang basin, wheat is seeded in October or early November after rice harvest, and is harvested in late May or early June. The winters are mild and relatively short. However, rainfall is often excessive during wheat maturation and disease is a severe problem. Nantai 2419 was the most popular soft wheat variety in 1962, having been grown on 4 million ha. This cultivar, originally the Italian variety Mentana, was introduced in 1932. It was extensively grown in Hupeh, Szechwan, Hunan, Kiangsu, and Anhwei provinces. Another soft wheat variety, Ailido, was also of Italian origin and had been named Ardito. It was grown on 400,000 ha in Kewichow, Kiangsu, Chekiang, Hunan, Hupeh, and Szechwan. Other soft wheat cultivars with significant production in 1962 included Sanyuehhuang, Chungnung, Gintai 2905, and Baipu (17). Introductions of newer cultivars have undoubtedly changed the picture.

The Szechwan basin is known for its warm climate and very mild winters. Seeding is done during the latter half of October and the wheat is harvested early in May. Precipitation is not as great as in the Ch'ang Chiang basin and the area sometimes undergoes a dry spell. Chungking, in the basin, has an annual rainfall of 1,090 mm, with June having the most at about 180 mm. The area often does not experience any frost during the year. Wheat, with rape and peas, serves as the second crop, and rice is the major crop in the area (17).

In the Hunan region, the climate is so warm that triple-cropping is often practiced: a double crop of rice and a single wheat crop. Wheat is sown during the latter part of November and is harvested in April. The crop often suffers from excessive rainfall during ripening; thus early maturing varieties are favored. Swatow, in the region, has 1,520 mm of rainfall and Nanning 1,320 mm per year, mostly in the summer months. Other areas in the region may average as much as 2,000 mm or more per year. Wheat is grown, though to a minor degree, in Kweichow, Yünnan, Kwantung, Fukien, and Kiangsi provinces, often as part of a double-cropping system (17).

In general, the wheat-growing region is east of an arc extending from the northern terminus of Burma to the northernmost point of the Amur river beyond the greater Khingan range. Within this region, the spring wheat area ranges north of a line from Peking to Lan-chou, (in Kansu), whereas the hard winter region extends south to the Huai Ho. This is also the line of rainfall demarcation, less than 1,000 mm to the north and more than 1,000 mm to the south, at a latitude of about 32 or 33° north.

China's wheat production was estimated at 35.9 million te in 1972, 36.0 million in 1973, 37.0 million in 1974, and 41.0 million te in 1975 (34). Soft winter wheats normally out-yield hard winter wheats, and the latter usually produce more than the spring wheats. From yield ratios of 1.00, 0.75, and 0.63[3] for the three classes, respectively, soft wheat production for 1975 has been estimated to be about 17.22 million te or 633 million bushels, based on a soft wheat area of 34.4% and winter wheat area of 90.1%. The spring wheat crop (area) for 1976 is said to be about 13% and that of winter wheat 87% (34). The effect of this alteration on the soft wheat production estimate is not known.

More recently a report (41) indicated the introduction of early maturing high-yielding fall-sown spring Mexican cultivars into areas in the south where little or no wheat was previously grown. These areas include provinces as far south as Kwangsi and Kwangtung. Wheat cultivars purchased for these tests included Potam, Tonori, and Saric.

TURKEY

The nation shares with Iran, Iraq, and Syria the distinction as the site of the first wheat domestication, reported as having evolved along the foothills of the Taurus and Zagros mountain ranges (59). Today wheat is cultivated in Turkey on the central Anatolean plateau and to the southeast and east, the Aegean and southern coastal areas, and in Thrace. The central, southern, and eastern parts of Anatolia, the principal wheat-producing area, have a continental climate; thus there we find the cold-tolerant hard winters and durum wheats, the latter mostly in the southeast. Much of the crop is rain-fed, so production somewhat depends on sufficiency and timeliness of rainfall (19). Among the hard winter (bread) wheats, high-yielding cultivars such as Bezostaya are gaining popularity. In Thrace, this variety had already occupied 60% of the wheat area in 1971 (20).

The Mediterranean climate along the coastal areas, with abundant winter rainfall, has favored the culture of spring wheats such as those of Mexican, French, and Italian origin. Although not a wheat-growing region until the recent introduction of high-yielding varieties, these coastal lands now account for 10% of the total wheat area of Turkey (5). The wheat produced there is normally consumed by the farmer or in small villages, and little finds its way into

[3]These ratios are based on yields reported for crops in the soft red winter, hard red winter, and hard red spring regions of the United States in 1974. Quite possibly the ratios have no relation to yields of the respective classes in the People's Republic of China, owing to differences in cultivars and cultural practices, but they provide a rough estimate of soft wheat production.

commercial marketing channels (50).

In 1973 about 5.7 million ha were seeded to *T. aestivum* and about 2.2 million ha to *T. durum*. Both spring and winter types were grown, the former accounting for about 1.3 million ha and the latter about 6.7 million ha. Production averaged 8.8 million te over the period 1969-1973 (33) and was about 13 million te in 1976 (57).

Five distinct types are grown in Turkey, according to Seeborg (50). They were, in order of commercial importance in 1974, the common white wheat, soft to semisoft in milling; compactum, a soft white similar to the club wheats of the United States; durum, an amber-colored winter-seeded hard wheat; hard red winter of the Bezostaya type; and the Mexican type of soft red wheat. The two white wheats provide much of the flour for Turkish bread and all for pastry and related goods. About 20% of the wheat production was in soft wheat in 1974, but the percentage is decreasing.

Turkish bread is distinctively different from the unleavened loaf in Arab countries but is similar in appearance to the solid, hearth baked, round loaves of eastern Europe (50). The per capita consumption of wheat in Turkey is among the highest in the world, more than 200 kg per year (33).

In 1972, about 88% of the wheat was used for bread production, 7% for bulgur, and about 2% for pasta. Cookie flour accounted for less than 0.1% of total flour production.

C. Europe

Three general climatic regimes characterize the European continent and influence wheat culture. Northwestern Europe, including the United Kingdom, Ireland, Scandinavia, and portions of France, West Germany, and Austria, come under the influence of the humid maritime climate. The southern tier of countries surrounding the Mediterranean Sea is subject to the Mediterranean climate, characterized by mild and rainy winters and dry summers. These climate conditions prevail in Italy, Greece, and portions of Spain, France, Turkey, and Yugoslavia. The cold winters and hot summers of a continental climate determine what type of wheat is grown in Eastern Europe, including portions of Czechoslovakia, Poland, Hungary, Romania, Yugoslavia, Bulgaria, and the U.S.S.R., as well as central Spain.

Wheat yields in northern and western Europe are the highest in the world, with several countries producing 4 to 5 te/ha, and yields of 8 to 10 te/ha are not uncommon. The use of high-yielding varieties, good water supply, efficient farming, and the use of nitrogen fertilizers all contribute (12).

Winter wheat is the most important type in Europe, but spring wheats are favored in the northern areas. Wheat production is assisted by a climate with mild winters, warm summers, precipitation sufficient (50 to 75 cm/year), and soils with good water-holding capacities.

In northern Europe, wheats are sensitive to day length (12), whereas varieties in southern France and the Mediterranean area are quite insensitive. Hot, dry summers such as those experienced in North America lead to rapid ripening and hard wheats of high protein content. The cooler summers of Europe lead to grain that is softer, better filled, and lower in protein. Frost damage does not generally occur in Europe, winters being considerably milder than those experienced in the U.S.S.R. and Canada.

Wet weather after the grain has matured can lead to sprouting and consequently flour with high α-amylase activity that reduces bread quality. At mill intake, such wheat is screened, generally by using the Hagberg Falling Number test. Wet weather at that time also leads to delayed harvest and to mold formation in beard and crease while the crop stands in the field, high microbiological counts occasionally ensue, with poor flour color owing to contamination with mold spores (22). Miscalculations of drying rates and temperatures can lead to damage of milling wheat. Wet harvest conditions tend to be greater in the United Kingdom than in France or West Germany, and hence the quality of U.K. grain is generally less reliable (36).

The climate has important implications for the disease situation in Western Europe and hence yield and quality of the wheat crop. The main diseases there are bunt (*Tilletia caries*), eyespot (*Cercosporella herpotrichoides*), take-all (*Gaeumannomyces graminis*), yellow rust (*Puccinia striiformis*), mildew (*Erisyphe graminis*), and septoria (*Septoria* spp.). These are also found in other soft-wheat regions and are described in Chapter 3, except for eyespot, a fungus carried on stubble; it overwinters on unburnt straw that carries spores and subsequently infects seedlings. The characteristic lesion at the base of the plant accounts for the name. The disease is most serious in high-yielding crops and where cereals are frequently grown. If damage is severe, the tillers are killed. More frequently, however, the plant remains alive, produces white-heads, and is susceptible to lodging. No chemical treatment is available, and countermeasures are limited to breeding for resistant cultivars.

In the pursuit of increasing wheat yields, seed companies in western Europe now offer farmers not only seed of high-yielding cultivars and the necessary fungicides for disease control but also further information on cultivation, crop plans such as the Laloux and Schleswig-Holstein systems, methods of cereal growing, and rates and timing of nitrogen applications.

FRANCE

As the largest producer of wheat in Europe (excluding the U.S.S.R.) and a central trading member of the European Economic Community (now the European Community [EC]), France occupies a key position in the wheat economy of Western Europe. Total wheat production for 1972-1976 is presented in Table II. In France, as elsewhere in Europe, commercial wheat is divided into soft (*T. aestivum*) and hard (*T. durum*) wheat. The former is grown throughout the country except in mountainous areas; the latter is concentrated in the warmer and drier south.

T. aestivum is not further subdivided in the market into soft and hard kernel types as it is in the United States, although millers and other processors recognize these differences. Although no official estimates are available, there is general agreement that the majority of French wheats have opaque kernels and are starchy, and lesser quantities of vitreous wheat are grown. These different types of wheat may have their basis in cultivars, but at the same time climatic variations in France may induce textural differences, attributable at least partially to protein content differences (A. Bourdet, personal communication).

In the last 80 years French wheat production has doubled and crop yields have increased 3.5 times, but the area sown to wheat has shrunk by 30 to 40% and now comprises 1/10 of France. The main growing area is in the Paris Basin, Aisne, Eure-et-Loire, and Aube. The Paris Basin and the northern areas produce high wheat yields. In the south, the yields are somewhat lower owing to hotter, drier conditions, causing premature ripening (28).

Although wheat is grown in every département of France, the 17 départements in the north supply 70% of exports so that the quality of French wheat bought within the EC is fairly well defined within a given year.

Because of the diversity of regions and climate, many different cultivars are grown throughout France. Some of the cultivars and the diseases to which they are resistant follow:

Disease Resistance	*Cultivar*
Black rust	Capitole, Etoile de Choisy, Hardi, Talent
Yellow rust	Remois, Hardi, Heima, Maris Huntsman, Talent
Fusarium and take-all	Atou, Cappelle, Joss, Remois
Mildew	Cappelle, Champlein, Maris Huntsman, Talent, Top

Very early maturing cultivars are Etoile de Choisy and Talent; Brennus, Capitole, Chrismar, and Top are early.

The winter cultivar Top was grown on 21.2% of the total wheat area in France in 1977, followed by Capitole on 16.2%, Hardi on 15.4%, Maris Huntsman on 7.1%, Champlein on 7.0%, and Talent on 6.9%.[4] Joss, grown on 11% in 1975, fell to 2.3%; it is being progressively replaced by Talent. Among the spring wheats Kolibri, Rex, Florence-Aurore, César, and Chrismar were the most popular in 1977.

Collection, storage, and drying of wheat are well advanced in France. The country is well supplied with machinery and equipment and with more than 7 million te of farm storage capacity and a further 17 million te provided by cooperatives and merchants (55). Collecting centers are equipped to clean and

[4]Source: Institut Technique des Céréales et des Fourrages, 1977.

TABLE II

Area, Yield, and Production of Wheat in France, 1972-1976

	1972	1973	1974	1975	1976
Area (1,000 ha)					
Soft wheat[a]	3,787	3,807	3,967	3,586	4,081
Hard wheat[b]	162	151	177	283	209
Yield (q[c]/ha)					
Soft wheat	45.7	45.1	46.2	38.9	37.6
Hard wheat	27.8	29.4	30.8	28.7	26.4
Production (1,000 te)					
Soft wheat	17,596	17,348	18,593	14,229	15,599
Hard wheat	450	444	548	812	551

[a]Soft wheat = *Triticum aestivum.*
[b]Hard wheat = *T. durum.*
[c]q = quintal (100 kg)
Source: 1977 World Wheat Statistics. 1977. International Wheat Council, London, England. June.

dry wheat. At worst, a fifth of the total crop needs to be dried and the available driers have the capacity to dry 12 million te within two months (3). The collection centers can store wheat shipments of different quality separately or blend them according to requirements.

About 37% of French wheat is used domestically in milling and for other food uses; about 20% stays on the farm with some sown; approximately 12%, mostly low-quality wheat, is fed to animals, and 31% is exported.

As the largest use of domestic ("soft") flour is in bread production, bread-baking potential is the prime quality consideration. In 1970, 87% of French flour was baked in bread, 3.8% used for *biscotte,* 4.3% for crackers and biscuits, 0.8% for pastry, and 4.1% for other uses (A. Bourdet, personal communication).

Almost one half of the exported wheat goes to EC member countries. In 1974-1975, France exported 7.0 million te, of which 3.1 million went to member nations, including Belgium, 0.78 million te; the Netherlands, 0.75; United Kingdom, 0.63; West Germany, 0.52; Italy, 0.24; Ireland, 0.17; and Denmark, 600 te. A total of 3.89 million te was shipped to third countries.

In the United Kingdom, French wheat is bought mainly on the basis of protein and Hagberg. Exclusion clauses referring to Huntsman and Clément, poor bread-making cultivars, are sometimes included.

In 1969, in order to encourage exports and reduce imports, the Office Nationale Interprofessionel des Céreáles (the French government agency that controls imports and exports) introduced a scheme of grading and classification to encourage the growing of quality wheat.

The classification for export was based on cultivar, protein content, and Zeleny sedimentation value according to Table III. In each class, wheat was graded according to its test weight, dockage, moisture content, and Hagberg Falling Number.

Although excellent in concept, this system seems to have found very little commercial acceptance. More recently, a hearth-type bread-baking test has been proposed.

WEST GERMANY

The division of Germany after the Second World War profoundly affected German agriculture. The industrial regions of West Germany used to depend for their food to a large extent on the farmlands of the present East Germany (and Poland), which contains 55% of the arable farmland of the country (14). The surplus produce of East Germany now passes to the Soviet Union, and faced by this loss, West Germany had to step up its own agricultural production. As a result of careful crop rotation and use of fertilizers, West Germany is now 80%

TABLE III
Classification of Wheat for Exportation from France

Class	Protein Content (14% mb)	Sedimentation Value
I	11.2	38
II	10.3	28
III	9.5	18
IV	8.6	13

self-sufficient in cereals.

Farm holdings are small; 25% in 1971 consisted of less than 2 ha. In 1973 there were 970,000 farms producing 21 million te of grain (7). Such small units make efficient farming difficult. Wheat is the most important crop and is grown where soil is suitable, but mainly in the loess lands at the foot of the central uplands. In 1974, wheat production was 7.1 million te, barley 6.6 million te, oats 3 million te, and rye 2.7 million te. Rye now tends to be used less as a bread grain. The area sown to wheat remains fairly stable.

German wheat production in 1978 was 8.12 million te at 11.8% protein db (11), 7.4 million te of winter and 0.72 million te of spring wheat being produced on 1.5 million ha, with winter and spring wheat yields of 5.14 and 4.05 te/ha, respectively.

Like other European countries, West Germany feeds to animals much of the wheat not suitable for bread production. In this respect, wheat competes with corn (maize) from a pricing standpoint. Corn is fed when its price is relatively lower, and wheat of lower baking quality is diverted to intervention (government price stabilization) stock.

The wheat grown in West Germany (predominantly *T. aestivum*) is not segregated into soft and hard textured types. As a member nation, West Germany observes the criteria for utilization quality established by the EC. In 1974, the quantity of wheat milled for human consumption was about 3.8 million te and approximately an equal quantity was fed to animals. Of the quantity milled for human use, about 74% was home-grown wheat (38). Intensive efforts are being made to improve the (bread) quality of new cultivars. However, although some of the newer releases have been very productive in the field, they are poor in technological (baking) quality (38).

Several organizations are commissioned annually to assess the quality of German wheat varieties. They are Bundesforschungsanstalt für Getreideverarbeitung, Detmold; Bayerische Landesanstalt für Bodenkultur und Pflanzenbau, Freising; and the Bundessortenamt, Hannover.

As a basis for assessment, the results of respective baking tests from previous years are employed, using samples from the Bundessortenamt and the various provinces (Länder). The classification of varieties is based on the results of baking using the "Rapid Mix" test, with special importance being attached to the surface and elasticity of dough, and final loaf volume.

Assessment of dough properties has been introduced recently owing to manufacturing problems encountered at the dough stage when certain high-yielding cultivars were used; hence, a "T-" subscript for varieties considered unsuitable for classification. Type A wheat includes a Q+ or Q− subscript denoting its effect on loaf volume when blended with other flours.

Table IV shows the definition for baking quality groups, and Table V shows how current varieties are placed.

GREECE

Climatically, Greece is similar to the coastal areas of Turkey and Yugoslavia and to Italy. Thus, the type of wheat grown also tends to be similar. From 1971 to 1974, approximately 19% of the wheat production was in durum varieties (E. Skorda, personal communication). Of the remaining 81%, most were in soft varieties rather than bread (hard) wheats, principally because of the higher field yields available from the former. However, soft wheat is extensively used for bread-baking purposes and for manufacture of biscuits and other pastry goods (E. Skorda, personal communication). Average wheat production for the years 1973-1975 was 1,960,000 te on 860,000 ha.

ITALY

With its Mediterranean climate, Italy is a major wheat-producing country. Wheat has a long history in Italy, having been grown and used since Roman times. The ruins of Pompeii (79 AD) revealed bakeries and even samples of bread. Much of the wheat today is grown in northern and central Italy, with a large portion in the Po valley or plain.

No distinction is made in Italy between hard and soft textured wheats within *T. aestivum*. All wheat of the species is termed "tenero" (soft) in contrast to the term "duro" (hard) that is applied to *T. durum*. Actually, much of the "soft" wheat grown in Italy is more similar to the soft wheats of the United States than to common hard winter and spring types. Breeding programs have as objectives the development of hard wheats of the types exported by the United States and

TABLE IV
Baking Quality Groups and Definitions for Wheat in West Germany

(These assessments presuppose that the wheat is sound and has for each relevant application an appropriately satisfactory content of protein.)

AI/Q+	Wheat varieties with "Eigenbackfähigkeit" (inherent baking quality) with very high yields by volume and normal qualities of the dough and of the baked goods. They feature a frequent and high degree of mixing ability.
AI/Q−	Wheat varieties with "Eigenbackfähigkeit" with very high yields by volume and normal qualities of the dough and of the baked goods. They feature a less frequent and a lower degree of mixing ability.
AII/Q+	Wheat varieties with "Eigenbackfähigkeit" with generally high yields by volume and normal qualities of the dough and of the baked goods. They feature a frequent and a high degree of mixing ability.
AII	Wheat varieties with "Eigenbackfähigkeit" with generally high yields by volume and normal qualities of the dough and of the baked goods. They do not feature mixing ability.
BI	Wheat varieties with "Eigenbackfähigkeit" with generally average yields by volume and normal qualities of the dough and of the baked goods. For obtaining commercial grades of flour, they generally need to be mixed with wheat of good quality.
BI/T−	Wheat varieties with "Eigenbackfähigkeit" with generally average yields by volume which, owing to their negative properties of the dough, are unsuitable for making bread if they are used on their own or if they are mixed in a proportion where they are a substantial constituent.
BII	Wheat varieties with "Eigenbackfähigkeit" with generally low yields by volume and with normal qualities of the dough and of the baked goods.
BII/T−	Wheat varieties with "Eigenbackfähigkeit" with generally low yields by volume which, owing to their negative properties of the dough are unsuitable for making bread if they are used on their own or if they are mixed in a proportion where they are a substantial constituent.
C	Wheat varieties with "Eigenbackfähigkeit" with very low yields by volume and normal qualities of the dough and of the baked goods.
C/T−	Wheat varieties with "Eigenbackfähigkeit" with very low yields by volume which, owing to their negative properties of the dough are unsuitable for making bread if they are used on their own or if they are mixed in a proportion where they are a substantial constituent.

Canada, but this end is difficult to attain owing to poor adaptability of hard wheat cultivars (C. F. Krull, personal communication).

Wheat production of the past several years averaged 9,200,000 te on 3,615,000 ha (23). About 70.4% was soft wheat (*T. aestivum*) (C. F. Krull, personal communication).

Distinctions between hard and soft wheats are based on species. Further classification with respect to utility for bread production may be based on test weight in some cases. In more advanced laboratories, the Berliner, Pelshenke, and alveograph tests are applied. Protein content is also a factor in the classification (C. F. Krull, personal communication).

The variety San Pastore was widely grown in Italy in 1971 (2). In quality tests of current Italian varieties, these classifications are used (13):

a) Varieties suitable for direct bread making: Jacometti 49, Lucera, Marzotto,

b) Varieties suitable for bread after supplementation with hard common wheat: Irnerio, Mara, Marimp 3, Farnese, Demar 4, Marco, Flavio, Livio,

c) Varieties with poor bread quality: Flamino, San Pastore, Curzio.

This three-tier classification system appears to be similar to that employed in Yugoslavia.

TABLE V
West German Classification of Wheat Varieties for Bread Quality

Cultivar	Quality Class	Distribution (%)	Protein Content (%)	Sedimentation Value	Rapid Mixing Test Volume (ml)
Winter Wheat					
Jubilar	B1	19.7	11.1	22	561
Caribo	B11	19.0	11.2	25	571
Diplomat	A11/Q+	14.2	12.0	35	645
Kormoran	A11/Q	13.7	11.5	42	636
Vuka	A11	8.4	12.3	42	669
Maris Huntsman	C/T−	2.7	11.9	22	...
Kranich	B11/T−	2.6	12.2	25	...
Carimulti	B11	1.6	12.0	24	582
Benno	B1/T−	1.4	11.9	21	...
Saturn	B11	1.4	11.7	31	598
Carisuper	A11/Q	1.3	12.9	31	642
Magnet	B1	1.1	11.2	22	562
Carstacht	A11/Q+	1.0	12.0	36	648
Topfit	B1	1.0	12.2	31	603
Schemauer	B1	0.9	10.8	21	553
Disponent	A11	0.8	12.0	26	618
Joss	B11	0.8	10.4	17	534
Monopol	A11/Q+	0.8	12.4	54	706
Florian	A11/Q	0.7	11.7	32	633
Clement	C/T−	0.6	12.2	16	...
Nimbus	B1	0.6	11.6	26	581
Spring Wheat					
Kolibri	A11/Q+	64.7	13.2	47	693
Janus	A11/Q	16.6	12.1	28	625
Arkas	A11/Q	5.9	12.0	35	645

UNITED KINGDOM

Wheat has been grown in the United Kingdom (England, Wales, Scotland, and Northern Ireland) for centuries. In 1977 wheat production was 5,274,000 te, representing 4.92 te/ha on 1,073,000 ha. About 97% of area and production were in England and Wales.

Wheat growing is well distributed throughout the United Kingdom: 52.5% in the east, including the eastern and east Midland region; 32.4% in the south, which includes the southeastern and southwestern regions, west Midlands, and Wales; the remaining 15.1% in the north, including the Yorkshire and Lancashire regions, the northern region, and Scotland. The maritime climate of the country is conducive to good yields for cultivars that are considered soft in quality. An intensive breeding and development program is underway to produce cultivars with improved bread-baking qualities.

Adopting the west European nomenclature, the United Kingdom classifies all of its wheat as soft wheat; this means that it does not produce durum wheat. However, if all the wheat were classified according to its value for milling and baking purposes, about 30% would be categorized as soft wheat (J. Wright, personal communication). In the United Kingdom soft wheat is defined as a cultivar that produces an amorphous, finely divided flour, tends not to produce a large or fine textured loaf of bread, gives a characteristic weak extensogram pattern, and has low water absorption.

The lack of strong topographic or ecogeological features in the major wheat-growing regions of England has influenced the varietal pattern, and single varieties such as Squarehead's Master, then Cappelle Desprez have dominated. Both winter and spring wheats are grown, with the former predominating. Leading winter cultivars in 1977 were Maris Huntsman, which accounted for 50,700 te or 35% of the winter wheat production; Bouquet, 17,800 te and 12%; Atou, 13,300 te and 9%; Flinor, 12,200 te and 8%; and Maris Freeman, 9,800 te and 7%. Sappo (5,400 te) contributed 66% of the total spring wheat production, followed by Maris Dove (1,700 te and 21%). All of the above-mentioned cultivars except Atou are considered to have hard-kernel texture, although they are all classed as soft wheat for marketing purposes (J. Wright, personal communication).

The relative scarcity of homegrown bread wheats has induced the importation of wheat of desired quality, since bread comprises by far the greatest end-use of wheat milling. It is estimated that 67% of flour is baked into bread, 11% into biscuits, 14% into cakes and similar goods, and 8% used as household flour (B. A. Stewart, FMBRA, England, personal communication).

A considerable portion of domestically grown wheat is fed to animals. In 1976-1977, 2,400,000 te was milled or otherwise used for human or industrial purposes; 2,035,000 te was fed to animals. The latter accounted for 42.6% of the 4,775,000 te produced in 1976.

The amount of U.K. wheat used in flour milling has increased over the last few years owing to the rising cost of third country wheat, and in 1975-1976 it represented 43% of the total wheat milled (6).

The largest wheat crop was harvested in 1978 with 6.59 million te recorded. The reasons were a 17.4% increase over 1976-1977 of the planting area, more sowings of high-yielding varieties, and favorable weather conditions at sowing time. Wheat growing in the United Kingdom is largely determined by climate since in northern areas the harvest becomes progressively later and in the west rainfall becomes excessive. For these reasons yields and quality are higher in the eastern part of the country. English wheat generally averages 10.0% protein content (at 14% moisture), but the hot, dry summer in 1976 produced grain with a much higher protein (13.9%) level.

According to seed sales, winter cultivars for the 1978 crop included Maris Huntsman, 31%; Hobbit, 19%; Flanders, 16%; Bouquet, 8%; Kinsman, 7%; and others, 19%. Sicco accounted for 60% of the spring wheat crop, Sappo 27%, and others, 13%. The poor baking quality wheat Maris Huntsman is being replaced by another feed-type wheat, Hobbit, a semidwarf with a protein level 10% below average.

OTHER WEST EUROPEAN COUNTRIES

Other members of the EC include Belgium, Luxembourg, Denmark, the Netherlands, and Eire.

Belgium grew almost 1 million te of soft wheat in 1977, mainly in the central plains in Brabant, Hainault, and Hesbaye provinces, and in Flanders in the west. Popular winter cultivars included Cama, Hardi, and Anouska, while César and Arabel were popular spring wheats. Millers in Belgium used 58% indigenous wheat, plus 29% EC, and 13% third country wheats.

Luxembourg millers used 15% homegrown wheat and 85% EC wheat in their grist.

The area devoted to wheat in Denmark is small, although agriculture is its largest industry. Much of its farm production is oriented toward its large dairy industry. Most of its 1 million te of wheat grown in 1978 was of the Solid and Sarah cultivars among the winters, and Sappo and Kolibri among the springs.

Wheat production in the Netherlands is decreasing, the tendency being to grow high-yielding cultivars for feed and to use other EC wheat for flour milling. Wheat is grown preferably on sea clay rather than sandy soil, such as the polder lands and the islands of Zeeland. Most of the farms are small (3 ha and less), but by intensive management, the crop yields are the highest in the world. Cultivars used in 1976 were: among winter wheat (109,000 ha), Caribo (24%), Lely (24%), Clement (17%), Manella (14%), Cyrano (5%), Norda (4%), Adamant (4%); and among the spring wheats (21,000 ha), Kasper (33%), Melchior (28%), Toro (22%), Bastion (11%).

The cultivars Caribo and Lely are increasing at the expense of Clement, which has fallen away due to yellow rust (the main disease problem in the Netherlands).

Crop yields are approximately 5 te/ha, the highest in Europe, giving a total wheat yield of 1.0 to 1.5 million te/year. Half of the wheat crop is used for feed, 15% for milling, and 30% for export.

Because yield is the main criterion for the farmer, the miller tends to use 10% homegrown, 45% EC, and 45% third country wheat. However, Dutch plant breeders, in line with other breeders in the EC, are aiming to produce wheat varieties with high yield and baking quality under pressure from government and farmers' organizations.

In 1978 Eire grew wheat on 49,000 ha, producing about 220,000 te of wheat. Wheat is grown mainly in the east of Eire in Leinster and in the south in Munster. Because of the generally warm and damp climate fostering fungal diseases in winter wheat, most wheat is spring sown. The proportions of spring to winter wheats can vary quite widely. The main varieties grown in 1977-1978 were Sappo, 45%; Kleiber, 35%; and Sicco, 15%. The preponderance of quality spring wheat makes life somewhat easier for the millers, although the climate can lead to sprouting and average protein levels are between 10 and 10.5% on a 14% moisture basis (G. Campbell, personal communication).

EAST EUROPEAN COUNTRIES

Bulgaria, Czechoslovakia, the German Democratic Republic, Hungary, Poland, Romania, and Yugoslavia are all major wheat-growing nations, with estimated productions in 1976 given in Table I. Little information is available regarding the extent to which soft-textured wheat as compared to hard wheat of the species *T. aestivum* is grown.

In Bulgaria, the percentages of the wheat crops consisting of soft wheat varieties were: for 1971, 5%; for 1972, 14%; for 1973, 24%; and for 1974, 8% (N. Donchev, personal communication). Soft wheat is usually used in bread production by blending with wheat of stronger gluten properties. Because of this necessity for supplementation in the use of soft wheat, its cultivation is being limited and the growing of hard wheat is being encouraged. In determining the quality of soft wheat, the choice of cultivars has been the principal criterion. Soft wheats include cultivars of poor baking quality, with less than 50% vitreous kernels, and 12% protein content (N. Donchev, personal communication).

Hron has reviewed the winter wheat situation in Eastern Europe (31). In the continental climate harder textured cultivars with greater bread potential are grown. However, the recent introduction of higher yielding cultivars has resulted in some quality loss, attributed by some to the negative correlation between quality and yield.

Yugoslavia grows cultivars that are classified as soft red and hard red winter. The percentage of soft wheat grown in the country is quite high, estimated at 70 to 80% (S. Borojevic, personal communication). The principal basis for classification are gluten quality as determined by the farinograph and by a bread-baking test. Cultivars belonging to groups A and B are considered hard wheats. Those in group C are categorized as soft wheat (S. Borojevic, personal communication). Wheat is grown throughout Yugoslavia, but the hard types are found in the eastern, less humid republics of Vojvodina and Serbia. Almost all of the national production is used in the form of bread. Several of the soft varieties of Yugoslavia are of Italian origin or derivatives of Italian varieties. They are not adapted to regions with more severe winters and do not yield well in the continental climate. Thus, the more winter-hardy varieties such as Bezostaya 1, Mironovskaya 808, and Rannaja 12 have been grown to a greater extent in the eastern region.

Bezostaya, a strong wheat with high productivity and good milling and bread-baking qualities, is grown exclusively in the Balkans. It is also grown in southeastern Poland and Czechoslovakia. Mironovskaya 808 is grown in Czechoslovakia, Poland, East Germany, and Hungary. Breeders have incorporated Bezostaya into new cultivars in Bulgaria, Hungary, Romania, and Yugoslavia (35).

In Poland, the varieties Grana, Jana, Aria, Dana, Luna, and Malwa comprised 70% of wheat production in 1973, with Grana alone accounting for about 40%. The qualities of these cultivars can only be inferred in the absence of firm data. Wolski stated (58) that the release of Grana, Luna, and Dana represented a backward step in baking (bread) quality, implying a weak gluten, a low gluten stability, and the other properties associated with soft wheat. Aria is a sister variety of Grana but is superior to Grana in grain quality. Malwa is described as having rather good grain quality, but apparently neither is the equal of Mironovskaya 808 or Kavkaz in (bread) quality. These Russian varieties comprised the bulk of the remaining 30% of Polish wheat production in 1973 (58).

The new Polish cultivars are not as winter-hardy as Mironovskaya 808. Perhaps they reflect their parentage, which includes soft wheat cultivars such as Ardito (Italy), Heine IV (Germany), Etoile de Choisy and Cappelle-Desprez (France)—all less hardy than the hard wheat cultivars (58). Based on these considerations, roughly 50% of the wheat in Poland can be regarded as soft.

The German Democratic Republic also does not distinguish between soft and hard wheats. The principal basis for classification lies in the capability of the variety for milling well and baking bread, with the poorer ones commonly recommended for livestock feeding. Among the cultivars grown in 1971-1975, the following were classified as having very favorable milling and (bread) baking properties: Alcedo, Aurora, Il'icevka, Kavkaz, Mironovskaya Jubilejnaja, and Mironovskaya 808. Those regarded as favorable included Delphin, Orlando, and Pluto among the winter wheats, and Carola, Hatri, and Ramses among the springs. Cultivars regarded as possessing modest milling and baking quality and recommended as feed for livestock included Eros, Fakir, Fanal, Pilot, Poros, Saladin, and Winnetou. Information was not available regarding the percentage of the wheat production attributable to the several varieties (C. Lehmann, personal communication).

Czechoslovakia has a variable agroecological wheat growing environment. Weather conditions in Bohemia are milder than in Slovakia, which has a continental climate (53). In 1974, approximately two thirds of the wheat production was of high-yielding varieties, chiefly Mironovskaya and Kafka from the U.S.S.R. Presumably these were found mostly in the eastern areas.

Approximately 40% of Hungary's wheat crop is fed to animals (16). Because it is unlikely that wheat suitable for bread would be thus used, this wheat probably consisted of the crop least suited for bread baking, presumably low-protein wheat or unsuited cultivars.

D. North America

Wheat is not indigenous to the western hemisphere. It was introduced by early colonists from Europe, and the present extensive culture of the crop is a result of early adaptation and breeding. Soft-textured wheats are grown extensively in the United States and to lesser extents in Canada and Mexico.

CANADA

Wheat in Canada is produced mainly in the prairie provinces of Saskatchewan, Alberta, and Manitoba, with some also in British Columbia and

Ontario. Most of the western wheat is of the hard red spring class. However, all wheat grown in Ontario is of the soft white winter class. Soft white winter wheat, spring types, and Alberta red are grown in small quantities in the west.

Total Canadian wheat production in 1977 exceeded 23.5 million te. Ontario's soft white winter yield was 822,000 te, and other soft wheat production brought the aggregate to 1.2 million te, or 5.1% of the Canadian harvest. Soft wheat is therefore not a major crop, but it is significant as the source of domestic confectionery flour; a small amount is exported.

In 1977, the leading soft white winter cultivar in Ontario was Fredrick, grown on an estimated 80 to 90% of the wheat area. The remainder was mostly Yorkstar. Minor quantities of soft spring wheats are grown in the maritime provinces. Alberta grew about 300,000 te of soft wheat in 1977, mostly reds (183,000 te); the remainder was soft white springs. Alberta red is a soft type, often blended with Ontario soft white or local soft white spring for milling to produce pastry flour.

Grain varieties are licensed in Canada. In addition to the two mentioned above, Talbot and Genesee were licensed for growing. Fredrick and Talbot were developed in Canada, whereas Genesee and Yorkstar are New York releases.

MEXICO

Recent wheat production in Mexico has exceeded 2 million te. Much of this is hard-textured grain, and presently only about 15% or less are soft-textured cultivars. There was a time during the last decade, however, when Lerma Rojo, a soft wheat, was by far the most widely grown cultivar, and the area devoted to soft wheat was estimated to be as much as 85% (R. Anderson, personal communication).

UNITED STATES

The United States ranks with France and the People's Republic of China as the largest soft wheat-producing nations of the world at about 16 to 17 million te (Table I). The major crop-growing areas are characterized by mild winters, adequate to abundant rainfall, and a humid climate, referred to variously as marine west coast, humid subtropic, and long summer humid continental by Heinzelman and Highsmith (29) and humid region by Blumenstock and Thornthwaite (10), for example. These environments appear to favor cultivars with soft-textured grain, although such types are not the only ones grown.

Wheat was brought to the United States by the earliest settlers. All were soft wheats (48), and it was not until the middle of the 19th century that hard wheats were introduced into the plains area of the country from Eastern Europe.

Table VI presents U.S. wheat and soft wheat production, and the percentages of soft wheat for the crop years 1975-1978, inclusive. The average soft wheat production for the period was 14,867,200 te, or 27.2% of the total. A little more than half (55.5%) of the soft wheat was red wheat.

According to the 1974 U.S. Department of Agriculture wheat class and varietal distribution survey (43), eastern soft wheat was grown on 4,285,200 ha, of which 3,899,100 ha were in soft red winter wheats and 386,100 ha in soft white winters. Leading eastern soft red winter wheat states were Ohio (639,400 ha), Illinois (598,100 ha), Indiana (582,800 ha), Missouri (472,200 ha), and Michigan

(389,600 ha). Significant production areas are also in Kentucky, Arkansas, Tennessee, Pennsylvania, North Carolina, and Virginia. Usually the cultivars grown in the North Central region differ from those in the south Atlantic seaboard and Gulf states. Those of the south are somewhat less winter-hardy and may have different photoperiod requirements than those adapted to northern states. Moreover, diseases hinder full productivity more in the South than in the North. In Michigan, the leading eastern soft white state, 297,300 ha were devoted to the crop, while New York grew 88,700 ha in 1974.

Soft red winter wheats grown in the United States in 1974 included Arthur 71, Monon, Blueboy, Redcoat, Abe, Logan, and Coker 68-15. All were grown on at least 50,000 ha in that order of area coverage. In all, 60 varieties were reported as grown in that year. Eastern soft white cultivars of importance were Genesee, Ionia, and Yorkstar, all of which were reported as being grown on more than 50,000 ha. Only seven cultivars were reported.

In the same year (1974), 1,700,200 ha of common white and 403,500 ha of club wheat were grown in 10 western states for a total western white wheat production of 2,103,700 ha. Most of the wheats were of winter habit, but some were spring-sown, and still others were facultative. In the three leading western wheat states of Washington (1,096,800 ha), Oregon (473,900 ha), and Idaho (388,800 ha), almost the entire crop was in common white and club wheats, and the three states accounted for 92.7% of the production of all western white wheats for 1974. The heaviest concentrations of wheat growing were in southeastern Washington, northern Oregon, and west-central Idaho. Smaller quantities of white wheat were grown in Arizona, California, Utah, Montana, Nevada, Colorado, and New Mexico, in that order. Wyoming did not report any white wheat for 1974. Soft-textured cultivars accounted for 98.5% of the white wheat grown in the region, and only 1.5% in semihard or hard white wheats. Cultivars widely grown in the western states included Twin, Springfield, Marfed, and Siete Cerros 66 (spring wheats), Nugaines, Hyslop, and Gaines (winter wheats), and Paha, Moro, and Omar (club wheats). All were grown on 50,000 ha or more (43).

TABLE VI
U.S. Wheat Production, All Classes (in 1,000 te), and Percent Soft Wheat, 1975-1978

	1975[a]	1976[a]	1977[b]	1978[b,c]
Soft red winter	8,880.5	9,162.2	9,532.4	5,502.4
White	6,972.6	6,738.6	5,884.4	6,795.6
Total, soft wheat	15,853.1	15,900.8	15,416.8	12,298.0
Hard red winter	28,804.2	26,565.8	27,017.8	22,711.2
Hard red spring	8,891.0	11,192.3	10,824.1	10,328.3
Durum	3,358.3	3,672.8	2,176.9	3,629.7
Total, wheat	56,906.6	57,331.7	55,435.7	48,967.2
Percent soft	27.9	27.7	27.8	25.1

Sources:
[a] Milling and Baking News 56(47):12, Dec. 27, 1977.
[b] Milling and Baking News 57(45):69, Dec. 19, 1978.
[c] Preliminary.

As noted in Table VI, U.S. wheat production data are subdivided into classes. The basis for this classification is cultivar, and the cultivar distinction between soft and hard types is clear-cut. Few cultivars of any importance could be said to be in the transitional area. Although the wheat class called "White" is not necessarily synonymous with or restricted to grain with soft texture, soft texture is a property common to practically all white wheat. The subclass "Hard White" accommodates wheat with harder texture.

The designation of soft red winter and white wheats as market classes serves a second purpose. These wheats are mostly used in flour production for cookies (biscuits), cakes, and similar confectionery products, whereas the hard red winter and spring wheats are used principally for breadbaking. Thus, grain texture has become associated with end-use.

In 1973 the per capita consumption of flour in the United States was 48.86 kg (21). The various types of flour consumed in recent years are given in Table VII. While the per capita consumption of bread has decreased somewhat in the past two decades, the decline in use of all-purpose flour and the steady increase in the use of soft wheat flour have been more notable.

As the United States for many years has produced much more wheat than is required for domestic consumption, the surplus is being exported all over the world, either through cash or concessional sales. Soft wheats are part of these exports. It is estimated that 85% of the western white wheat is exported, mostly to Asian countries. However, export of eastern soft wheat is much more variable, increasing when the crop is large and declining when it is low. An adequate supply is available for domestic demand, but carryover to the next crop year is not large in relation to that of wheat of other classes.

E. Oceania

Commercial quantities of wheat are grown in Australia and New Zealand. New Zealand's wheat production totaled 456,000 te in 1976, of which 4% was soft wheat. Most varieties are semihard, but a soft variety, Arawa, is grown mainly in South Canterbury. The soft wheat is blended with others in milling for bread flour, but limited quantities are used for special purposes (R. W. Cawley, personal communication).

AUSTRALIA

Australia, with western Europe, the United States, and the People's Republic of China, is one of the major soft wheat-producing nations of the world. Unlike

TABLE VII
Consumption of Flour in the United States, 1959-1973 (in kg/year)

Type	1959-1960	1965-1966	1973
White pan bread	23.00	20.14	20.37
All-purpose	17.60	13.43	9.75
Soft	9.26	11.75	13.02
Whole wheat	3.31	2.95	2.86
Total	53.36	50.45	48.86

Source: USDA-ERS Wheat Situation WS-233, Aug. 1975.

Europe and the People's Republic of China, but similar to the United States regarding wheat classification, Australia makes a clear distinction between hard and soft textured wheats within the species *T. aestivum*. This distinction is made on the basis of cultivar (56). More often than not, an association is found among certain cultivars, vitreous kernels, and hard kernel texture on one hand, and among other cultivars, mealy kernels, and soft texture on the other. However, vitreosity is not as important as cultivar in determining texture, which is measured by a grinding test (9) such as a particle size index test (56). A distinction between hard and soft wheat cultivars is also based on suitability of wheat for end-product usage, i.e., hard wheat for bread and related products and soft wheat for biscuits and other confections.

Wheat is grown in the southern region of Australia and may be found in all states except the Northern Territory. Until about 1936, all Australian wheat was soft. Since then, however, breeding efforts have brought about a significant shift in wheat class, such that for the period 1975-1976, when 57.4% of the Australian wheat crop of 11,982,000 te was classed as hard and 42.6% as soft (Table VIII) (C. Wright, personal communication).

Queensland grows little or no soft wheat; most of the crop in Western Australia is in soft wheat cultivars. In New South Wales, South Australia, and Victoria the percentage of soft wheat has steadily decreased to the extent that in New South Wales and South Australia soft cultivars presently comprise a distinct minority. In 1973, the leading varieties were Gamenya, Falcon, Insignia, and Heron, of which only Falcon is classed as hard wheat. In Victoria, Halberd, Olympic, Summit, and Insignia were popular. All but Halberd are soft wheats. Olympic, Heron, and Insignia were also grown in New South Wales and South Australia (4). Tasmania grows little wheat, and all of it is soft.

Varieties grown in Australia have been identified as hard or soft according to their response to particle size index and pearling resistance tests (56), and classification of varieties into one of five classes is given by Wrigley and Shepherd (60). Variety is used in at least one state (New South Wales) as a basis for establishing quality standards. To qualify as Special Agency Soft Wheat, grain must be of the Egret, Olympic, Pinnacle, Summit, or Teal variety.

TABLE VIII
Percent of Wheat-Growing Area Sown to Soft Wheat Varieties in Australia, 1973-1976
(shown as percent of total area sown to specified varieties)

Season	Percentage Production of	New South Wales	Victoria	Queensland	South Australia	West Australia	Australia
1971-1972	State	44.1	92.8	0.3	57.8	81.8	
	Australia	14.9	13.5	0.0	8.5	23.4	60.3
1972-1973	State	41.8	81.5	0.0	28.0	77.3	
	Australia	14.4	11.7	0.0	3.6	24.8	54.5
1973-1974	State	38.9	60.8	0.0	16.8	76.6	
	Australia	12.5	8.6	0.0	2.7	25.5	49.3
1974-1975	State	31.0	51.0	0.0	11.1	76.6	
	Australia	9.9	7.0	0.0	1.6	25.9	44.4
1975-1976	State	21.5	49.8	0.0	9.7	76.4	
	Australia	7.0	6.2	0.0	1.1	28.3	42.6

Source: C. W. Wrigley, personal communication.

Australia's concern about wheat classification into soft and hard stems from the nation's reliance for foreign exchange through exports. Designation of wheat by texture, often related to product end-use, can be of value in determining the quality of the wheat and its probable utilization. Most of Australia's wheat is exported to Asian nations, especially Japan. A clear understanding of quality designation and market grades is mutually advantageous in international trade.

Recent breeding efforts in Australia appear to be directed to maintaining and enhancing its position in foreign sales. Because the Australian government controls wheat sales through the Australian Wheat Board, steps can be taken to promote uniformity in types of wheat to be grown and marketed.

F. South America

Argentina, Brazil, and Chile are the largest wheat-producing nations on the continent. Argentina has exported wheat (all hard or bread types) for a number of years; Brazil is expanding its wheat area in the humid southern region; the soft wheat region of Chile has a climate similar to that of western Europe, from where a number of cultivars have been introduced.

ARGENTINA

Argentina is the premier wheat-growing country in South America and one of the leading wheat-exporting nations of the world. In 1975-1976, total production was approximately 8,000,000 te, of which about 3,600,000 were exported (46). Wheat is grown in Santa Fé to the north, La Pampa to the west, and in the coastal area, between the cities of Mar del Plata and Bahia Blanca (46).

Varieties currently grown include Klein Toledo, Klein Atlas, Buck Atlantico, Tacuari INTA, Buck Cencerro, DeKalb Tala, DeKalb Lapacho, DeKalb Urunday, and others classed as "duro," hard, whereas Pergamino Gaboto and Precoz Parana INTA are "semiduro." Argentina does not grow soft wheat (R. J. E. Agrasar, personal communication).

BRAZIL

Brazil, long an importer of wheat, has seen its domestic production increase dramatically since 1970. In 1976, Brazil harvested 3,100,000 te, compared with about 360,000 te in 1967. Much of this wheat is found in the northeastern area of the state of Rio Grande do Sul, where about 65% of the nation's crop is grown. In this region rainfall is adequate (1,700 mm per year). The area farther south is more likely to experience drought and a Mediterranean type of climate. The crop is almost exclusively of the soft spring type; about 60% are presently of the cultivars Nobre and IAS 54 (F. W. Bergmann, personal communication), but have fair to good protein content (45). In Rio Grande do Sul, seed is planted in May or June; the plants head in August or September and are harvested in October-November. Early spring cultivars, resistant to the high aluminum content of the soils, are used. Double cropping with soybeans is often practiced; hence early maturing cultivars are preferred (8,45).

The western portion of Santa Catarina and southwestern Paraná comprise extensions of the agroecological condition of Rio Grande do Sul but are warmer and less humid in the spring. However, the same (soft-kerneled) cultivars grown in the south are planted here (45).

In northern Paraná, southwestern São Paulo, and southern Mato Grosso, wheat growing is expanding rapidly. These areas lend themselves climatically to the early spring Mexican wheats or their derivatives (45).

Rocha (44) reported wheat production in 1974 for Rio Grande do Sul at 1,600,000 te; Santa Catarina, 64,000; Paraná, 930,000; São Paulo, 120,000; and Mato Grosso, 121,000; and a total of 2,835,000 te. If the production in the first three states is regarded as consisting entirely of soft-textured wheats and partially so in the others, as much as 95% of Brazilian wheat could be classed as soft.

Wheat marketing in Brazil is controlled by the government, which allocates the supply to mills and other outlets, but a small quantity of locally grown wheat is often also available. Approximately 80% of the wheat is used for bread production, and the remainder is milled for confectionery and pastry products. The poorer grain is used as animal feed (F. W. Bergmann, personal communication).

CHILE

Chile ranks third (behind Argentina and Brazil) in South American wheat production, growing about 1,250,000 te in 1976. The wheat-growing region extends from the province of Tarapoca in the north, with its continental climate, to Chiloe in the south, where the environment is almost maritime and rainfall is adequate to plentiful. Corresponding to the range of climates are the types of wheat grown and cultural practices. Hard spring and durum wheats predominate in the irrigated lands of the north. In the central-south zone, with its moderate rainfall averaging 1,000 mm annually, wheat is grown mostly under rain-fed conditions. Both spring and winter wheats are grown, as well as facultative types. The region extending from Malleco to Chiloe provinces constitutes the southern zone where winter wheat, mostly soft-textured, predominates. Irrigation is absent in this region. The north usually accounts for 34.4% of the production, and the central-south and south produce 26.9% and 38.7%, respectively (1).

Most of the winter cultivars were introduced from northern Europe. Those popular in Chile in the early 1970s included Vilmorin 29, Cappelle Desprez, Hesbignon, Manella, Lilifen, and Intermedio. Most of these are soft-textured; and, indeed, it is estimated that 55% of the Chilean crop consists of soft wheat cultivars. In spite of textural differences among cultivars, the preponderant portion of the crop is used in bread baking. The softer varieties are used for biscuit, cake, and tart production (H. Wulf, personal communication). A significant portion of the crop (182,000 te in 1972) is fed to livestock (42).

G. The Union of Soviet Socialist Republics

The U.S.S.R. grows more wheat than any other nation in the world; more than 65 million ha are sown. About 45 million ha are devoted to spring wheats and 20 million ha to winter wheats. Much of the winter wheat area is in the European U.S.S.R., mostly in the Ukraine, the southern Federated S.S.R. in the Caucasus region, the central chernozem, central, and in the southern Kazakh S.S.R. Spring wheats are grown in the eastern regions, including the Volga, western edge of the Urals, northern Kazakhstan, and southwest Siberia.

The varieties Bezostaya and Mironovskaya 808 occupied 85% of the total winter wheat acreage in the U.S.S.R. in the early 1970s (35). Others released more recently are Aurora and Kavkaz. All are hard-textured bread wheats. There appears to be no breeding program to develop low-protein, soft-textured varieties—either winter or spring—similar to those used in the Western nations and Japan for the production of flour confectionery products. Such confections are usually made from durum flour in the U.S.S.R. (L. W. Briggle, personal communication).

In view of the above and without further information regarding breeding, the extent of area with climatic and environmental conditions favoring the culture of soft-textured wheats, we can only conclude that the culture of soft kernel-textured wheat varieties is minimal and can be considered to be available, at most, only in trace quantities.

IV. INTERNATIONAL TRADE IN SOFT WHEAT

Wheat is grown in many countries, but few produce enough for both domestic use and export. Much of the need of grain-deficit nations is for wheat that can be used to bake bread or to blend with available wheats to improve bread quality. Wheat is also needed by some nations to supplement domestic supply for products normally baked by them. Still others import grain to produce new and diverse goods in response to consumer demand.

Economic and nutritional considerations also affect international trade in

TABLE IX
Total and Soft Wheat (and Flour) Exports by Major Exporting Nations, 1973-1976 (in 1,000 te)

	1973-1974	1974-1975	1975-1976
Australia, total[a]	7,418	8,550	8,233
Fair Average Quality or Australian Standard White[b]	5,285	6,156	5,603
Canada, total[c]	10,816	10,225	11,635
Eastern	26	279	327
United States, total[d]	29,823	27,048	30,553
Soft red winter	657	3,762	4,266
White	3,164	5,224	5,595
Argentina, total[e]	1,106	2,178	3,111
Soft	0	0	0
European Community[e,f]			
Total	5,467	7,122	7,729
Soft	5,467	7,122	7,729
Sweden, total[e]	394	1,054	738
Soft	394	1,054	738
Total wheat	55,024	56,177	61,999
Total soft	14,993	23,597	24,258
Percent soft	27.2	42.0	39.1

[a] Annual report 1975-1976. 1977. Australian Wheat Board, Melbourne. Appendix 7.
[b] *Ibid.* Appendix 6.
[c] 1977 World Wheat Statistics. 1977. International Wheat Council, London. Appendix 5b.
[d] U.S. 1973-1974 data from Wheat Situation. 1974. USDA/ERS, Washington, DC, W-229, Table 6. U.S. 1974-1975 data from Wheat Situation. 1975. USDA/ERS, Washington, DC, WS-233, Table 5. U.S. 1975-1976 data from Wheat Situation. 1976. USDA/ERS, Washington, DC, WS-237, Table 6.
[e] 1977 World Wheat Situation. 1977. International Wheat Council, London. Appendix 3.
[f] Excludes intra-EC trade.

wheat. Price differentials exist for wheats of various classes, which may not differ greatly in nutritional value. Price per calorie may be more important to some importing nations than other factors.

From the quantitative standpoint, soft wheat is less important than hard wheat in world trade. As Table IX indicates, soft wheat averaged about 36% of total wheat traded by the six major wheat-exporting groups in the period 1973-1977. The nations did not participate equally in the soft wheat trade. Argentina grows no soft wheat, and Canada relatively little. On the other hand, much of the wheat

TABLE X
Major Importing Nations of Soft Wheat, 1975-1976

Importing Nation	Exporting Nations (in 1,000 te)			
	Australia[a]	United States[b]	EC[c]	Sweden[c]
Algeria			531	44
Bangladesh	51	542	232	
China (People's Republic)	947			
China (Taiwan)	79	103		
Egypt	950	1,057	1,482	11
India	930	3,528	925	
Indonesia	486	195	47	
Iran	56	95	13	
Iraq	314			
Italy	136	236		
Japan	1,011	1,078	4	
Jordan			106	
Korea (South)		909		
Kuwait	130			
Libya			329	
Malaysia	329			
Morocco		87	874	
Netherlands		107		15
Norway			1	125
Pakistan	127	749	142	
Philippines		116	7	
Poland		340	138	287
Portugal		46	101	5
Romania		111	247	
Saudi Arabia	79		187	
Senegal			108	
Singapore	129	4	3	
Sri Lanka	128		398	
Switzerland			117	
Syria			131	
Tunisia		75	256	18
United Kingdom	103			
U.S.S.R.	1,289		3	
Yemen	258		91	
Others	450	483	1,256	233
Total	7,962[d]	9,861	7,729	738

[a] Annual Report. 1975-1976, 1977. Australian Wheat Board, Melbourne.
[b] Wheat Situation. 1977. USDA, Economic Research Service, Washington, DC. WS-241, Aug.
[c] 1977 World Wheat Statistics. 1977. International Wheat Council, London, June.
[d] Includes 680,000 and 1,160,000 te respectively, of Australian Prime Hard and Australian Hard, destinations of which were not available.

exported from Australia under the classification Fair Average Quality or, more recently, Australian Standard White could be considered as soft or semisoft. Almost all wheat exported to other nations by the EC and Sweden may be considered as soft wheat. Wheat shipments within the EC are not recorded in Table IX.

Shellenberger (51) studied the qualitative aspects of wheat exported to Europe in 1955-1956. Almost all wheat from Argentina graded Hard by U.S. marketing standards except two (of 60) that graded Soft Red; most wheat from Australia and Uruguay graded Soft; Canadian wheats were almost entirely Hard Red Spring except for one sample each of Soft Red and Soft White. Tests (including bread baking) on flours milled from export wheat samples showed that most grading Soft had low protein content and fair to very poor bread quality, except those from Argentina and Uruguay, which had higher protein content and good bread quality.

Because the U.S. wheat marketing system permits the classification of soft red winter wheat as such for inspection purposes and because practically all of the exported white wheat (originating in either the western or eastern region) is soft, the portion of exported wheat that can be classed as soft is readily ascertained. Thus, in 1975-1976, the United States exported 9.86 million te of soft wheat to 45 countries and 9.503 million te to 49 nations in 1976-1977. These quantities constituted 32.3 and 39.3% of total U.S. exports in the respective periods.

Major soft wheat-importing nations in 1975-1976 and nations supplying the wheat, are shown in Table X. The trading patterns are not the same each year; they vary with the size of the crop of some importing nations (demand), with the size of the crop of exporting nations (availability), price (soft wheat is usually less costly than hard or bread wheat), specific wheat type needs, and other factors.

All soft wheats exported by Australia and Canada are white wheats, whereas those of the EC and Sweden are red. Most of the soft red winter exports from the United States have gone to North Africa and Europe and to South and Central America. Nations in the Far East (and some in western and central Asia) have tended to prefer white wheats. Seed coat color is a quality factor in the wheat utilization for chapati or similar bread; white is preferred. If the imported wheat is milled to flour before food preparation, the inherent endosperm properties (such as protein content, physical properties of gluten, and particle size distribution in flour) may be more important than the external features of the kernel.

ACKNOWLEDGMENTS

The authors acknowledge the contributions, mostly in the form of personal communications, from the following in the preparation of this chapter: R. J. E. Agrasar, DeKalb Argentina, Buenos Aires, Argentina; R. G. Anderson, CIMMYT, Mexico City, Mexico; M. A. Aziz, FAO, UNDP, Baghdad, Iraq; R. K. Bequette, DeKalb Ag Research, Wichita, Kansas, U.S.A.; F. W. Bergmann, Centro de Experimentacaoe Pesquisa, Cruz Alta, R. S., Brazil; S. Borojevic, University of Novi Sad, Novi Sad, Yugoslavia; A. Bourdet, National Institute of Agronomic Research, Paris, France; L. W. Briggle, USDA, Beltsville, Maryland, U.S.A.; G. Campbell, Ranks (Ireland), Ltd., Limerick, Eire; R. W. Cawley, Wheat Research Institute, Christchurch, New Zealand; M. H. Chaudhary, Agricultural Research Institute, Bahawalpur, Pakistan; I. de la Roche, Agriculture Canada, Ottawa, Canada; N. Donchev, Dobrudja Agronomic Research

Institute, Tolbuhin, Bulgaria; C. V. Fulton, Agriculture Canada, Ottawa, Ontario, Canada; T. Gotoh, Tohoku National Agricultural Experiment Station, Morioka, Japan; M. S. Jalalyar, Ministry of Agriculture, Kabul, Afghanistan; V. A. Johnson, USDA, Lincoln, Nebraska, U.S.A.; C. F. Krull, DeKalb Italiana, Chrarano, Italy; H. N. Lafever, Ohio Agricultural Research and Development Center, Wooster, Ohio, U.S.A.; C. Lehmann, Akademie der Wissenshlaften, Gatersleben, GDR; J. MacKey, Royal Agricultural College of Sweden, Uppsala, Sweden; T. Mahlooji, Ministry of Agriculture, Teheran, Iran; P. C. J. Maree, Department of Agricultural Technical Services, Pretoria, Republic of South Africa; S. Nagao, Nisshin Flour Milling Co., Ltd., Tokyo, Japan; M. V. Rao, Indian Agricultural Research Institute, New Delhi, India; L. P. Reitz (retired), Sun City, Arizona, U.S.A.; G. Rubenthaler, USDA, Pullman, Washington, U.S.A.; D. R. Sampson, Agriculture Canada, Ottawa, Ontario, Canada; W. Saurer, Swiss Federal Research Station of Agronomy, Zurich, Switzerland; E. Skorda, Institute of Cereals, Thessalonika, Greece; A. J. Tulett, International Wheat Council, London, England; K. Waltl, State Institute for Plant Research and Development, Vienna, Austria; J. Wright, Home Grown Cereals Authority, London, England; C. W. Wrigley, CSIRO, North Ryde, NSW, Australia; H. Wulf, Institute of Agricultural Investigations, Santiago, Chile.

LITERATURE CITED

1. Acevedo A., J., Ramirez A., I., and Barnier B., R. 1972. Winter wheat improvement in Chile. Proc. 1st Int. Winter Wheat Conf., Ankara, Turkey, pp. 279-288.
2. Alessandroni, A., and Scalfati, M. C. 1972. Results of LWWPN in 1969 and 1970 at S. Angelo Lodigiani. Proc. 1st Int. Winter Wheat Conf., Ankara, Turkey, pp. 227-243.
3. Anonymous. 1966. The Practice of Flour Milling. (Two vol.) Nat. Joint Ind. Council for Flour Milling Industry, London. Northern Publ. Co., Ltd.
4. Anonymous. 1975. The wheat industry, 1973-74. Commonwealth Bureau of Census and Statistics, Canberra, Australia.
5. Anonymous. 1976. Turkey harvests large wheat crop. For. Agric. 14(19):15.
6. Anonymous. 1977. NABIM statistics. Inc. Nat. Assoc. Br. Irish Millers, London.
7. Becher, M. 1974. West Germany's flour milling and animal feed industries. J. Flour Anim. Feed Milling 156(6):28.
8. Bekenkamp, R. L. 1976. Field crops take lead in south Brazil's rich farmland. For. Agric. 14(23):2-3.
9. Berg, S. O. 1951. Is wheat grittiness a varietal characteristic? Cereal Chem. 24:274-283.
10. Blumenstock, D. I., and Thornthwaite, C. W. 1941. Climate and the world pattern. In: Yearbook of Agriculture, U.S. Dept. Agric., Washington, DC, pp. 98-121.
11. Bölling, H. 1978. Die Qualität der deutschen Weizenernte 1978. Mühle u. Mischfuttertechnik, 19 Oct.
12. Bommer, D., and Dambroth, M. 1974, 1976. Climate and wheat production. Agrometeorology of the wheat crop. Proc. World Meteorological Office Symposium, Braunschweig, W. Germany, 22-27 Oct. 1973. (Two vol.) Deutscher Wetterdienst, Offenbach, W. Germany.
13. Borghi, B., Testoni, A., Cattaneo, M., and Corbellini, M. 1975. Effects of genetic-environment interaction of different tests for bread making quality evaluation. Proc. 2nd Int. Winter Wheat Conf., Zagreb, Yugoslavia, pp. 423-434.
14. Brannigan, J. J. 1974. Europe. New Certificate Geography Series, 2nd ed. MacDonald & Co., Ltd., London.
15. Briggle, L. W., and Reitz, L. P. 1963. Classification of *Triticum* species and of wheat varieties grown in the United States. USDA Tech. Bull. 1278.
16. Champeau, H. C. 1972. Romania and Hungary expand feed and livestock output. For. Agric. 10(6):6.
17. Chin, S., and Liu, D. 1962. Wheat cultivars in China. Agricultural Book Publ. Corp., Peking, China. (Original not seen; translation of portions supplied by T. Gotoh).
18. Cutler, G. H., and Brinson, G. A. 1935. The granulation of whole wheat meal and a method of expressing it numerically. Cereal Chem. 12:120-129.
19. Demirlicakmak, A. 1972. Problems and opportunities for increased winter wheat production in Turkey. Proc. 1st Int.

Winter Wheat Conf., Ankara, Turkey, p. 3.
20. Demirlicakmak, A., Ergin, Y., and Yakar, K. 1972. Results of third IWWPN grown at Ankara, Turkey. Proc. 1st Int. Winter Wheat Conf., Ankara, Turkey, p. 259.
21. Economic Research Service. 1975. Wheat situation. USDA ERS, Publ. WS-233.
22. Farrand, E. A. 1972. Potential milling and baking value of home grown wheats. J. Nat. Inst. Agric. Bot. 12:464-470.
23. Food and Agriculture Organization of the United Nations. 1976. Production yearbook, vol. 29. FAO, Rome, Italy, pp. 60-61.
24. Freeman, G. F. 1918. Producing breadmaking wheats for warm climates. J. Hered. 9:211-226.
25. Freeman, T. R., Jr. 1969. New wheats power bigger Pakistan harvests. For. Agric. 7(14):12-13.
26. Gotoh, T. 1975. Variation in the vernalization requirements of winter wheat cultivars. Proc. 2nd. Int. Winter Wheat Conf., Zagreb, Yugoslavia, pp. 292-297.
27. Hafiz, A. 1972. Winter wheat improvement problems in the Near East and North Africa. Proc. 1st Int. Winter Wheat Conf., Ankara, Turkey, pp. 22-24.
28. Hébert, J. 1975. Wheat Agronomy in France. In: Bread. A. Spicer, ed. Appl. Sci. Publishers, Ltd., London.
29. Heinzelman, O. H., and Highsmith, R. M., Jr. 1967. World Regional Geography, 3rd ed., Prentice Hall, Inc., Englewood Cliffs, NJ.
30. Hoseney, R. C., and Seib. P. A. 1973. Structural differences in hard and soft wheat. Bakers Dig. 47:26-28, 56.
31. Hron, R. 1972. Winter wheat improvement problems in eastern Europe. Proc. 1st Int. Winter Wheat Conf., Ankara, Turkey, pp. 8-12.
32. Kurtzig, M. E. 1975. Iran's imports of U.S. products soar in fiscal year 75. For. Agric. 13(14):2.
33. Kurtzig, M. E., and Parker, J. B., Jr. 1974. Middle East will buy more wheat as hopes for big harvests dim. For. Agric. 12(38):5.
34. Larson, M. R. 1976. China's crop outlook dims. For. Agric. 14(27):2-4.
35. Lukyanenko, P. P. 1972. Some results of winter wheat breeding in the U.S.S.R. and the trends of its progress. Proc. 1st Int. Winter Wheat Conf., Ankara, Turkey, pp. 13-21.
36. Lupton, F. G. H., and Pushman, F. M. 1975. The crop, the environment and the genotype. In: Bread. A. Spicer, ed. Appl. Sci. Publishers, Ltd., London.
37. Mehmood, A. 1975. Pakistani wheat output up; sufficiency not imminent. For. Agric. 13(51):10.
38. Mesdag, J. 1975. Quality improvement of European wheats. Proc. 2nd Int. Winter Wheat Conf., Zagreb, Yugoslavia, pp. 341-350.
39. Moss, H. J., Goodchild, N. A., and Edwards, C. S. 1973. Selection of cultivars for soft wheat quality. I. Tests for softness in flour. Aust. J. Exp. Agric. Anim. Husb. 13:292.
40. Murray, K. L. 1976. North Africa anticipating near record wheat crop. For. Agric. 14(27):5, 12.
41. National Academy of Sciences. 1975. Plant studies in the People's Republic of China. National Academy of Sciences, Washington, DC, pp. 54-64.
42. Ramírez A., I., and Hewstone M., C. 1974. La produccion de trigo en Chile. Proc. Latin Am. Wheat Conf., Porto Alegre, Brazil, pp. 21-27.
43. Reitz, L. P., and Hamlin, W. G. 1978. Distribution of the varieties and classes of wheat in the United States in 1974. USDA ARS Stat. Bull. 604.
44. Rocha, M. A. B. 1974. Relatório sôbre a producão de trigo no Brasil e seus problemas. Proc. Latin Am. Wheat Conf., Porto Alegre, Brazil, pp. 15-20.
45. Rocha, M. A. B., and Schlehuber, A. M. 1972. Some problems of wheat production in Brazil and the role the International White Wheat Performance Nursery may play in their solution. Proc. 1st Int. Winter Wheat Conf., Ankara, Turkey, pp. 272-278.
46. Rudbeck, J. P. 1976. Argentina's wheat exports may be highest in decade. For. Agric. 14(2):6.
47. Schlehuber, A. M., and Tucker, B. B. 1967. Ch. 4, Culture of wheat. In: Wheat and Wheat Improvement. K. S. Quisenberry and L. P. Reitz, eds. Monograph 13, Am. Soc. Agron., Madison, WI.
48. Schmidt, J. W. 1977. Wheat—Its role in America's heritage. In: Agronomists and Food: Contributions and Challenges. M. D. Thorne, ed., ASA Spec. Publ. 30, pp. 45-52.
49. Seeborg, E. F. 1971. U.S. wheats can bolster Iran's short grain supply. For. Agric. 9(47):8.
50. Seeborg, E. F. 1974. Turkey's wheat outlook clouded by weather. For. Agric. 12(25):2.

51. Shellenberger, J. A. 1958. Survey of the quality of European wheat imports. Kans. Agric. Exp. Stn. Bull. 396.
52. Simmonds, D. H. 1974. Chemical basis of hardness and vitreosity in the wheat kernel. Bakers Dig. 48:16-18,20,22,24, 26-29,63.
53. Spaldon, E. 1972. The IWWPN in Czechoslovakia. Proc. 1st Int. Winter Wheat Conf., Ankara, Turkey, pp. 246-251.
54. Steiner, H. H. 1975. Maghreb grain crops decline. For. Agric. 13(32):2.
55. Stewart, B. A. 1973. French wheat. FMBRA Bull. 5, Oct.
56. Symes, K. J. 1961. Classification of Australian wheats based on the granularity of their whole meal. Aust. J. Exp. Agric. Anim. Husb. 1:18-23.
57. U.S. Department of Agriculture. 1977. Agricultural Statistics. U.S. Government Printing Office, Washington, DC.
58. Wolski, T. 1975. The actual state and main problems of winter wheat breeding in Dankow and Rogaczewo, Poland. Proc. 2nd Int. Winter Wheat Conf., Zagreb, Yugoslavia, pp. 114-124.
59. Wright, H. E., Jr. 1976. The environmental setting for plant domestication in the Near East. Science 194:385-389.
60. Wrigley, C. W., and Shepherd, K. W. 1974. Identification of Australian wheat cultivars by laboratory procedure: Examination of pure samples of grain. Aust. J. Exp. Agric. Anim. Husb. 14:796-804.

CHAPTER 2

SOFT WHEAT BREEDING IN THE UNITED STATES

F. L. PATTERSON
Department of Agronomy
Purdue University, West Lafayette, Indiana, U.S.A.

R. E. ALLAN
United States Department of Agriculture
Science and Education Administration
Wheat Breeding and Production Research
Pullman, Washington, U.S.A.

PART ONE: CULTURE AND BREEDING IN THE EASTERN UNITED STATES

I. INTRODUCTION

A. Adaptation

Soft wheats (*Triticum aestivum* L. em Thell.) historically have been grown in the areas east of a line of about 762 mm annual precipitation. This line extends in a northern and northeastern direction from eastern north-central Texas through the eastern edge of Oklahoma and the northwestern corner of Missouri, bisects Iowa, crosses through extreme southeastern Minnesota, and bisects Wisconsin (127,135). Generally, wheat ranks second or third in crop acreage on individual farms. Soft wheats are of either red or white grain color. The white wheats have been grown principally in the states of New York, Michigan, and adjacent areas and in Ontario, Canada. The white wheat area is fairly homogeneous, with most cultivars generally adapted to the entire area. The eastern soft red winter wheat region may be divided for convenience of discussion into four production areas: a) the extended Ohio river valley, b) northern soft wheat areas of Illinois and Wisconsin, c) northeastern United States and d) southern and Atlantic coastal areas. The largest production is in states in the Ohio River Valley. Cultivars adapted here have about a 50 to 65 day vernalization requirement (winter period) and have good winter-hardiness to both low temperatures and to frost heaving. Older cultivars grown here had long day length requirements. More recently developed cultivars, such as Knox, Monon, or Arthur, are more day neutral and are adapted to a wider area (19). In the northern area, e.g., northern Illinois and southern Wisconsin, cultivars require greater resistance to low temperatures and to smothering under ice. Adapted cultivars also have a longer vernalization requirement. In the northeastern United States, cultivars require high resistance to low temperature, resistance to frost heaving and to smothering under deep snow cover. In the southern and Atlantic coastal area, cultivars have shorter vernalization requirements, only moderate resistance to low temperatures, and shorter day photoperiod response.

Soft wheats are adapted to a wide range of climates and soils. In the United States they are grown from eastern Texas to southern Wisconsin and eastward to the Atlantic coast. In general, the crop is not adapted to poor sandy or poorly drained soils. Wheats grown on the more acid soils must also be tolerant to aluminum. It is interesting to note that cultivars developed at the Purdue University Agricultural Experiment Station are not tolerant to aluminum, whereas those developed at the Ohio Agricultural Research and Development Center, on the high aluminum soils at Wooster, Ohio, are all tolerant. The crop is grown without irrigation.

B. Culture

Traditionally, soft wheats were grown in a rotation with clover or other legumes seeded in the growing wheat in early spring. Wheat was grown only one year and followed oats, corn, soybeans or other crops in the rotation. Beginning in about the 1940s, legumes were frequently omitted from rotations and increased levels of commercial fertilizers, including nitrogen, phosphorus, and potassium were applied. In the Ohio river valley, the acreage of soybeans greatly increased beginning in the 1940s. Wheat was generally seeded following soybeans in the fall, immediately after harvest of the soybeans. The seedbed was generally prepared by discing (or plowing). Fertilizer was applied with the seed drill or in rotation. Additional nitrogen was applied as a top dressing. The seedbed following soybeans is usually excellent and a good soybean crop may fix an estimated 45 kg/ha of nitrogen for use by the wheat crop. Nearly all wheat grown receives commercial fertilizer. In 1973, the average for the states of Indiana, Illinois, Ohio, Michigan, and Missouri was 62 kg N, 65 kg P_2O_5, and 65 kg K_2O per hectare (118). Current average fertilizer rates are similar. Rates 50% higher than these are common on the more productive farms. Nitrogen is generally applied both at seeding and as a top dressing in late winter or early spring. Manganese may also be required on some soils. Soybeans are excellent indicators for manganese deficiency. Manganese should be applied for wheat on soils where soybeans show deficiency symptoms.

Growing wheat followed immediately by soybeans in the summer (double cropping) has been popular since the 1960s after good cultivars of soybeans were developed for the southern Ohio valley and for the southern and coastal areas. Early maturing cultivars of wheat are used. Several methods of planting the soybeans have been devised. Some of the more successful farmers plant directly in the wheat stubble to prevent soil moisture loss. Chemical control of weeds is usually necessary. Many systems of planting soybeans in growing wheat sometime in the two-month period before harvest are also being tried.

Wheat is normally seeded at about 0.7 to 1.4 q/ha. The higher rate is used when seeding is delayed. Seeding rates lower than 0.7 q/ha generally result in lower yields. In much of the area, a normal seeding rate is essential for good winter survival.

The preferred seeding time is soon after the Hessian fly-free date which was established to avoid losses from the fall brood of Hessian fly; it corresponds to the expected beginning of light frosts as the plants emerge in the fall. If seeding is delayed three or four weeks in the northern half of the eastern soft wheat area, severe winter killing of the wheat can occur. Several diseases have been observed recently to be more severe on early seeded wheat. These include the barley yellow dwarf virus disease with aphids as the vectors for the virus; the spindle streak mosaic soilborne virus disease; take-all (*Ophiolobus graminis* Sacc.) root rot; and the Septoria leaf blotch (*Septoria tritici* Rob. em Desm). Even with cultivars resistant to Hessian fly, seeding after the fly-free date is generally more profitable. A delay of one week in seeding date at Lafayette, Indiana, results in a two to two and one-half day delay in flowering in the spring. When double cropping is planned, timely fall seeding is essential to early harvest of the wheat.

With the recent increased use of herbicides in the production of corn, soybeans, and other crops, there is a growing trend toward more limited rotations among crops. The shorter rotations involving wheat have greatly increased foot and root rot disease problems.

Drilling is the preferred method of seeding. Recently the trend has been to broadcast fertilizer from a bulk truck, and in some cases, wheat is seeded mixed in the fertilizer. This is followed by shallow discing. Seeding rates 50% higher than normal are advised to get a good stand of wheat and seeding depth is variable. Plants with shallow crowns are more subject to frost heaving.

In the southern half of the eastern soft wheat region, wheat is often grazed during the winter to provide both pasture and grain. Removal of excess vegetative growth by grazing is desirable for producing a good grain crop. Additional fertilizer is necessary to provide for the two crops.

C. Harvest and Marketing

The wheat crop is harvested by direct combining, dried by forced, heated (or unheated) air if necessary, and generally marketed at or near harvesttime. The harvest is frequently delayed by wet weather after the wheat is ripe and germination may be initiated in the head. White wheats lack a postripening dormancy that is present in red wheats for about three weeks or more. There were no cases of dormancy in white wheats (109). Catechin-tannin could be related to the formation of red seed color and to germination inhibition through a reversible chemical reaction (108). There are differences in degree of dormancy among red wheat cultivars. In three weeks there was a loss of 67% of the catechin-tannin present in red wheats at harvest (109). Sometimes the harvest may be delayed so long that red wheats are no longer dormant.

Another problem of delayed harvest is weed growth. Wheat competes quite well with weeds until leaves senesce and no longer shade the weeds. Then weeds grow rapidly and may overtop the shorter cultivars and increase harvest losses. This may be a major problem with semidwarf cultivars in the eastern United States.

Wild garlic (*Allium oleraceum* L.) is by far the worst weed for reduction of wheat quality and for market discounts in the southern half of the eastern soft wheat region. Planting wheat on relatively garlic-free land helps avoid the problem. A well-fertilized, vigorous crop of wheat will compete better against wild garlic and reduce the latter's aerial bulblet formation. The herbicides 2,4-D or dicamba may be applied for garlic control in wheat in the spring when wheat is 4 to 8 in. tall and fully tillered but before jointing (92). Rates should be those approved and labeled for use on wheat.

Garlic produces aerial bulblets about the same height as wheat. The use of earlier maturing wheats has not solved the problem. If garlic occurs in ripe wheat, the combine should be adjusted to remove as many of the bulblets as possible.

Wheat is graded light garlicky (two to six green bulblets/1,000 g) or garlicky (more than six green bulblets/1,000 g). In grading, three dry bulblets equal one green bulblet (53). Price discounts for garlicky wheat are not uniform and are determined by each buyer (53). An average of about 15% of the soft red winter crop grades garlicky each year. The miller is aware of the garlicky grade and may or may not elect to tolerate a small amount of garlic. In most years discounts have been in the range of 6 to 12 cents per bushel (53). Wheat with high amounts of garlic is often used as a feed grain.

In 1974, garlic counts were very high in wheat with 30% of the soft red winter wheat crop graded as garlicky (59) and discounts ranged as high as $2.00/bu (53). The 1974 spring was cold and wet, nitrification was slow, and many fields were too muddy for ground application of nitrogen topdressing or chemical weed control; thus garlic competed greatly with wheat.

In flour milling garlic smashes and forms a sticky coating on the break rolls, reducing the efficiency of milling and increasing flour ash (141). Garlic also imparts an odor to flour. It is necessary to stop the mill frequently and scrape and wash off the break rolls (141,155) or to use a different washing technique while the mill is running (155).

Garlic may be removed from the grain before milling, but success is not complete. One method is to dry the grain extensively and after several months remove the dry garlic by a combination of aspiration and cleaning. Dry garlic does not gum up the break rolls, but some flour odor problems still occur (141).

II. SOFT WHEAT IMPROVEMENT BEFORE 1925

A. Introduction and Evaluation

Wheat is not native to the Americas. Columbus brought wheat from Spain on his 1494 voyage to the West Indies (137). Early colonists brought wheat seed from Europe and settlers carried these cultivars westward. Congressmen distributed small packets of wheat obtained by ambassadors and counsels in other lands. After about 1875, agricultural societies, State Agricultural Experiment Stations (AES), and the U.S. Department of Agriculture (USDA) developed programs for the introduction of new wheat cultivars into the eastern United States.

One of the earliest activities in the state experiment stations were performance comparisons of wheat cultivars grown by farmers in the area. These were cultivars introduced from Europe or were plant selections made by farmers from these introductions or possibly from chance hybrids between cultivars. Latta (96) summarized 19 years of testing of wheat cultivars in Indiana, begun in 1880, the first year after the Indiana Experiment Station was established. Wiancko and Fisher (175) reported protein content of the grain of their 1905 test plots. The Michigan Amber check yielded 21.7 q/ha with a protein content of 13.08%. The researchers noted that protein quantity was a fair index of value of any lot of wheat for bread making and that cultivars retained a relative ranking in protein year to year. They also noted the low protein of the white wheats, Dawson's Golden Chaff, Gold Coin, and Tennessee Fultz.

B. Cultivar Distribution

Wheat cultivar surveys have been conducted every five years since 1919 by the USDA. The distribution of cultivars and classes of wheat in the eastern United States in 1919 and 1924 were summarized by Clark et al (46). The area of soft winter wheats was 8,880,200 ha or 30% of all wheat in 1919, a war production year, and about 4,539,500 ha or 22% of all wheat in 1924, a more normal production year. In 1924, the 10 most widely grown cultivars accounted for 76% of the eastern soft wheat area. Tables I and II present the areas occupied by the four most widely grown soft red winter and eastern soft white winter cultivars, respectively. Certain soft white cultivars such as Goldcoin and Martin were grown in the western United States as well. Cultivar percentages are based on the total area of soft white wheats in the east only.

Most of the early cultivars reported in 1919 and 1924 were developed before 1900 (Table III). The history of these cultivars is often unknown or confused. A cultivar was also often known by several names, thus adding to the difficulty of identification (45).

Differences in commercial quality were noted by millers and bakers. Red Wave was one of the first to be recognized as being of poor quality for bread (44).

C. Breeding

Wheat breeding began in the early 1900s in the USDA and in state AESs. W. M. Hays of Minnesota was practicing mass selection, pure line selection, and controlled hybridization before 1900. A wheat breeding project was initiated at Purdue University in 1904 and in many other eastern states at about this time.

Winter survival, in addition to quality, was an important objective of early wheat breeders. Cutler and Mulvey (56) listed the annual percentages of abandonment of the winter wheat crop in Indiana for 1900 to 1951. The period 1900 to 1925 is characterized by several severe winters for wheat survival.

A number of wheat cultivars that were important in more limited regions were developed in 1900 to 1925. Two cultivars, developed at Purdue University from the cross Michigan Amber × Malakof, were excellent in winterhardiness but too high in gluten strength for good commercial acceptance. They were Michikof (1920) and Purkof (1924). Three cultivars originated as selections from Fulcaster: Forward (1920), by Cornell AES; VPI 131 (1915), by Virginia AES; and Nittany (1918), by the Pennsylvania AES. Selections from Fultz included Fulhio (1920) and Trumbull (1916), from the Ohio AES. Redhart (1920) adapted to the southern coastal area, was selected from Flint or Red May by Coker's

TABLE I

Most Widely Grown Soft Red Winter Wheat Cultivars in the United States (Cultivars Grown on More Than 1.0% of Crop Area, 1919-1974, in 1,000 ha)[a]

	Importance of Cultivar													
	First			Second			Third			Fourth			Cultivars (no.)	
Year	Cultivar	Area	Percent of Crop	Cultivar	Area	Percent of Crop	Cultivar	Area	Percent of Crop	Cultivar	Area	Percent of Crop	Reported	Grown on 1%
1919	Fultz	1,943	23.8	Fulcaster	1,043	12.8	Mediterranean	1,036	12.7	Poole	993	12.2	50	15
1924	Fulcaster	735	17.3	Fultz	723	17.1	Poole	425	10.0	Mediterranean	243	5.7	58	20
1929	Fultz	586	14.2	Fulcaster	567	13.7	Trumbull	365	8.9	Red May	323	7.8	63	20
1934	Fultz	757	15.4	Fulcaster	565	11.5	Trumbull	460	9.4	Red May	396	8.1	74	21
1939	Fultz	589	12.2	Trumbull	520	10.8	Fulcaster	495	10.3	Kawvale	494	10.2	64	21
1944	Thorne	643	13.7	Fultz	491	10.4	Clarkan	365	7.8	Fulcaster	330	7.0	70	20
1949	Thorne	1,395	33.2	Clarkan	380	9.0	Fairfield	280	6.7	Redhart	245	5.8	67	22
1954	Vigo	737	24.8	Thorne	664	22.3	Seneca	384	12.9	Butler	160	5.4	68	15
1959	Knox	710	23.8	Seneca	476	22.3	Dual	281	9.4	Vermillion	277	9.3	69	12
1964	Monon	1,088	36.0	Knox	335	11.1	Vermillion	226	7.5	Redcoat	211	7.0	65	17
1969	Monon	952	37.7	Redcoat	228	9.1	Benhur	217	8.6	Blueboy	193	7.6	63	14
1974	Arthur	1,871	49.5	Arthur 71	753	19.9	Monon	174	4.6	Blueboy	138	3.7	61	12

[a] References for values: 44–46, 48–51, 129–132, 136.

TABLE II

Most Widely Grown Eastern Soft White Winter Wheat Cultivars in the United States (Cultivars Grown on More Than 1.0% of Crop Area, 1919-1974, in 1,000 ha)[a]

	Importance of Cultivar													
	First			Second			Third			Fourth			Cultivars (no.)	
Year	Cultivar	Area	Percent of Crop	Cultivar	Area	Percent of Crop	Cultivar	Area	Percent of Crop	Cultivar	Area	Percent of Crop	Reported	Grown on 1%
1919[b]	Goldcoin	183.3	74.4	Dawson	50.8	20.6	Martin	4.6	1.9	Democrat	2.1	1.0	9	4
1924[c]	Goldcoin	163.7	83.1	Dawson	24.9	12.7	Greeson	4.5	2.3	Honor	1.9	1.0	6	3
1929[d]	Goldcoin	191.3	86.3	Dawson	17.2	7.8	Honor	7.0	3.2	Greeson	4.0	1.8	9	4
1934[e]	Dawson	144.1	52.0	Goldcoin	98.4	35.5	Honor	27.9	10.1	Greeson	5.8	2.1	8	4
1939[f]	Dawson	153.6	55.2	Goldcoin	55.7	20.0	Yorkwin	49.5	17.8	Honor	15.0	5.4	5	5
1944[g]	Dawson	186.5	46.8	Yorkwin	183.2	46.0	Goldcoin	19.5	4.9	Greeson	5.4	1.4	6	4
1949[h]	Yorkwin	448.2	73.1	Cornell 595	135.9	22.2	Dawson	23.6	3.9	Goldcoin	4.7	0.8	5	3
1954	Yorkwin	296.0	62.4	Cornell 595	105.6	22.3	Genesee	70.9	15.0	Dawson	1.2	0.3	8	3
1959	Genesee	406.1	77.5	Yorkwin	73.5	14.0	Cornell 595	43.0	8.2	Dawbul	0.7	0.1	7	3
1964	Genesee	294.9	72.6	Avon	88.1	21.7	Yorkwin	16.7	4.1	Cornell 595	5.4	1.3	6	4
1969	Genesee	188.8	63.5	Avon	74.3	25.0	Yorkstar	29.8	10.0	Yorkwin	5.6	1.8	7	5
1974	Genesee	129.8	34.8	Ionia	105.5	28.3	Yorkstar	101.5	27.2	Avon	25.6	6.9	7	6

[a] References for values: 44–46, 48–51, 129–132, 136.

[b] Goldcoin, area in U.S. 383,300 ha; Martin, 15,300 ha; other cultivars grown exclusively in eastern United States.

[c] Goldcoin, area in U.S. 271,500 ha; other cultivars grown exclusively in eastern United States.

[d] Goldcoin, area in U.S. 361,100 ha; other cultivars grown exclusively in eastern United States.

[e] Goldcoin, area in U.S. 177,100 ha; other cultivars grown exclusively in eastern United States.

[f] Goldcoin, area in U.S. 108,300 ha; other cultivars grown exclusively in eastern United States.

[g] Goldcoin, area in U.S. 175,800 ha; other cultivars grown exclusively in eastern United States.

[h] Goldcoin, area in U.S. 145,600 ha; other cultivars grown exclusively in eastern United States.

Pedigreed Seed Co., South Carolina. Honor (1920) was selected from Dawson's Golden Chaff by the Cornell AES and the USDA.

By 1925, older wheat cultivars had been tested extensively. New cultivars were being developed by plant selections from the better older cultivars or by crosses of two adapted types. Cultivars had developed a reputation for good, average, or poor commercial value. Good progress was made in this period for yield, straw strength, and winter survival.

III. SOFT WHEAT IMPROVEMENT, 1925-1949

A. Breeding Objectives

This period of wheat breeding is characterized by an increase in emphasis on breeding for improved commercial quality. Parental cultivars were selected for known commercial acceptability. Hybridization, followed by pedigree selection, was the most used breeding method. Progenies were carefully studied for quality and other characteristics. Treatments were replicated and statistical analyses applied to the data.

The severe winter of 1928 resulted in a 60% abandonment of the Indiana crop and focused attention on the need for increased winter-hardiness. Breeding for disease resistance was initiated also. The soilborne mosaic virus disease discovered in 1919 in Kansas (135) was soon widespread in the eastern soft wheat region. Other diseases considered of major importance were leaf rust, stem rust, scab, loose smut, and flag smut. Purdue No. 1 wheat released in 1934 was very susceptible to the loose smut disease incited by *Ustilago tritici* (Pers.) Rostr. and to soilborne mosaic.

B. Breeding for Quality

With the introduction of hard wheat into the Great Plains and their acceptance for bread making (47), soft wheat, formerly used for all baking purposes,

TABLE III
Origin of Some of the Most Widely Grown Wheats in 1919 and 1924[a]

Cultivar	Year	Selection or Hybrid	Origin
Fulcaster	1886	Fultz/Lancaster	S. M. Schindel, MD
Fultz	1862	Selected from Lancaster	A. Fultz, PA
Poole	before 1880	Unknown	Unknown
Mediterranean	about 1840	Introduction	Mediterranean area
Trumbull	1916	Selected from Fultz	Ohio AES
Leap	1905	Selected from Mediterranean	J. S. Leap, VA
Red Wave	1906	Early Red Clawson/unnamed	A. N. Jones, NY
Harvest Queen	1897	Selected some other cultivar	E. S. Marshall, KS
Red May	1845	Unknown	Unknown
Red Rock	1940	Selected from Plymouth Rock	Michigan AES
Purplestraw	before 1822	Unknown	Unknown
Flint	before 1852	Unknown	Unknown
Goldcoin (white)	1865	Selected from Fultz	G. Clawson, NY

[a]Summarized from Clark and Bayles (43).

assumed a dual purpose role. It was often blended with hard wheat in the production of bread flour (42) and won continuing recognition for its unique suitability for the baking of confectionery products (99).

Although flour quality requirements for the two types of products were not clearly delineated, there was some suspicion of mutual exclusivity (11). Quality tests applied to soft wheat at the time were those originally developed for bread quality evaluation (13). These included the pearling (163) and granularity (55) tests to distinguish between hard and soft textured cultivars (15). Other tests included the wheat meal fermentation time test (57,58,122,138), the acid viscosity test (110,150), and the mixograph test (161,162). Although there may have been some relationship between results of these tests and bread quality, no specific association between them and soft wheat product quality had been developed. Bayfield and Shiple (13) attributed the confusion of the period regarding soft wheat quality to the multiplicity of uses for soft wheat, the paucity of research on this wheat class, and the large number of commercially grown soft wheat cultivars with its attendant diversity in milling and baking quality (12). Organized research was initiated in 1929 with the founding of the Tri-State Wheat Improvement Association (10). A few years later the Federal Soft Wheat Laboratory (later renamed the Soft Wheat Quality Laboratory) was established cooperatively in Wooster, Ohio, by the USDA and the Ohio AES.

C. Cultivar Distribution (1929-1949)

SOFT RED WINTER WHEATS

During the period, an average of more than 67 cultivars of soft red winter wheat were reported in the five-year surveys, and of these, 21 were reported as being grown on more than 1% of the soft red wheat crop area (Table I). Fultz and Fulcaster, old line cultivars, dominated in the earlier years. Records indicate that the utilization quality of these two cultivars was similar to those of Poole and Trumbull, also popular with farmers. The general quality level of all four cultivars would be regarded as subnormal today.

Direct combine harvesting became established and attention was given to shorter and stiffer straw and to resistance to grain shatter after ripening. Thorne (released in 1937) became one of the most widely grown cultivars. It had better straw strength, shorter height, and higher yield than most cultivars. Moreover, the quality level of the cultivar constituted a significant improvement over earlier cultivars.

A related cultivar, Seneca (1949), similar to Thorne but higher in test weight, replaced much of the Thorne area. Blackhawk CI 12218 (1944), developed at the Wisconsin AES, combined good winterhardiness from Minhardi with the good quality from Fultz. This provided a good soft red winter cultivar for this northern soft wheat area. Clarkan (1934), selected by Earl Clark, Kansas farmer, occupied important areas, particularly in Kansas and Missouri. Kawvale (1932), distributed by the Kansas AES, was grown mainly in Kansas and Missouri. Kawvale had poor soft wheat quality but was important as a parent for resistance to Hessian fly, loose smut, and leaf rust. Atlas 50 (1948) and Atlas 66 (1948) were important new higher protein wheats for the southeastern area. A number of other new cultivars of this period were important in more local areas.

The period 1926 to 1949 is characterized by breeding by selection from older cultivars or from crosses between two adapted types. Yield advance over older cultivars was quite modest. A large number of cultivars was grown to some extent but eight or fewer cultivars were each grown on 202,350 ha in any survey year.

EASTERN SOFT WHITE WINTER WHEATS

White wheats are produced in New York and Michigan primarily, in the eastern soft wheat area, with smaller areas in adjacent states. As recorded in the first survey, New York has produced predominantly white wheats (Table IV). Michigan, however, grew mostly soft red wheat until 1939, when the area of white wheats exceeded that of the reds for the first time. The percentage of the wheat crop seeded to white wheat steadily increased until today it is stabilized at about 75 to 80%.

Goldcoin and Dawson, two old cultivars, dominated the eastern soft white area until 1944 when Yorkwin (1935), a Cornell University release, was second in area (Table II). Much of the increase in white wheats between 1934 and 1949 can be attributed to two outstanding cultivars developed at Cornell University, Yorkwin and Cornell 595 (1942). Both were excellent in straw strength, in winter-hardiness, and in commercial quality. Their use increased rapidly. The favorable reputation that Michigan and New York soft wheats enjoyed as a source of cookie flour was well deserved. Soft white wheats were also in demand as raw material for breakfast foods because light seed coat color led to attractive products.

D. Breeding for Disease Resistance

Breeding for disease resistance was initiated at Purdue University in the early 1920s. Resistance was sometimes found among the many cultivars that were adapted to the eastern United States and that had good commercial quality. In other cases, resistance had to be transferred from unadapted or poor quality parents of common wheat, from other species, and from other genera to common wheat.

TABLE IV
Soft Winter Wheat Distribution in New York and Michigan, 1919-1974

	New York			Michigan			
	White		Red	White		Red	
Year	(%)	(1,000 ha)	(1,000 ha)	(%)	(1,000 ha)	(1,000 ha)	Ref.
1919	70	132.1	55.7	27	96.1	262.2	45
1924	90	113.8	13.2	27	86.4	232.4	46
1929	83	79.7	14.8	47	149.9	167.1	48
1934	80	84.8	19.6	53	182.1	159.5	49
1939	80	90.9	22.0	51	159.0	151.0	50
1944	92	137.6	10.6	83	258.6	133.7	51
1949	97	168.0	4.3	87	456.7	59.9	44
1954	98	134.4	3.0	82	334.7	68.9	136
1959	96	112.3	3.5	83	405.3	65.4	129
1964	97	85.1	2.6	80	329.1	80.9	130
1969	98	78.1	1.3	83	227.4	47.2	132
1974	100	88.7	0.4	76	291.9	92.5	131

Mains and Jackson, at Purdue University in 1926, identified 12 physiologic races of *Puccinia recondita* Rob. ex Desm. f. sp. *tritici,* inciting leaf rust of wheat, on seedling plants of 11 cultivars falling into eight differential groups of wheat (103). They noted that cultivars were sometimes resistant to one race and susceptible to another. They began breeding work by 1926 to combine the resistances of two cultivars in order to build up resistance to more races. They recognized also that some cultivars, such as Kanred, were resistant in the field but not in the seedling stage. From genetic studies, Mains, Leighty, and Johnson (104) concluded that resistance to leaf rust was inherited as a dominant, intermediate, or sometimes recessive character.

From these early experiments, a number of cultivars were used to breed improved soft red winter wheats. Research was continued by Caldwell and Compton at Purdue University to identify sources of resistance to leaf rust in the adult plant among cultivars and introductions. Important among these were several Fultz selections, including Wabash CI 11384 and Wea CI 11512.

Vigo, developed at Purdue University AES and released in 1946, is an example of a program planned to combine the leaf rust resistance of Wea with the quality and high yield of Trumbull. Progenies were selected for leaf rust resistance and quality. The Fultz type resistance was also utilized in the cultivar Blackhawk (1944).

Many other crosses were made to combine leaf rust resistance with good quality and good agronomic type. Most of these did not become cultivars as such but were parents for the next step in recombination of resistances to several diseases and to Hessian fly.

The significance of the procedure is that once resistances are all in good quality types, and when these improved types are recombined, segregation for good quality will occur often. In the 1925 to 1949 period, many sources of resistances to disease were assembled in excellent quality types (Table V).

TABLE V
Some Sources Used in Breeding Disease-Resistant Soft Red Winter Wheats, 1926-1949

Source	CI Number	Type
Malakof	4898	Hard red winter
Mediterranean	3332	Soft red winter
Hussar	4843	Hard red winter
Hungarian	4830	Hard red winter
Webster	3780	Hard red spring
Hope	8178	Hard red spring
Frondoso	106504[a]	Hard red spring
Fronteira	103832[a]	Hard red spring
Chinese	6223	Hard red spring
Fultz selection	11512	Soft red winter
Wabash	11384	Soft red winter
Transfer	13296	Hard red spring
Wisconsin 245	12633	Hard red spring
Kenya Farmer	12880	Hard red spring
Supresa	103833[a]	Hard red spring
Trumbull	5657	Soft red winter
Kawvale	8180	Semihard red winter

[a]Plant introduction number.

Resistances transferred from other species included resistances to stem rust, loose smut, and leaf rust in the cultivar Hope. Hope was derived by McFadden from a cross involving Yaroslav emmer (105). Resistances from *Agropyron elongatum* Host. were transferred to common wheats (Agrus and other selections) by Caldwell and Compton at Purdue University. Resistance to leaf rust was transferred by Sears from *Aegilops umbellulata* Zhuk. by irradiation into a wheat line, which was named Transfer (142). Resistance to the leaf rust, stem rust, and powdery mildew diseases were transferred from *Triticum timopheevi* Zhuk. to wheat in Wis. 245 line (and in the Timwin cultivar) by Shands (144,145). The spring wheats Frondoso, Supresa, Chinese, and Kenya Farmer were among those used for resistance to the rusts and to powdery mildew.

The good resistance to leaf rust from Frondoso is associated (presumed linked) with higher protein. Two soft wheat cultivars, Atlas 50 and Atlas 66, were developed with this resistance and released in North Carolina in 1948. These represented a "new type" of wheat quality for the southeastern United States.

Unpublished research by Caldwell and Compton, followed by research by Bever (16), indicated that a number of races of the *Ustilago tritici* incite the loose smut disease. Many cultivars are resistant to one or more races. Trumbull, Hope, and Kawvale have been most useful as parents in breeding for resistance.

Resistance to the soilborne wheat mosaic virus was found in many current cultivars of wheat. Commonly grown cultivars that proved susceptible included Purdue No. 1 (1934) with the rosette symptom and Michigan Amber with the mottle symptom.

E. Breeding for Insect Resistance

Breeding for Hessian fly (*Mayetiola destructor* Say) resistance was initiated by Purdue University and the USDA about 1937. Three sources of resistance were discovered that were effective in the eastern soft wheat region. These included: W38 selection from Illinois No. 1 spring wheat (H_3 gene) (32), PI 94587 spring durum (32) including the H_6 gene (5), and Ribeiro PI 56306-8 spring wheat (H_5 gene) (146). Kawvale was also noted to have resistance. Resistance was incorporated into winter types of good soft wheat quality, but none was available as a cultivar before 1955. Methods of testing for Hessian fly resistance in the greenhouse were developed by Cartwright and LaHue (40) and helped greatly in breeding for resistance.

IV. SOFT WHEAT IMPROVEMENT, 1950 TO PRESENT

A. Breeding for Quality

DEVELOPING QUALITY TESTS

In recent years, soft wheat has been increasingly used in the production of cookies, cakes, crackers, family trade flour, and similar products, and less for bread baking. This trend has emphasized the evaluation of breeding lines for their utility in baking pastry products and reduced the previous ambivalence about quality objectives in the development of new cultivars in the eastern United States. Moreover, it has stimulated research into wheat and flour factors contributing to, or responsible for, quality in soft wheat products.

In spite of the availability of chemical and physicochemical tests for quality (20), the baking of a product remains the ultimate quality test. The diversity of products from soft wheat, the quality factors of each of which may differ from each other and from bread quality, complicates the evaluation. However, the American Association of Cereal Chemists at the present time recognizes as official two baking tests for soft wheat quality. These are the sugar-snap cookie test (8) and the high-ratio white layer cake test (8).

It is thus only natural that these two tests, or modifications thereof, be incorporated insofar as possible as quality tests for soft wheat breeding lines. Modeled after Alexander's cookie test formula proposal (2) and following Hanson's adoption of the test for flour quality evaluation (75), a laboratory "micro" test for cookie quality was developed by Finney et al (64) for evaluating the baking potential of soft wheat breeding lines. They also investigated the effects of variable quantities of certain ingredients on the test (65).

Studies of factors associated with cookie quality as expressed by spread resulted in the development of an alkaline water retention capacity (AWRC) test (180). Essentially it is a measure of the quantity of sodium bicarbonate solution held by flour against centrifugal force. In a cookie dough the very limited quantity of water is partitioned mostly between sugar and flour. A flour with greater AWRC value holds relatively more water, making less available to dissolve sugar to form syrup. A dough with less syrup is not as fluid during baking, hence resists the expansive effect of leavening, and sets earlier and at lower temperature during the bake, resulting in cookies of smaller spread (181). The AWRC test has been reduced in scale and applied to sifted meals (185).

Cakes comprise another large end-use of soft wheat flour. The cake quality of a breeding line is evaluated first by making a patent grade milling. The flour's particle size is then reduced by passage through a pin mill, and the flour is treated with chlorine gas. These postmilling treatments parallel those applied to the production of commercial cake flour. A formula sometimes used to evaluate the cake potential of breeding lines is that of Kissell (89). A distinction of this test procedure is that no egg white or milk is included in this fairly high-ratio (130%) white layer cake, thus forcing the flour to make the major contribution to cake structure. The relative simplicity of the formulation makes possible the study of ingredient contribution to cake volume and internal structure (90). A finding of such a study was that the formula as used represented very nearly the best balance of ingredients.

More recently, a study showed that an inverse relationship exists between flour particle size expressed as mass median diameter (MMD) and its cake volume potential (184). Related work has shown that chlorination is a random process and that the reaction between chlorine and flour is surface-dependent (179). Because flour MMD, a surface parameter, is inversely correlated with the results of the wheat particle size index test (182,184), the results of the latter test or a modification of it may be used as an index of cake potential.

Kernel texture is important in soft wheat quality also from the standpoint of its relation to milling. Patent and break flour yields are functions of kernel fracturing properties. These milling parameters, as well as total flour yield, are important aspects of soft wheat quality (20).

EVALUATING BREEDING LINES

The association between results of the cookie and cake tests on one hand and physicochemical tests on the other have been used in "early generation screening evaluations" whereby quality may be appraised on as little as 20 g of grain (185). Three tests are done—protein content, meal AWRC, and kernel texture. Wheat (and therefore flour) protein content of soft wheat is considered important because of the present belief in industry that pastry products made from lower protein content flours are more tender than those from higher protein content flours.

The same tests have been used to determine inheritance of grain quality from single crosses. In a seven-parent diallel cross including soft red, soft white, and hard red winter cultivars, quality of the 21 hybrids was generally near the average of the two parents (72). Hard × soft hybrids were not suitable in soft wheat quality. Many other tests of hybrid wheats have supported these conclusions.

The three quality factors also show high heritability from generation to generation over years in other studies (22). Good progress in selection for improved quality is expected from data of the microtests in early generations.

Breeding lines in the advanced nursery stage are available in larger quantities and are tested for microtest weight, wheat moisture, and protein, pearling index (163) and particle size index (182). An appropriate quantity of conditioned wheat is milled by a tandem Quadrumat-Bühler milling procedure to obtain straight-grade flour, which is analyzed for moisture, protein, and ash contents. A viscosity test is applied to an acidulated flour suspension, a micro-AWRC test is run, and cookies are also baked. Although flour is not available for a cake test, the particle size index value may serve as an index of cake potential.

Entries in still later stages of development, such as drill plot material or those in uniform nursery trials, and available in still larger quantities, are evaluated more rigorously than those above. Following wheat tests, each sublot is milled on an Allis Chalmers laboratory mill to obtain a straight-grade flour, and where conditions permit, a cake patent is milled as well. The latter is pin milled, treated with chlorine, and baked into cakes. Flour tests applied to advanced nursery entries are also run on these flour samples. Because more sample is available, all tests, especially baking tests, can be run in duplicate.

In all of these evaluations, performance data of test lines are compared with those of standard or released cultivars that have been grown under conditions identical to those of lines being tested. Because inherited quality traits can be modified by environment, comparisons of this type tend to lessen the intrusion of factors extraneous to evaluation of heritable traits. One would expect the quality potential of a cultivar to be maintained over the years since wheat is self-pollinating.

Practically all soft red and eastern soft white cultivars released since 1945 had been previously evaluated for quality. The release of Knox (1953), Vermillion (1955) (36), and Dual (1955) (34) provided wheat types with softer kernel texture than that of older cultivars such as Trumbull. These releases had high break flour yields and excellent soft wheat quality for cakes and cookies. Knox remains a standard of excellence in quality.

Air classification has been used as a postmilling process to fractionate flour for special uses (151). Recent research has been done at Purdue University with a high protein selection of soft wheat (named Key, formerly Purdue 4930A6-28-2-1). The hypothesis was that a high protein fraction of the flour could be removed for independent sale and that the remaining flour would be suitable for soft wheat products. Key derives its high protein character from Frondoso, as does Atlas 66. Both have been tested internationally and maintain high protein levels in diverse environments (158). Lysine level of the protein is not improved (158). Currently, research is in progress at Purdue to determine the value of Key for breakfast food products.

A wheat market quality factor is test weight. Low test weight in the past was associated with low flour yield because of the presence of shriveled grain. However, normal grains of certain cultivars have inherently low test weight and yet give normal flour yields (9,183). Blueboy and Yorkstar, both released in 1966, represent cultivars with low test weight but with good milling yield (183).

One deleterious association was found, in unpublished research, of the leaf rust resistance from the parent Agatha with yellow flour in soft red winter wheats. Agatha derives resistance from Agrus, and Agrus from *Agropyron elongatum*, a wheat relative.

In hybrid wheat, if fertility is not completely restored in all florets, one may expect increased occurrence of the scab and ergot diseases that effect grain quality.

MAINTAINING HIGH-QUALITY SOFT WHEATS

In commercial practice soft wheat flours are milled to meet bakery specifications under the most economic conditions. This can often mean the use of certain selected mill streams for one application and the remaining streams for others. This procedure, called split milling (113), is commonly used where cake flour is one of the mill products. In the laboratory evaluation of soft wheat breeding lines, it is not possible to produce special flours for all the different products. The practice has therefore been to prepare flours for only two—sugar-snap cookies and high-ratio white layer cake. The former is baked from a straight-grade untreated flour; the latter requires a highly refined flour, followed by postmilling treatments. If a test line bakes both products with the same quality as those baked from a standard or control cultivar, it is presumed to have the same soft wheat quality potential.

In spite of attempts to follow the above principle, it is not always possible to control the quality of cultivar releases within narrow limits. Variations in quality are usually less than the calculated least significant difference values for the various tests. It is possible, however, that with each new release, the quality level could be slightly lower than that of its predecessor. If enough of such quality lowering took place, the quality level could eventually be signficantly lower than in earlier releases (quality erosion).

The quality level of current soft wheat cultivars is assumed to be satisfactory. The principal concern in the release of new cultivars is that it not be significantly different. A major change in quality of new release could alter the general soft wheat quality level, especially if it were grown extensively. In commerce, uniformity in wheat and flour quality on a continuing basis is perhaps equal in importance to actual quality level. The intent of the present laboratory quality evaluation program is therefore to ensure that drastic quality changes do not occur with each new cultivar release and that quality is maintained at its present high level.

B. Cultivar Distribution (1954 to Present)

The genetic diversity in a wheat production area has great bearing on crop vulnerability to disease, insects, or unusual weather (54). Since 1954 unique and diverse new cultivars have been developed; farmers have tended to choose a few for their major plantings. Nevertheless, seed of a large number of cultivars is available for the next crop if the leading cultivars should become vulnerable.

Following the extensive cultivation of Thorne, Vigo became the leading cultivar in 1954 (Table I), accounting for 24.8% of the soft red wheat area. The high protein cultivars Atlas 50 and Atlas 66 reached their maximum with 100,800 ha in 1954, then decreased to 13,000 ha in 1969.

Knox was grown on 710,000 ha in 1959 to become the leading cultivar. Monon, with 36.0 and 37.7% of the soft red wheat acreages in 1964 and 1969, respectively, was then the leading cultivar. However, its acreage decreased sharply by 1974, not because of any inadequacies of the cultivar but because of the great adaptability and acceptance of Arthur (1968), which was grown on 1,871,000 ha and accounted for almost half (49.5%) of all soft red wheat plantings in the 1974 survey (131).

UNIQUE NEW CULTIVARS

In this period, a great deal more diverse germ plasm was used in cultivar breeding. This came about by breeding for resistance to several diseases and insects or by using more than one source of resistance to a single disease. Some unique cultivars were developed.

The first example is Knox (1953) and Vermillion (1955), developed from the same series of crosses (36). They were derived from seven original parents in a series of six crosses; the first in 1927 and the final one in 1941. Chinese spring wheat was the source of mature plant resistance to leaf rust and a source of earliness. Michigan Amber was used for adaptation and winterhardiness. Purplestraw was chosen for earliness, and Minhardi for winterhardiness. Wabash and Wea were used for resistance to leaf rust, adaptation, and good quality. Trumbull provided resistance to loose smut, good grain quality, and adaptation. These were the start of early maturing wheats for the Ohio river valley. Knox provided a more day neutral heading response, shorter vernalization requirement, and yet good winterhardiness, a rare combination in cultivars of the world (19). Knox provided about a 25% advance in yield, a new level of excellence in quality, the germ plasm for early maturing wheats, and resistance to leaf rust in the commercial crop for about 12 years until races of the inciting organism changed. Perhaps of greater long-range value were the slow rusting and slow mildew development characters of Knox, discovered more recently (115,133,147).

Another example is Dual (1955), which was selected following a series of seven crosses among eight original parents (34). Resistances to leaf rust from four parents were combined with resistance to Hessian fly from W38 in a high yielding type with good soft wheat quality.

Redcoat (1960) is a unique cultivar for excellence of straw strength, disease resistance, and tolerance to high rates of nitrogen application. Its adaptation was in the northern Ohio River Valley and Pennsylvania. Its characters were combined from nine parents in a series of nine crosses from 1929 to 1948 with the final selection in 1951 (33).

Monon (35) was developed by combining the earliness, adaptation, and disease resistance of Knox with resistance to Hessian fly of the W38 (H_3) type. Monon also has excellent winterhardiness and became one of the most widely grown cultivars (Table I). Unfortunately, time proved that Monon did not inherit the slow mildewing and slow rusting characters of Knox and it is now extremely susceptible to powdery mildew and to leaf rust in Indiana.

Blueboy (1966) is a unique semidwarf wheat developed at the North Carolina AES using Brevor-Norin 10 as the source of dwarfing (111). It has excellent standing ability and a very high yield potential under optimal conditions. Test weight is somewhat lower, but milling yields of plump grain are excellent.

Other semidwarf cultivars are McNair 4823 (1971) and Timwin (1967). McNair 4823, developed by the South Carolina AES and the McNair Seed Co., is a short cultivar with stiff straw and excellent yield potential under optimal conditions (107). Timwin (145) has semidwarfness derived from Brevor-Norin 10 and disease resistance from *Triticum timopheevi*. Timwin is adapted to the northern soft wheat area but is less winterhardy than Blackhawk (1944), Racine (1959), or Kenosha (1968), all adapted to the same area.

Arthur (1968) is another cultivar with resistance to several diseases and a breakthrough in yielding ability. Arthur (120) was derived from eight original parents in a series of crosses developed over a 35-year period. The final cross was between two lines, neither of which was superior in yield to Arthur.

Genesee (1950) is a superior white wheat developed at Cornell University (80). It is excellent in yield and straw strength, and by 1959 was the predominant cultivar in the eastern soft white wheat area. Yorkstar (1968) is a shorter white cultivar with shortness derived from Brevor-Norin 10 (86). It was grown on large areas of the eastern white winter wheat region in 1969.

BREEDING METHODS

Two cultivars of soft red winter wheat were derived by irradiation of lines selected from crosses. They are Lewis (1965) and Statler (1965), both developed at the Missouri AES (76,77).

Several cultivars were developed by backcrossing to add disease or insect resistance to proven genotypes. These include Knox 62 (1962), derived from Knox; Riley 67 (1967), from Riley; Blueboy II (1971), from Blueboy; Arthur 71 (1971); Abe (1972), and Beau (1976), from Arthur; Oasis (1973) and Downy (1976), from Arthur 71; and Ionia (1969), from Genesee.

Theoretical considerations by Hanson (73,74) indicated that normal breeding procedures may not be adequate to break up wheat chromosomes to achieve maximual recombination of genetic factors from the parents. He suggested using four or more parents and maintaining cross pollination of the resulting plants for three or more generations before selfing to homozygosity. In view of possible serious linkage restrictions to recombination in wheat, an experiment was designed to determine if detrimental linkages for yield and quality were associated with factors for disease and insect resistances in current parental materials. Five linkage blocks were examined in Knox background, singly and in combinations of two or three (121). There were only very small linkage block effects for yield. Research is in progress to determine influences on milling and baking qualities.

A breeding method is under way at Purdue University to maintain cross-pollination for several generations after crossing several parents. In unpublished research of F. L. Patterson, eight parents were crossed in all combinations of two. These F_1s were grown in the field in isolation and several hundred spikes were scissor-emasculated and exposed to wind-blown pollen of the additional F_1s of the 28 crosses. If necessary for an adequate volume of pollen, the F_2 generation was sometimes used as pollinator. After three or four generations of forced cross-pollination, plants were selfed several generations and plant selection was begun. Selected lines are now in yield trials and there is evidence for transgressive segregation for several traits.

C. Hybrid Wheat

Commercial hybrid wheat is being produced in the hard red winter wheat belt of the United States and is expected to be available soon for the eastern soft wheat region. The principles and breeding techniques will be discussed only briefly with illustrative literature citations.

The mechanism in hybrid wheat is similar to that for hybrid sorghum. The female parent has a cytoplasmic type of sterility control (termed A-line). The male parent (R-line) for the hybrid must have genetic restorer genes that interact with the sterile cytoplasm to produce fertile pollen and selfed seed set in the commercial single cross hybrid. The sterile female parent is produced by crossing to its similar male counterpart except for fertile cytoplasm (termed B-line) and the resultant normal pollen formation.

Three lines are required to produce hybrid wheat. The A-line is used as the male-sterile female parent and is pollinated by the R-line. The seed from this single cross is sown for the commercial production. To maintain seed of the A-line it is used as the female and crossed with the B-line. Seed from this cross is sown as the A-line parent. The B- and R-lines are self-fertile and are multiplied alone.

The cytoplasm of *Triticum timopheevi*, a wheat relative, is currently used as the source of male-sterile cytoplasm for hybrid wheat. Wilson and Ross (178) developed male-sterile Bison wheat now used by most breeders in developing A-lines. Using this as female and a selected cultivar as male and backcrossing several generations, an A-line of the selected cultivar is obtained. The original cultivar is the B-line. In general, there have been three types of A-lines from this procedure. There are A-lines hard to sterilize but easy to restore, as with the cultivar Reed. A second type, as Timwin, is easy to sterilize and easy to restore. The third type, as Arthur, is easy to sterilize and difficult to restore. The first R-line was developed by J. A. Wilson by transferring two genetic factors from *T. timopheevi* to the cultivar Marquis (102). Generally, at least one additional genetic factor, or more, is necessary for a satisfactory R-line. For convenience in selecting the restorer factors, the sterile cytoplasm is generally used in the R-line (87). Other sources of sterile cytoplasm and restorer systems have been proposed but none are in use commercially. A major limiting factor in hybrid wheat is few good R-lines.

Wilson has discussed some of the problems of hybrid wheat breeding and production (177). Wheat produces a relatively small amount of pollen, which is airborne over relatively short distances (17,18). Cultivars differ in pollen production and dispersal. A hybrid production field ratio of two or three farm drill widths of A-line to one drill width of R-line might be expected to produce 50 to 70% seed set. Most pollen is produced during a four to five day period, so the A-line and R-line must be matched in maturity.

According to studies of the amount of hybrid vigor in wheat, increases of 15 to 20% above the average of the two parents occur quite frequently in soft red winter wheats (18,24,66,72). If both parents of the hybrid are high yielding, the hybrid frequently exceeds the better parent. In breeding hybrid wheats, the development of high yielding cultivars for use as parents still needs strong emphasis.

In much of the eastern soft wheat region, relatively large amounts of seed are required per hectare (0.7 to 1.4 q) for good winter survival. Hybrids seeded at lower rates are generally no more productive than cultivars seeded at normal rates. The extra cost of hybrid seed must be recovered through extra yield. If the seed industry is efficient and the price of commercial wheat is high enough, hybrid wheat will be grown on major areas in the eastern soft wheat region.

D. Breeding for Disease Resistance

The losses from diseases of wheat were greatly reduced in the 1950-1976 period by the breeding and use of resistant cultivars. Leaf rust, considered the most damaging disease of wheat, seldom caused serious losses because of the use of resistant cultivars for the majority of the crop. Losses to stem rust were minor because of the use of resistant cultivars or early maturing cultivars that escaped the disease. Cultivars resistant to powdery mildew were developed and grown on a considerable area, reducing losses. Resistance to the Septoria leaf blotch was available for the first time in Oasis (1973).

Programs of breeding for resistance to other diseases were initiated. For the most part, resistance was based on a specific genetic factor or a combination of two or more factors. Protection from these genes is usually of limited duration because of the evolution of new races of the disease-inciting organism. Nevertheless, good protection of the commercial crop was achieved by using several sources of resistance and changing resistant sources when necessary. An important development in this period was the concept of general resistance (30,31,37) that is long lasting and not subject to race change in the disease-inciting organism. General resistance was recognized in certain cultivars, and programs for breeding for general resistance were initiated.

Breeding for resistance is the most practical method for the control of wheat diseases. This food crop is produced almost entirely without the use of pesticides.

LEAF RUST

This rust has been one of the most serious diseases of wheat in the eastern United States. A number of sources of resistance have been used singly or in combination to provide protection. The use of resistance before 1955 was reviewed by Caldwell et al (38). Mature plant resistance of Fultz selection CI

11512 (Wea) was utilized in Vigo (1946), Blackhawk (1944), and Racine (1956) and in combination with resistances of Hope and Hussar in Dual (1955). Resistances from Frondoso or Fronteria have been used in Atlas 50 (1948), Atlas 66 (1948), Coastal (1949), Coker 47-27 (1950), Anderson (1951), Taylor (1953), and Key (1976). Resistance from Chinese has been used in Knox (1953), Vermillion (1955), Monon (1959), Knox 62 (1962), and Riley (1965). Resistance from *T. timopheevi* has been used in Timwin (1968), Kenosha (1968), and Beau (1976), and also in combination with resistance from Transfer in Arthur 71 (1971), Abe (1972), and Oasis (1973). Resistance from *Agropyron* via Agent has been used in Blueboy II (1971).

Highly important in this period was the development of concepts of breeding for general resistance (30,31,37), recognition of the slow rust character in Knox (115), and initiation of breeding for this general type of resistance.

STEM RUST

This rust, incited by *Puccinia graminis* Pers. f. sp. *tritici* Eriks. & E. Henn., caused only minor losses in the eastern soft wheat area. Hope, Frondoso, Frontiera, Supresa, and *T. timopheevi* have been used as sources of resistance in cultivars discussed for resistance to leaf rust from these sources. Arthur and backcross derivatives Arthur 71, Abe, Oasis, Beau (1976), and Downy (1976) derive resistance from three sources: Hope, *T. timopheevi,* and Kenya Farmer.

The development and use of early maturing cultivars was important as an escape mechanism. They mature about one rust cycle earlier than previous cultivars, enough to avoid serious losses in the 1950-1976 period.

POWDERY MILDEW

With the use of higher rates of nitrogen fertilizer, this disease has become much more serious. Sources of resistance included Frondoso in Atlas 66 and other cultivars, Supresa in Redcoat, and *T. timopheevi* in Timwin, Kenosha, Arthur, and cultivars derived from Arthur. A number of biotypes of *Erysiphe graminis* DC. f. sp. *tritici* E. Marchal. occur.

Knox has a valuable slow-mildewing resistance, and breeding for this type of general resistance has been initiated (133,147).

LOOSE SMUT

This disease caused losses particularly in the 1950s. A number of races of *U. tritici* have occurred and few cultivars are resistant to all races (16). The resistance of Trumbull in Vigo, Knox, and Vermillion was overcome by new races. The resistance of Kawvale was utilized in Riley (1965) and Riley 67 (1967). Resistance from Hope occurs in many cultivars and is often associated with glume necrosis as in Benhur (1966) and Arthur (1968). Arthur has been immune to most races. Many cultivars have resistance to some races of the organism. Chemical seed treatment is now available to control the disease if necessary.

BUNT

Covered smut (bunt) has been controlled by seed treatment. Little work has been done in developing cultivars resistant to bunt in the eastern United States. Avon (80), developed at Cornell University, is resistant to dwarf bunt. Timwin

(1967) is resistant to common bunt. Bunt is an infrequent problem in grain quality in the eastern United States.

SEPTORIA LEAF BLOTCH

Losses to this disease have increased recently. Cultivars differ in resistance to *Septoria tritici*. Early maturing cultivars and short cultivars are generally more damaged. Wheats planted early are usually more heavily infected than those planted later. In Indiana the leaf blotch is nearly always present in the young wheat in March and April. If weather is favorable, the disease spreads up the plant, kills leaves, and may progress to and kill the flag leaf before the kernels have filled. Shaner et al reviewed the expression and effectiveness of resistance to the Septoria leaf blotch (148). Oasis (1973) was developed with resistance from Bulgaria 88. Resistance is conditioned by a single dominant gene (148), and is expressed as a reduction in fruiting of *Septoria tritici*, thereby limiting the buildup of the organism. In epidemics of leaf blotch, Oasis has yielded 15% more than Arthur 71 (148). These two cultivars are similar in yield in the absence of the disease, alike in maturity, and related in genetic background.

In addition to the specific resistance of Oasis, a general resistance, termed the "green-leaf" type, was observed in Indiana during the severe epidemics of 1973 and 1974 (148). Cultivars with this kind of resistance have reduced leaf necrosis without reduced fruiting of the organism. These types are distinguishable in severe epidemics but not in light infections. Beau (1976) is a cultivar with this general resistance. Control of this disease by breeding resistant cultivars is most promising.

SEPTORIA GLUME BLOTCH

Caused by *Septoria nodorum* Berk., this disease is considered serious in the southeastern United States (114). Glumes and awns on infected wheat heads are gray to black and the head dies prematurely, yielding shriveled grain or none at all. Early maturing cultivars are generally more heavily damaged than later maturing ones. Shorter cultivars, particularly semidwarf heights, heads on lower tillers, and heads near the ground from lodging are more seriously diseased. Cultivars adapted to the southeastern United States vary in degree of resistance (114).

A number of cultivars have field resistance or tolerance in some areas of the world, but efforts to establish resistance on a precise genetic basis have been unsuccessful (152). Nevertheless, breeding for resistance or tolerance to the disease offers the best approach to reducing yield and quality losses.

VIRUS DISEASES

There are three major virus diseases of wheat in the eastern United States. The soilborne wheat mosaic virus is widespread in soils of the eastern soft wheat region. All cultivars must be resistant. Susceptibility is often introduced into crosses with foreign germ plasm of specific other values. Plots maintained in continuous wheat are used to test for resistance to the disease. Planting at the fly-free date or earlier in the fall, contributes to fall infection. Symptoms of susceptibility are evident in early spring.

The wheat spindle streak virus disease is also soilborne. It was reported in

Canada by Slykhuis (153) and was widespread in Michigan and Pennsylvania in 1969 (176).

Spindle streak mosaic was observed in Indiana in 1967, 1969, 1973, and 1974. Infection occurs in the fall and symptoms appear in early spring. Cultivar differences were distinct at Lafayette, Indiana, in 1974 (85). Monon was resistant; Arthur, and derived Arthur types, were intermediate; and Redcoat was susceptible. Some breeding lines were more susceptible than Redcoat and stands were thinned by the disease. Differences between cultivars were not noted in Michigan (176).

In the spring of 1974 the disease appeared throughout Indiana; entire fields showed symptoms. The disease was more severe in early-planted fields and in fields where wheat followed wheat rather than following soybeans or corn in rotation. Breeding for resistance is the most promising means for control of the disease.

The barley yellow dwarf virus disease occurs on oats, barley, wheat, and some grasses. The disease has been recognized since about 1952. Aphids are vectors of the virus, and the bird-cherry oat aphid has been a principal vector (106).

The symptoms of the disease in wheat are generally less striking than those in oats and barley and may be confused with poor nutrition. The disease is a problem when the young wheat is exposed to a prolonged period of warm weather in the fall when aphids are active. Wheat sown early is generally damaged more than that sown later. In Indiana the disease has been prevalent on spring oats, and viruliferous aphids are often abundant in early autumn for transmission of the virus to wheat.

Unpublished work of Henry Jedlinski in Illinois, cited by McKinney (106) indicated that yield reductions greater than 80% could occur from the disease. Losses may be reduced by control of the aphid vectors with systemic insecticides that persist in the wheat plant for a few weeks. To date there has been little work in breeding resistance to this disease in wheat. This is the most promising path to control.

In summary, the three virus diseases of wheat are much more damaging when fall weather is warmer than normal after the seeding of wheat. In such years, losses from the diseases may be substantial.

FOOT AND ROOT ROT DISEASES

Several are important in wheat. Bruehl (25) reviewed their importance in limiting wheat production. In the past, these diseases were considered less important in the eastern United States where wheat was grown in rotation with other crops. In recent years, losses to root rots have been severe in Indiana and adjacent states, particularly where wheat follows wheat.

The take-all disease is probably the most serious. Infection of the roots occurs in the fall, particularly on early-planted wheat. Early fall seeding and nitrogen stress predispose wheat to severe take-all. Stunting, root necrosis, irregular height, tiller death, and sterile white heads before crop ripening are all symptoms. Ample nutrition, particularly nitrogen, in the spring is important in reducing losses. Take-all is generally more serious on light, porous soils where leaching of nitrogen occurs.

Unpublished research of D. M. Huber at Purdue University has shown the value of stabilizing fall-applied nitrogen with nitrapyrin, which inhibits nitrification to a leachable form. Huber and associates are breeding for improved tolerance to the disease and for a positive response to spring-applied nitrogen by infected plants. The progress in breeding is encouraging.

E. Breeding for Insect Resistance

Breeding new cultivars for resistance to the Hessian fly (*Mayetiola destructor* Say) and the cereal leaf beetle (*Oulema melanopus* L.) has been successful. Resistant cultivars provide protection to both the fall and spring broods of the Hessian fly. Delayed planting in the fall until after the fly-free date has been used as a measure to escape damage from the fall brood on susceptible cultivars. The fly-free date is based on the average temperatures when the wheat emerges. If the fall is warmer than average, the Hessian fly is still active and infestations occur.

The interactions between the host wheat plant and the Hessian fly are highly specific. The young larvae cannot survive on resistant plants. During the period 1950-1976, five races of the Hessian fly, A to E, have developed in the eastern soft wheat region. The H_3 gene from W38 was introduced into cultivars beginning with Dual (1955) and followed in Redcoat (1960), Monon (1959), and several cultivars. In about 1962 a new race (B) was found that could overcome this resistance. The H_3 resistance gave protection for about 10 years. The H_6 gene for resistance to races A and B, from durum wheat PI 94587, was introduced in the new cultivars Knox 62 (1962), Benhur (1966), and others. This resistance is still quite effective in Indiana, but two new races, C and D, have developed. Resistance of the H_5 type from Ribeiro, effective against races A, B, C, D, and E was introduced in cultivars Arthur 71 (1971), Abe (1972), Oasis (1973), Beau (1976), and Downy (1976). Resistance to the new race E in Georgia was found in cultivars Ga. 1123 and Seneca (H_7H_7, H_8H_8) (119). Gallun and Reitz (69) estimated, based on the 1969 cultivar survey (132), that resistant cultivars were grown on 1,619,000 ha or about one third of the entire soft wheat acreage of the eastern United States in 1969.

Breeding for resistance is a constant challenge in this highly dynamic host-insect relationship. Research continues to discover the genetic potential of the insect to form new virulent races. Research also continues to find new sources of resistance in wheat.

Resistance to the cereal leaf beetle is based on dense leaf pubescense (70). Pubescent wheats deter oviposition and larval feeding. Levels of about 100 hairs (trichomes) per square millimeter give almost complete control. The insect was first found in Michigan and Indiana in 1962 where extensive research efforts were made to find resistant germ plasm and to breed resistant cultivars. The first commercial cultivar, Downy, was released by the Purdue University AES and USDA in 1976. Downy derived pubescense from a Russian spring wheat, CI 9321. Downy is a soft red winter wheat of the Arthur type with good commercial quality and good resistance to diseases and to Hessian fly.

Since the resistance to the cereal leaf beetle is of a mechanical nature, new virulent races of the insect are not expected to evolve. Control of this insect by a resistant cultivar only 14 years after discovery of the insect in the United States is a remarkable achievement.

F. Progress in Breeding

A general improvement in yielding ability of new wheat cultivars was accomplished in the 1950-1976 period. Data from Lafayette, Indiana, illustrate the advances for this location (Table VI). Trumbull represents the long-time yield level check and the quality most acceptable before 1950. Vigo, released in 1946, provided similar quality to Trumbull, resistance to leaf rust, and better straw strength but little gain in yield potential. The two early cultivars, Knox 62 (representing Knox, 1953) and Monon (1959), are somewhat low in this comparison since they are now susceptible to leaf rust, whereas they were resistant when released. The gain for Knox was estimated at 25% over Vigo in 1958. Redcoat represents a cultivar intermediate in maturity and with a straw strength that permitted use of higher rates of nitrogen. It was adapted to the northern part of the Ohio river valley and the northeastern United States. Blueboy (1966), released in North Carolina, showed exceptional yield potential there and good yield potential in Indiana. Logan (1968) released from Ohio, showed improved yield potential in areas requiring a high degree of winterhardiness and tolerance to aluminum. Arthur (1968) and Arthur backcross derivatives Arthur 71 (1971), Abe (1972), and Oasis (1973) have established the highest level of productivity for Indiana; this was estimated from different summations of data as a 50 to 60% increase over cultivars available in 1950. Similar advances have been made for other areas in the eastern United States.

Increasing the productivity of cultivars is very important in the eastern United States, where wheat competes with other crops for an acreage large enough to satisfy domestic needs for soft red and soft white wheats.

A number of high yielding cultivars are available for production immediately ahead. In the white wheat area, Genesee, Yorkstar, Avon, Ionia, Ticonderoga, Arrow, and Tecumseh are expected to predominate. In the extended Ohio River

TABLE VI
Yield Improvement from Breeding Soft Red Winter Wheats, Field Plot Trials, Lafayette, Indiana[a]

		Yield (q/ha)[b]		
Cultivar	**Year**	**5-yr. avg.**	**4-yr. avg.**	**3-yr. avg.**
Trumbull	1916	30.6	30.1	34.1
Vigo	1946	31.0	30.6	33.9
Knox 62[c]	1953	36.2	35.1	41.5
Monon	1959	37.0	36.4	41.5
Redcoat	1960	39.9	40.7	43.1
Blueboy	1966	...	...	47.0
Logan	1968	43.6	43.3	46.9
Arthur	1968	47.7	46.6	52.9
Arthur 71	1971	45.4	44.1	50.4
Abe	1972	48.6	47.4	54.0
Oasis	1973	...	46.6	...

[a] Data from Purdue University Agric. Exp. Stn.: Station Bull. 104 (1975) and 56 (1975); Research Bull. 896 (1973), 883 (1972), and 882 (1971), and data for unreleased lines in the same tests.
[b] Five-year average 1970-1972 and 1974-1975. Trumbull was not included in 1973 tests; four-year average 1971-1972 and 1974-1975; and three-year average 1970-1972.
[c] Knox 62, a backcross derivative, represents the yield level of Knox (1953).

Valley, the Arthur types (Arthur, Arthur 71, Abe, Oasis, Doublecrop, Beau, and Downy) are expected to predominate on nonacid soils and Logan, Lucas, Ruler, and Redcoat in other areas. In southern Wisconsin and northern Illinois, the hardy cultivars Timwin, Kenosha, and Racine are expected to continue being important and in the southern and southeastern United States, they will probably include Blueboy, Coker 68-15, Coker 68-19, Coker 68-20, Georgia 1123, Holley, McNair 4823, and McNair 701.

PART TWO: CULTURE AND BREEDING IN THE WESTERN UNITED STATES

I. INTRODUCTION

Soft wheat (*Triticum aestivum* L. em Thell) is grown in 11 western states, but the main production areas are the Columbia and Snake River drainage systems of Washington, Oregon, and Idaho, the Willamette Valley of Oregon, and scattered, principally irrigated areas of California, Arizona, and Utah. Smaller amounts are produced in Colorado, Montana, Wyoming, New Mexico, and Nevada. Of the 4.9 to 6.5 million ha of wheat sown yearly in the west, between 1.4 and 2.1 million ha are planted to soft wheat.

Climatic, physiographic, and adaphic features of these areas vary dramatically (1,128,135). Many different and distinct climates occur in the west. The area is large and extends through 1,900 km of latitude and 1,800 km of longitude. Annual mean temperatures decrease by about 0.90 C for each 100 km from the Mexican to the Canadian borders. Mean temperatures also decrease by about 0.50 C for each 100 m rise in elevation. The climate can change drastically within a short distance owing to the varied relief. Most of the region is relatively dry with less than 50 cm of precipitation annually, but it is not unusual for annual precipitation to change by 120 cm within a distance of 30 km as terrain changes (1).

The ranges of some climatic factors for this region are: growing season, 30 to 320 frostfree days; snow cover, 0 to 150 days; average precipitation, 13 to 200 cm; mean daily relative humidity for the growing season, less than 20 to 80%. Soil types are diverse in the West, and soft wheat is grown on 12 of the 13 soil groups of the region (1).

Four main soft wheat management systems are practiced: dryland annual crop, dryland wheat-fallow, irrigation, and double-crop. Double-cropping is relatively minor. About 50, 32, and 28% of the present soft wheat area is grown under wheat-fallow, dryland annual crop, and irrigated management, respectively, in the western states.

Irrigation is gaining rapidly. The irrigated area increased about 15% from 1967 to 1974, and it is expected to grow 25 to 50% in the next 10 to 15 years (61). In the Pacific Northwest many farmers now provide supplemental water of 7 to 13 cm once or twice to their wheat before anthesis.

Great diversity also occurs in crop rotations. Specific rotations depend on water availability, potential for soil erosion, prevalent weeds and pests, access to markets, need to diversify, and other special economic considerations. The main common soft wheat rotations are: 1) fallow/winter wheat/fallow/winter wheat, 2) fallow/winter wheat/spring grain/fallow, 3) peas/wheat or spring grain/wheat; 4) irrigation, with specialty crop/wheat/specialty crop/wheat, and 5) forage (1 to 3 years)/wheat/forage (1 to 3 years).

Fertilizer requirements also vary owing to the diversity of soil types, climate, moisture availability, and specific management practices (98). The light arid, semiarid, and desert soils often require complete fertilization, particularly with irrigation. The moderately dark, dark semiarid, and subhumid soils usually need applications of nitrogen and phosphorus. Annual crops frequently require additional sulfur, and wheat sown on the clay knolls of the Pacific Northwest Palouse area respond to added potassium. Several trace elements (Zn, Mo, B, Mn) may be limiting in specific areas of the region. Nitrogen is the most critical soil nutrient. In the Pacific Northwest soft wheats require about 1 kg of nitrogen per 22 kg of grain produced (97). Dryland farmers in areas with rainfall of 25, 38, and more than 38 cm may typically need to add 45, 67, and 90 kg/ha of nitrogen (98). Under irrigation, rates in excess of 225 kg/ha are not uncommon. Applications of phosphorus and potassium above 34 and 56 kg/ha each are rarely needed for dryland, but much higher rates are often used under irrigation. Soil pH also varies throughout the western region, but generally only the humid forest soils below a pH of 5.5 need lime for wheat production.

Seeding dates of fall-sown wheat also differ markedly. In the major soft winter wheat areas of the Pacific Northwest, seeding is done from August 1 to November 15. Early seeding is favored on fallow, because early stand establishment offers high yield potential and inhibits wind and water erosion of soil. Extremely early seedings, however, waste soil nutrients and moisture and increase vulnerability to several diseases (128). Seeding dates of annual crop or recrop dryland are dictated by moisture availability and weed control practices, but early stand establishment is desired. In the Northwest, it is not uncommon to sow in dry seed beds. Emergence then occurs in the winter after precipitation begins.

Seeding dates of irrigated soft winter wheats are more flexible. In the Columbia Basin irrigated semidwarf cultivars may be successfully sown from early August until mid-February. Rotation requirements and length of the growing season primarily determine seeding dates under irrigation. Seeding dates for soft spring wheat on dryland also vary widely; the optimal dates are about mid-March to mid-April in Washington and Idaho. Irrigated spring wheat may be sown over an even longer time span. In fact, a common practice in Arizona and California is to seed spring wheat in November or December for spring harvest (128).

Wheat has been successfully double-cropped with vegetables, sorghum, corn, soybeans, and annual forage in specific irrigation systems of the West. Each of these rotations often requires particular seeding dates to fit each situation.

Seeding rates of soft wheat in the West are generally lower than those of the East. Normal fall seeding rates for dryland fallow, dryland annual crop, and irrigated wheat are 45 to 67 kg/ha, 67 to 101 kg/ha, and 67 to 135 kg/ha, respectively. Seeding rate, however, varies with management, moisture level, seeding date, seed bed, and the cultivar. Rates are usually increased for late seedings, trashy or rough seed beds, suboptimal moisture conditions, and low-tillering cultivars such as the clubs. Spring seeding rates vary from 67 to 135 kg/ha and depend on individual management and local environmental conditions.

II. SOFT WHEAT BREEDING

A. Early Wheat Improvement in the West

Soft wheat production in the West began with wheat cultivars introduced from several areas. Cultivars from Chile, Mexico, Australia, England, and the eastern United States had significant impact on early production on the Pacific Slope. Wheat culture in the West actually predated the time of the Louisiana Purchase, and history is sketchy about the specific dates, places, and kinds of wheat that were first grown.

The Catholic missions fostered early western wheat culture. The old soft white common spring wheat, Sonora, was grown at the Magdalena Mission of Sonora, Mexico, as early as 1770 and arrived in our Southwest about 1820. Sonora could be easily ground by hand or with the early stone mills (29).

Club-type wheat was also grown early, principally around Spanish missions. The old club wheats were probably brought to California from Chile and perhaps also Australia and France (39). Archaeological finds suggest that a wheat identical to Little Club was used in adobe bricks of California missions erected between 1701 and 1845 (43). Little Club was introduced from California to the Pacific Northwest, where it became the progenitor of most of the club wheats developed later. Carleton (39) praised club wheats for their adaptability to the Pacific Northwest because they had excellent cracker quality, high yield, stiff straw, earliness, and drought and shatter resistance.

The introduction of Australian cultivars had great impact on soft wheat improvement in the West. Pacific Bluestem arrived from Australia in the 1850s. A common white wheat with soft to semihard kernels, it became very popular because it could be seeded in the fall or spring in mild climates. This cultivar milled satisfactorily and had good cookie and cake flour characteristics (52,63). White Australian, a popular early wheat in California, is almost identical to Pacific Bluestem (43).

Baart, introduced about 1900 from Australia, was well adapted to the drier areas of California and the Pacific Northwest. It was purported to be drought-resistant (41). Baart came to Australia from South Africa but probably originated in western Europe (43). Although the grain of Baart often graded hard white, it milled well and produced excellent cake flour and good bread flour at high protein levels (63). Baart and Pacific Bluestem were the first true dual-purpose pastry and bread wheats grown on the western slope.

Fortyfold or Goldcoin from New York state was an early soft white common winter wheat that became widely grown in the Pacific Northwest and served as an important parent for western soft wheats. It ranked among the best of the old wheats for flour yield and cookie and cake flour properties (63).

The most important soft red winter wheat in the West before 1900 was Jones Fife. Selected by a New York farmer in 1889, it became a major wheat in the Pacific Northwest and also served as an important parent of subsequent cultivars.

Other less important early-grown soft wheat cultivars were: the common whites—Propo (Prosper), Dawson, Dicklow, Surprise, White Winter, and Defiance; the common reds—Red Russian and Lofthouse; and the clubs—Jenkin, Big Club (Chile Club), Red Chaff, Blue Club. But many other introduced wheats also were grown in the West before 1900. Carleton (39) listed 60 selections and cultivars of soft kernel texture that originated in eight foreign countries and other parts of the United States and that showed promise for culture in the West.

Unlike other U.S. wheat regions, the Pacific Northwest did not significantly exploit selections from originally introduced land cultivars. The club wheats Genro, Hood, Bluechaff, Coppei, and Mayview were selected from introduced cultivars or farmer fields of other wheats. Big Club was derived from Little Club and grew on more than 8,500 ha in 1919. Dicklow, Propo, Foisy, and Silvercoin were common white wheat field selections (43). Most of these cultivars had limited merits compared with those of their progenitors.

Selection within the older, soft wheat land cultivars received little attention after W. J. Spillman of Washington State College began his classic wheat hybridization studies, which quickly gave superior types. Spillman (154) was aware as early as 1895 that Little Club, Red Chaff, and Pacific Bluestem cultivars had their own specific adaption to eastern Washington. All three were spring types, however, and lacked winterhardiness. Because no winter-hardy types were adapted, he decided to breed new cultivars by crossing the best winter wheats with those three extensively grown spring wheats. Aside from his famous independent rediscovery of Mendel's Laws of Genetics, Spillman's effort led to several high yielding, shatter-resistant, stiff strawed, winter-hardy cultivars. They included Hybrids 60, 128 (winter); 108, 143 (intermediate); 63, 123 (spring). The most successful cultivar was Hybrid 128, derived from a Jones Fife × Little Club cross. Released in 1907, it was grown on nearly 50% of the club wheat area between 1924 and 1929. Hybrid 128 excelled both in yield and quality in tests conducted from 1911 to 1914 and yielded about 6% more grain than older wheats (139). Hybrids 108 and 123 had high yields but milled poorly, had red kernels, and lacked popularity. About 1924, Spillman hybrids began to decline because of their common bunt susceptibility and because Federation, a common soft white spring wheat, became popular.

B. Early Quality Studies and Observations

The importance of soft wheat quality and cultivar differences was recognized as early as 1907 in the West. Thatcher (164) observed that nearly three fourths of Washington's wheat crop was marketed elsewhere. He believed that wheat grown in the state should be at least equal in quality to wheats with which they competed, and he listed five important wheat quality considerations: 1) chemical composition, 2) percentage each of several mill products, 3) distribution of chemical constituents, 4) flour quality, and 5) flour color.

Tests of 19 cultivars grown at several localities showed notable differences in chemical composition among cultivars, but differences were even greater within cultivars; wheats grown in the driest sections produced the largest and best loaves (164). White-seeded types yielded slightly more flour than their red-seeded sister lines (165). He stated that flours of Little Club, Goldcoin, and Red Russian were poor for bread baking but adapted to the manufacture of special pastry uses.

Clark et al (52) found that Baart satisfied early demands for a better milling wheat and noted that millers paid premiums for Pacific Bluestem grown in the Columbia Basin because of its superior milling quality. Their data showed that Sonora milled very poorly, had low ash flour, and was unsuited for bread but valuable for specialty purposes. The major advantages of Australian wheats were their superior milling and flour qualities; White Federation ranked the best.

III. SOFT WHEAT IMPROVEMENT, 1919-1944

A. Cultivar Distribution

During this time more than 110 cultivars with soft wheat properties, including common soft red winter, spring and winter club, and soft and semihard spring and winter common white were grown in the West (45,46,48–51). In 1919 about 54% of the wheat area of the 11 western states was soft wheat. By 1924 the proportion dropped to 33% owing to the sharp increase in popularity of Turkey-type hard red winter and hard red spring cultivars. Soft wheat comprised 31 to 35% of the area from 1924 to 1944. Soft red winter cultivars consistently declined during that time (Table VII). They represented about 5% of the area in 1919 but decreased to less than 1% by 1944. Club wheat also declined from 11% of the wheat area in 1919 to only 4% in 1944 (Table VII). The common white wheat area, composed of both spring and winter types, varied from 38% of all wheat sown in the West in 1919 to 19% in 1924; and in 1944 this class comprised 30% of the total. For the years 1919, 1924, 1929, 1934, 1939, and 1944, 44, 56, 61, 69, 65, and 56 different cultivars with soft wheat properties were reported to be grown in the West, respectively (45,46,48–51). Table VIII lists the more important cultivars of the four major soft wheat classes and gives their percentage of the total area sown to wheat in the West. Table IX shows the actual area devoted to the principal soft white winter, soft white spring, club, and soft red winter cultivars per period.

TABLE VII
Area and Percent of Total Soft Wheat Grown by Class from 1919 to 1974 in the 11 Western States[a]

	Common White		Club		Soft Red Winter		All Soft		Total
Year	(ha)	(%)	(ha)	(%)	(ha)	(%)	(ha)	(%)	(ha)
1919	1,464,205	38	417,069[b]	11	197,795	5	2,079,069	54	3,839,679
1924	682,868	19	338,961[b]	10	138,949	4	1,160,778	33	3,550,700
1929	1,293,749	26	293,712[b]	6	135,580	3	1,723,041	35	4,944,335
1934	1,030,756	25	281,649[b]	7	84,447	2	1,396,852	34	4,139,066
1939	1,188,364	26	166,569[b]	4	56,358	1	1,411,291	31	4,547,340
1944	1,440,204	30	188,849[b]	4	36,386	<1	1,665,439	34	4,867,290
1949	1,497,299	21	508,900	7	27,256	<1	2,033,454	28	7,161,615
1954	870,733	15	720,260	13	7,460	<1	1,598,452	28	5,689,440
1959	807,775	16	745,655	14	4,011	<1	1,557,441	30	5,187,240
1964	1,228,894	25	213,450	4	933	Tr	1,443,277	29	4,909,410
1969	1,376,913	26	285,834	5	217	Tr	1,662,963	31	5,314,046
1974	1,700,207	26	403,453	6	Tr	Tr	2,103,660	32	6,514,830

[a]References for values: 44–46,48–51,129–132,136.

[b]Includes club cultivars with red seed (Dale, Coppei, Mayview, Hybrids 123 and 108).

TABLE VIII
Cultivars with Soft Wheat Properties Most Extensively Grown in the West and their Percentages of Total Western Wheat Area at Five-Year Intervals[a]

	Percentage of Total Western Wheat Area—Five-Year Intervals											
Cultivar and Class	**1919**	**1924**	**1929**	**1934**	**1939**	**1944**	**1949**	**1954**	**1959**	**1964**	**1969**	**1974**
Common white winter												
Goldcoin	5.2	3.0	3.4	1.9	1.2	3.2	2.0	trace	trace	0.4		
Golden					0.3	1.4	1.3	1.1	0.8	trace	0.2	0.1
Rex					3.3	3.7	2.4	1.1	trace	trace	trace	
Brevor								1.7	1.7	1.1	trace	
Gaines										13.2	7.9	3.3
Nugaines											11.1	10.3
Hyslop												4.2
Yamhill												0.7
Common white spring												
Baart[b]	5.3	5.5	6.3	7.8	7.9	8.3	3.7	2.8	1.2	0.3	0.2	
Sonora	2.8	1.0	0.8	0.5	0.2	0.1	trace	trace	trace			
Pacific Bluestem[b]	14.4	4.2	3.0	1.6	1.2	0.5	0.2	0.1	0.1	0.1	trace	
Defiance	2.0	0.4	0.3	0.3	0.1	trace	trace	trace	trace			
Dicklow	1.7	1.3	2.1	1.7	1.2	0.7	0.2	0.1	trace	trace		
Bunyip		0.3	1.0	0.7	0.8	0.4	0.1	0.1	trace	trace		
Federation[b]		0.5	7.0	8.0	7.4	7.9	4.8	2.3	1.8	0.8	0.5	0.1
Idaed[b]					trace	0.4	0.9	1.0	0.6	0.6	0.8	0.2
Lemhi[b]					trace	0.9	1.1	1.4	2.2	0.6	1.2	0.1
Ramona[b]					0.1	0.1	1.1	1.2	2.1	2.3	0.8	0.1

Marfed							0.2	0.4	0.8	0.8	1.6	1.0
Orfed							1.0	0.2	0.1	0.1	trace	
Twin												2.0
Springfield												2.0
Siete Cerros[c]												1.0
Club												
Little Club	1.1	0.3	0.1	0.3		trace						
Hybrid 128	3.0	4.8	2.9	1.4	0.4	0.9	0.4	trace	trace			
Jenkin	0.7	1.3	0.8	0.5	0.1	0.1	0.1	trace				
Albit			0.6	3.8	1.1	0.2	trace	trace				
Hymar					1.1	1.7	1.5	0.3				
Elgin							4.7	2.0	1.2	0.7	trace	
Elmar								10.1	1.0	0.1	trace	
Omar									12.0	3.5	0.8	0.8
Moro											4.5	2.4
Paha												3.0
Soft red winter												
Red Russian	1.6	0.6	0.5	0.2	0.1	0.2	0.1	0.1	trace			
Jones Fife	2.8	.9	0.7	0.5	0.2	0.1	trace	trace				
Triplet		.1	1.4	1.2	0.8	0.4	0.3	trace				

[a] References for values: 44–46,48–51,129–132 136.
[b] Includes backcross prototypes, and for Federation also White Federation and Hard Federation.
[c] Siete Cerros is classed as a hard white spring but has some soft wheat uses when sown in the fall as in California.

TABLE IX
Area of the Three Most Widely Grown Cultivars for Each Class or Subclass of Soft Wheat in the 11 Western States (1919-1974)

		Soft White Winter		Soft White Spring		Club		Soft Red Winter	
Year	Rank	Cultivar	1,000 ha	Cultivar	1,000 ha	Cultivar	1,000 ha	Cultivar	1,000 ha
1919	1	Goldcoin	201.4	Pacific Bluestem	552.2	Hybrid 128	117.1	Jones Fife	108.6
	2	White Winter	21.3	Baart	202.5	Little Club	43.0	Red Russian	62.7
	3	Martin	10.7	Sonora	106.8	Jenkin	26.9	Odessa	10.5
1924	1	Goldcoin	107.8	Baart	196.5	Hybrid 128	168.7	Jones Fife	68.7
	2	White Winter	12.0	Pacific Bluestem	150.4	Jenkin	45.4	Triplet	40.8
	3	Prohibition	6.3	Dicklow	47.0	Hybrid 123	21.0	Red Russian	20.5
1929	1	Goldcoin	169.9	Federation[a]	345.5	Hybrid 128	144.5	Triplet	68.0
	2	White Winter	10.8	Baart	310.5	Jenkin	37.3	Jones Fife	35.1
	3	Wilhelmina	9.3	Pacific Bluestem	147.4	Albit	31.7	Red Russian	23.3
1934	1	Goldcoin	78.8	Federation[a]	330.5	Albit	159.0	Triplet	49.9
	2	Wilhelmina	15.2	Baart	322.1	Hybrid 128	57.8	Jones Fife	19.9
	3	White Winter	6.6	Dicklow	70.1	Jenkin	19.4	Red Russian	9.9
1939	1	Rex	149.9	Baart[a]	360.2	Hymar	51.4	Triplet	38.0
	2	Goldcoin	52.6	Federation[a]	338.3	Albit	50.9	Jones Fife	11.1
	3	Golden	15.6	Dicklow	56.6	Hybrid 128	18.8	Red Russian	4.6
1944	1	Rex	182.2	Baart	403.9	Hymar	82.9	Triplet	17.8

	2	Goldcoin	156.4	Federation	385.0	Hybrid 128	43.2	Red Russian	8.9
	3	Golden	66.8	Lemhi	43.9	Alicel	25.4	Jones Fife	5.9
1949	1	Rex	173.1	Federation[a]	342.6	Elgin	336.2	Triplet	17.9
	2	Goldcoin	141.0	Baart[a]	266.6	Hymar	109.3	Red Russian	3.7
	3	Golden	94.9	Lemhi	81.9	Hybrid 128	31.5	Mediterranean	2.2
1954	1	Brevor	94.8	Baart[a]	160.5	Elmar	573.9	Red Russian	3.0
	2	Rex	62.7	Federation[a]	128.7	Elgin	112.4	Triplet	2.0
	3	Golden	62.5	Lemhi	79.0	Hymar	19.1	Jones Fife	0.9
1959	1	Brevor	87.8	Lemhi[a]	112.5	Omar	622.6	Red Russian	1.3
	2	Golden	40.3	Ramona[a]	109.6	Elgin	64.3	Triplet	1.2
	3	Alba	15.8	Federation[a]	91.6	Elmar	50.0	Jones Fife	0.8
1964	1	Gaines	646.8	Ramona[a]	112.5	Omar	172.0	Only trace area	
	2	Brevor	53.3	Marfed	39.6	Elgin	34.7		
	3	Goldcoin	17.3	Federation[a]	38.8	Elmar	4.9		
1969	1	Nugaines	589.4	Marfed	86.5	Moro	237.9	Only trace area	
	2	Gaines	422.1	Lemhi[a]	62.7	Omar	42.3		
	3	Druchamp	12.4	Idaed[a]	43.7	Big Club 60	1.0		
1974	1	Nugaines	567.9	Twin	132.8	Paha	194.1	Only trace area	
	2	Hyslop	275.5	Springfield	127.3	Moro	158.0		
	3	Gaines	215.9	Marfed	66.6	Omar	50.3		

[a] Includes all forms of the cultivar.

The five common white spring wheats Baart, Sonora, Pacific Bluestem, Defiance, and Dicklow dominated during 1919 (Table VIII). Pacific Bluestem and Defiance declined notably by 1924 and were replaced by hard red winter and hard red spring wheats. The area of Baart and Dicklow remained rather stable throughout the period. These cultivars were well adapted to drier regions on the Columbia and Snake River plateaus and were desired by industry. Sonora also enjoyed long popularity, especially in Arizona and southern California. Bunyip, bred by William Farrer of New South Wales, Australia, was important in California from 1929 to 1939 (43).

One of the most important common soft white spring wheats from 1929 to 1944 was Federation (Tables VIII, IX). Farrer bred Federation from a cross Purplestraw × Improved Fife × Etewah (14). Heralded as this famous breeder's greatest crossbreeding achievement, Federation had a major impact on western soft wheat culture. It contributed genes for early maturity, short stiff straw, shatter resistance, dual-purpose end product use, and facultative growth habit to many subsequent soft wheats. Federation was tested at Moro, Oregon, and released in 1920. By 1929 it was grown on 7% of the western wheat area (Table VIII). Federation and Baart were the two most extensively grown soft wheat cultivars from 1929 to 1944 (Table IX). They complemented each other, because Federation was adapted to heavy soil and moist areas, and Baart grew best in light soil and dry areas.

Three popular common soft white winter wheat cultivars during this time were Goldcoin (Fortyfold), Golden, and Rex (Table IX). Goldcoin was well liked because it was more winter-hardy, matured earlier, and was less subject to smut than many other early winter wheats (139). Golden was a selection made from Goldcoin at Moro, Oregon, in 1923; it exceeded Goldcoin for shatter resistance, straw strength, and yield (135). Less important soft white winter cultivars grown from 1919 to 1944 included White Winter, Martin, and Wilhelmina (Table IX).

Rex was the first extensively grown soft white common winter wheat with resistance to common bunt. Selected from the cross Hard Federation × White Odessa in 1926 at Moro, Oregon, Rex possessed the desired combination of shatter resistance, short straw, lodging resistance, and high yielding ability and was grown extensively throughout the Northwest after 1944. Notoriously poor milling quality became the cause of its downfall.

Red Russian, Jones Fife, and Triplet were the major soft red winter wheats grown from 1919 to 1944 (45,51) (Tables VIII, IX). Odessa was important early in the period but declined rapidly in popularity (Table IX). Red Russian and Jones Fife were the two most important winter wheats of the soft type in the West before 1919. Jones Fife's strong points were adaptation to dry regions and high yield, but it lodged and shattered badly and was very susceptible to common bunt. Red Russian was adapted to heavy soil and high rainfall, had strong straw, competed with weeds, matured late, and had "dilute resistance" to common bunt (67,139). Triplet was developed in 1910 at Pullman, Washington, from a compound cross. It had shorter, stiffer straw, shattered less than Jones Fife, and averaged about 30% higher yields than Jones Fife or Red Russian (139). Triplet was the last important soft red winter wheat to be grown in the West and was ultimately replaced by improved common white winter cultivars.

Five club wheats were grown extensively between 1919 and 1944 (Table VIII).

The old white, spring-sown Little Club was phased out of production after 1934. Jenkin grew on a sizable area between 1919 and 1935 mainly because of its superior flour quality traits (Table IX), but it was replaced by Hybrid 128 and later Albit because of its poor yield, low test weight, weak straw, and lack of adaptation to fall seeding (169). Shatter resistance, winterhardiness, straw strength, and improved yield favored Hybrid 128 over Red Russian, Jenkin, and Little Club. It was the best yielding fall-sown wheat from 1911 to 1914 in Washington tests, but it was highly susceptible to common bunt (139).

Albit was the first club cultivar with resistance to common bunt and was the leading club wheat by 1934 (Tables VIII, IX). Selected from a cross of Hybrid 128 × White Odessa in 1923 at Pullman, Washington, it carried the Bt_1 gene for bunt resistance and was more winterhardy than Jones Fife or Hybrid 128 (169). Hymar had common bunt resistance similar to Albit's and was selected at Pullman from a cross of Hybrid 128 and Martin. Hymar was very winter-hardy, had high test weight, and resisted shattering. However, its yield equalled that of Hybrid 128 and it produced too much straw, so it never became too popular (169) (Tables VIII, IX). Hybrid 123 and Alicel were also grown in significant quantities for rather short periods during that time (Table IX).

The club wheats Dale, Coppei, Mayview, Hybrid 123, and Hybrid 108 actually had red seeds with many properties similar to soft red winter wheats. They were grown on 4,300 to 75,600 ha between 1919 and 1944 but later became unimportant.

B. Cultivar Quality Traits

During this time the wheat quality characteristics of the region were formulated. More than 100 major and community mills were located in the West by about 1918 (157). Before then, many more mills were operating in the area (157). The region's millers and bakers quickly identified the specific desirable or undesirable quality traits of local cultivars. Furthermore, with the diversity of growing conditions and classes of wheat found in the West, complex marketing and utilization problems were encountered early (135). The quality traits have been described for the wheats grown most extensively during this period.

The old club white wheat, Jenkin, rated excellent for flour yield and very good for cake quality (63). Jenkin's cookie quality was outstanding. It produced yellow, medium ash flour. Hybrid 128 gave good flour yields but less than those of Jenkin. Its flour rated very good for cookies, but it was inferior to Jenkin for cakes and was high in ash (63). Albit, with flour yields lower than those of Hybrid 128, was low in ash but equal to Hybrid 128 for cake and cookie quality (63). The club, Hymar, milled slightly better than Hybrid 128, but its flour quality varied from season to season (63).

The old soft white common wheat, Goldcoin (Fortyfold), milled comparably to Hybrid 128 and produced good cookie and cake flour. Golden was similar to the quality of Goldcoin except that it was less desirable as a cake flour (63).

Pacific Bluestem often graded hard white, but produced a flour valued for its soft wheat properties. Its flour yield was less than that of Jenkin, but the flour made excellent cakes, good cookies, and fair bread (63). Like Pacific Bluestem, the common white spring, Baart, often graded hard white but had several desirable soft wheat quality traits. Baart produced a soft, velvety flour, low in ash and carotenoid content, that was excellent for cake and fair for cookies (63). Like Pacific Bluestem, Baart was considered a dual-purpose wheat. Processors in Idaho and Utah liked the common white spring wheat, Dicklow. It produced a very white, excellent cake and cookie flour but had flour yields below those of Federation (63). Fall-sown Federation was recognized as one of the better soft white cultivars for pastry purposes and was in demand by the trade. Industry rated it good for milling, family flour, crackers, and cake but only fair for cookies (156). Idaed was very similar to Federation, but it rated lower as a cracker flour.

The old soft red winter wheat, Jones Fife, produced flour yields similar to those of Pacific Bluestem but had lower flour quality; Red Russian was rated even poorer than Jones Fife for quality (165). The flour yields and pastry flour quality of these two cultivars were inferior to those of Goldcoin. The soft red winter wheat, Triplet, milled well, made fair cookies and excellent cake, but produced poor family or cracker flour (156).

The most infamous variety of this period was Rex, a common soft white winter wheat with outstanding grain yield. Rex milled extremely poorly and produced a fine, poor-bolting flour. These properties reduced mill capacity and flour yield; hence, it was despised by the trade. Its creamy yellow flour was rated poor for family use, fair for crackers and cake, and good for cookies (63).

C. Identification of Adaptive Traits for the Region

In a region with such diverse environments as the West, it is not surprising that adaptive traits for wheat varied widely from one production area to another. Some traits were common to nearly all areas, however. Because many of the areas were geographically isolated and early transportation was both poor and expensive, wheats with dual-purpose uses were popular. Logically, cultivars with both bread and pastry qualities were initially sought.

The important adaptive traits were recognized nearly from the onset of wheat farming in the West. Spillman (154) polled eastern Washington farmers in the 1890s on the desired traits for a wheat cultivar. The farmers consistently mentioned three things: shatter resistance, stiff straw, and winterhardiness. Because the three major wheats grown at that time (Jenkin, Little Club, and Pacific Bluestem) were fall-sown spring types, winterhardiness was considered the most important trait.

Vogel and Barbee (169) summarized the important eastern Washington adaptive features for fall and spring wheat cultivars during the period 1931-1943. Fall seeding outyielded spring seeding by 7 to 21%. Shatter resistance was a particularly important adaptive trait. The soft red winter wheat Mosida had high yield potential but limited usefulness because of shattering. Oro was described as prone to yet another form of preharvest loss, termed "head snapping." Cultivars that matured early to midseason generally had higher yields and test weights. The shortest wheats resisted lodging. Cultivars with rapid spring recovery competed more favorably with weeds than did the highly dormant, slow-recovery types. Semidormant to highly dormant types such as Rex and Hybrid 128 were best adapted to cold areas. The only disease mentioned during this time was common bunt (169).

D. Breeding to Control Common Bunt

Common bunt, or stinking smut, was identified early as the major disease of wheat in the West. Heald and Woolman (78) stated in 1915 that common bunt was clearly the most destructive wheat disease in the Pacific Northwest with estimated annual losses in the millions of dollars. The losses occurred several ways: increased costs of production owing to seed treatment and disease-reducing cultural practices, yield loss, lowering of grade or quality, and fire hazard. Gaines (67) reported in 1923 that losses to common bunt in the Pacific Northwest had increased steadily for the previous 25 years. Infection most commonly occurs from seedborne spores, but in the Pacific Northwest airborne spores deposited on fallow land survived the summer to germinate and infect seedlings in the fall. According to Holton (83), the Pacific Northwest's environment and farming practices provided nearly ideal conditions for common bunt.

The disease is most severe on fall seedings, because the pathogen apparently cannot overwinter in the soil to infect spring wheat. Yet early tests showed that maximum yields in the Palouse region were best achieved by fall-sown wheats. In 1923, 60% of the Washington crop was sown to winter wheats (140), reaching 88% by 1944. Because available seed treatment methods did not control soilborne infection, host resistance received high priority.

As early as 1915, Heald and Woolman (78) predicted the ultimate development of highly resistant cultivars. Their early tests showed that most club wheats were highly susceptible, as was Jones Fife; but Goldcoin was less susceptible and Turkey had the least smut. Gaines (67) began screening cultivars for bunt reaction in 1914, and by 1916 he had tested over 500 named wheats. Turkey, Alaska, Florence, and Marquis rated resistant (67). He recorded which wheats produced partially smutted versus completely smutted spikes, a trait that five decades later was associated with nonspecific resistance. Gaines (67) noted that Fortyfold, Red Russian, and Marquis each had different "dilute resistances" that reduced smut and that, when added together, gave a more concentrated resistance.

In a multiteam approach in three western states, Tisdale et al (166) tested most of the commercial wheats in the United States for common bunt. As a class, the hard red winter types were most resistant, soft red winters and hard red springs were frequently intermediately susceptible and clubs were highly susceptible. Hussar and Martin were identified as useful parents in the development of bunt-resistant wheats. They optimistically stated (166) that the resistance of Sherman, a Turkey selection, was such that it could be safely sown without seed treatment.

In 1927, Stephens, then at Moro, Oregon, noted that previously resistant selections of Turkey wheat had become highly susceptible (68). Ridit, Albit, Martin, and White Odessa also showed increased smut. In a pioneering set of experiments, Gaines (68) identified at least five forms of the pathogen. He urged increased effort in breeding resistance to the new forms but recommended that immediate emphasis be placed on seed treatment studies and cultural control practices.

Common bunt research next focused on the search for new and different sources of resistance, complemented with genetic studies on the inheritance of resistance and identification of number, distribution, and presence of physiologic races that attack the various host-gene combinations. By 1945 Rodenhiser and Holton (134) identified 31 races. Knowledge of the extent of physiologic specialization served as a useful breeding tool and facilitated selection of resistant types in the Pacific Northwest.

In a series of detailed genetic studies based primarily on a single race of the pathogen, Briggs and Holton identified four strong and two weak host genes for resistance (23). In 1950, optimism prevailed about the possibility of final control of bunt through breeding. At that time Briggs and Holton (23) suggested that combining the Martin (Bt_1) and either the Turkey (Bt_4) or Rio (Bt_6) genes would give resistance to all known bunt races. They stated that a breeder need only work with two major genes and screen his populations against only two races to detect these genes. Unfortunately, this simplistic solution did not stand the test of coevolution of host and pathogen. Race T18, which could attack the Bt_1, Bt_4, or Bt_6 combination was found by 1961 (88). In 1976, Hoffmann and Metzger (81) reported a common bunt race capable of attacking as many as five host genes.

E. California Backcross Wheat Cultivars

From 1919 to 1941, wheat workers at Davis, California, began their comprehensive studies of the backcross breeding method. Their work contributed much useful basic information on the inheritance of several traits, provided a basis for modern backcross theory, and led to the release of several cultivars improved for various traits, including awn expression and resistance to common bunt, stem rust, leaf rust, and Hessian fly. By repeated backcrossing, these rather simply inherited traits were added to Pacific Bluestem, Federation, White Federation, Ramona, Onas, Baart, Big Club, Sonora, and Poso. Pacific Bluestem 37, Big Club 37, Sonora 37, Federation 41, and Onas 41 were improved for a single trait wherein all received the Bt_1 gene for common bunt resistance (160). Otherwise they were very similar to their recurrent parents.

Other backcross-derived wheats were improved for two or more traits. Baart 38, White Federation 38, and Ramona 44 resisted several races of common bunt and stem rust. Poso 48 and Big Club 43 surpassed their progenitors for resistance to common bunt, stem rust, and Hessian fly (160).

A strong argument for the backcross approach was that it would assure retention of the desired traits of the old cultivar in the improved type. This advantage would have particular significance for quality attributes because processors could expect the new improved types to behave almost exactly as their progenitor cultivar in physiochemical quality traits. The evidence usually affirmed this expectation. Suneson (160) reported no significant differences among six prototype cultivars and their six progenitor cultivars in comparisons of 26 quality criteria. One exception occurred, however, in that Baart 38 had bread properties slightly inferior to those of Baart. Backcross breeding was nevertheless established as the best way to retain desired quality traits and yet effect improvements for simple genetic traits. Conservative estimates indicate the California backcross cultivars added 81,600 tons of wheat to the state's production each year (135).

F. Relative Changes in Yield

Significant progress in yielding ability was made during this period. Obviously, increases cannot be attributed only to improved cultivars. Improved management, mechanization, and disease control also contributed significantly. The average wheat yields in the Pacific Northwest increased from 1,171 kg/ha in 1919 to 1,817 kg/ha in 1944 (173). The relative yield improvement made by certain soft winter wheats can be estimated from Vogel and Barbee's data (169) comprising 33 to 49 station-years (Table X). Comparisons were made to Kharkof, a long-term hard red winter check, and to the soft wheat Goldcoin. When compared to Goldcoin, rather remarkable yield increases were achieved. The club wheats yielded 15 to 21% more than Goldcoin, with Hymar the most productive. Rex gave the highest yields of any cultivar then popular, about 25% more than Goldcoin. The soft red winter wheat Triplet produced about 24% more than Goldcoin.

Spring wheat yield improved slightly less during the same period, according to similar eastern Washington data (169) (Table X). The common white spring wheats Dicklow, Onas, Fedawa, Idaed, and Federation exceeded the yields of Pacific Bluestem by 22, 21, 19, 19, and 12%, respectively. Baart and Jenkin exceeded the yield of Pacific Bluestem by 9 and 5%, respectively. When Baart was used as the base cultivar, improved yield potential was more modest, representing only 10 to 12% for the higher yielding types.

Yield improvements made in western wheat-producing regions other than the Pacific Northwest were probably more modest because of their generally less favorable environment, the secondary importance of wheat to these areas, and their different research priorities.

IV. SOFT WHEAT IMPROVEMENT, 1945-1959

A. Cultivar Distribution

This period was characterized by a further drop in soft wheat production in the 11 western states, from about 36% in 1919-1944 to about 29% in 1945-1959 (Table VII). The soft red winter wheat area declined to less than 1% of all

TABLE X
Relative Yields (Percent) of Certain Cultivars Grown in 1919-1944 Compared with Check Cultivars from Tests in Eastern Washington in 1931-1943

	Average Percent of			Average Percent of	
Winter Cultivar	**Kharkof**	**Goldcoin**	**Spring Cultivar**	**Baart**	**Pacific Bluestem**
Hybrid 128	99	118	Dicklow	112	122
Albit	97	115	Onas	111	121
Hymar	102	121	Idaed	110	119
Red Russian	83	99	Fedawa	110	119
Triplet	105	124	Federation	103	112
Goldcoin	84	100	White Federation	102	111
Golden	98	117	Baart	100	109
Rex	105	125	Jenkin	97	105

production and the most dramatic change occurred in the club wheat area, which rose from about 4% in 1944 to 14% by 1959. Club wheat increased at the expense of common white wheat, which dropped from 30% to 15%. The numbers of cultivars with soft wheat properties reportedly grown were reduced somewhat from the previous period. During 1949, 1954, and 1959, 49, 48, and 54 soft wheats were grown, but only 8 to 10 cultivars occupied significant areas (1% or more) when sampled at these five-year intervals (44,129,136).

Soft red winter cultivars occupied only 4,000 to 27,000 ha in the West during this period; the most important were Triplet, Red Russian, Jones Fife, and Mediterranean (Table IX).

The seven most important common white spring wheats were Baart, Federation, Ramona, Idaed, Lemhi, Orfed, and Marfed (Tables VIII, IX). Areas of Baart, Federation, and their backcross prototypes declined sharply after 1944, as farmers switched to club wheats, the newer soft white spring wheats, and improved hard red winter types. Ramona and its backcross types predominated in California and comprised 34 to 43% of the state's wheat crop between 1954 and 1959. Liked by the trade and resistant to several diseases, Ramona 44 and related types remained popular until semidwarfs were introduced in California. Idaed was released in 1938 and became popular in Idaho and Washington. It had short stiff straw and high test weight and matured 7 to 10 days earlier than most other white spring wheats. Lemhi was released in Idaho in 1939 and became popular in the irrigated Snake River plateau. Lemhi combined the short stiff straw of Federation with the white flour color of Dicklow; it was similar to Federation in other quality factors. Orfed was never grown extensively, but it became an important parent of many later cultivars. Bred at Pullman, Washington, Orfed was a short, stiff-strawed, shatter- and common bunt-resistant cultivar that yielded well, particularly from fall seedings, in eastern Washington. Orfed outyielded Triplet, the next most productive winter wheat, by 4% in 23 station-years of eastern Washington tests (171). Marfed was adapted to the 30-cm rainfall areas of eastern Washington and to irrigation in the Columbia Basin.

The five most important common soft white winter wheats grown in this period were Goldcoin, Golden, Rex, Brevor, and Alba (Tables VIII, IX). Goldcoin, the oldest cultivar of the subclass, disappeared from production about 1954 after having been grown in scattered areas of the West for more than 70 years (Table VIII). Golden occupied between 62,500 and 94,900 ha until 1959 when its area dropped by nearly one-half (Table IX). The popularity of Rex, with extremely poor milling quality, waned after 1949, when new, more bunt-resistant wheats became available.

Brevor established a new high level of common bunt resistance and was extensively grown in the Pacific Northwest by 1954. It was selected from a complex cross involving Turkey, Florence, Goldcoin, Federation, and Oro at Pullman, Washington. Brevor had at least three genes for common bunt resistance (Bt_1, Bt_3, and Bt_4) and had the shortest and stiffest straw of all other Pacific Northwest winter wheats (172). Because of its resistance to lodging, applications of medium to high levels of nitrogen were possible, which raised yields even further. However, Brevor had poor milling, low threshability, and mediocre seedling vigor and never was grown as extensively as Goldcoin or Rex.

By 1959, Alba became popular in western Oregon. It lacked adaptation to other major soft wheat production areas of the West (129) (Table IX).

Important club wheats grown from 1945 to 1959 were Elgin, Elmar, and Omar (Tables VIII, IX). Elgin became the leading club wheat and was grown on 4.7% of all western wheat land by 1949. Elgin was selected from Alicel at Pendleton, Oregon, in 1932. It established a new level of yield in the Pacific Northwest, being the most productive of 27 fall-sown cultivars tested in eastern Washington from 1931 to 1943 (169). In addition to high yield potential, Elgin had short stiff straw, resisted lodging and shattering, and responded very well to commercial fertilizer. Despite being very susceptible to common bunt, Elgin rapidly replaced Hymar. Pacific Northwest farmers came to expect all cultivars to have the desirable straw characteristics of Elgin (135).

By 1954, the major switch to fall-sown club wheat occurred and clubs made up 48% of the area sown to wheats with soft-wheat properties. The major club wheat was Elmar, a backcross-derived selection from the cross Hymar × 3*Elgin, with Hymar smut resistance and Elgin plant type. Because it yielded similarly to Elgin, stood well, resisted shattering, and had desirable quality, Elmar was widely grown, and by 1954 it occupied 10.1% of the western wheat area (Table VIII).

Common bunt caused a major change in cultivar distribution by 1959. Elmar, which had only the Bt_1 gene for resistance, was severely attacked by several previously minor races and planting dropped to only 1.0% of the area (Table VIII). Fortunately, a new white winter club, Omar, was resistant to the then prevalent races of common bunt (88). Omar immediately became popular and was grown on 12.0% of the western wheat area by 1959. It was particularly adapted in the Pacific Northwest, where it represented 47% of the area in Oregon and Washington in 1959. In addition to common bunt resistance, Omar outyielded Elmar and set new high quality standards for club wheat.

B. Establishment of the Western Wheat Quality Laboratory

During this period the complexity of quality problems of western-grown wheats was of greatest concern. Before 1946, publicly supported cereal quality laboratories in other regions conducted quality evaluations of western-grown wheats. The West grew every kind of market class of wheat, which compounded quality problems (135). By 1946, quality differences caused by environmental factors had also become very complex production considerations in the West. Cultivars such as Onas 41, Baart 38, Requa, Marfed, Mosida and the bread types Oro, Rio, Wasatch, and Cache had less than satisfactory milling or flour quality (116). After the cultivar Rex, with notoriously poor milling quality, was released, the industry unified to support a federal quality laboratory.

In mid-1946 the Western Wheat Quality Laboratory was established at Pullman, Washington, and by 1948 it was staffed with a cereal technologist, milling technologist, two cereal chemists, and a baking technologist. The two primary purposes of the laboratory were cultivar characterization and research. The first was needed to enable breeders to make accurate, intelligent decisions on the value of new wheats for processing. Research was needed to determine the fundamental factors responsible for the desirable and undesirable traits of wheat and flour (M. A. Barmore, personal communication).

C. Cultivar Quality Traits

The extensively grown club wheats Elmar and Omar had quality attributes very similar to those of Elgin. Western Wheat Quality Laboratory tests over several years indicated that Elmar was somewhat lower than Elgin for milling score, flour yield, cookie diameter, and sedimentation and slightly higher for absorption and viscosity (M. A. Barmore, personal communication). Omar was almost identical to Elgin in every respect for milling and physiochemical flour quality traits. During the era of Omar, the superior milling and pastry flour quality characteristics of white club wheat became firmly established, and a strong export market for the low-protein clubs was secured with several Pacific rim countries.

Industry rated Idaed good for milling performance, family flour, and pastry, cake, and cookie quality but poor to fair for cracker use (156). It rated Marfed only fair for milling performance, family flour, crackers, and cake and poor for pastry and cookies (156). In addition to industry comparisons, the Western Wheat Quality Laboratory demonstrated that Idaed milled better than Federation and that Marfed and Lemhi definitely ranked below Federation for milling score and flour yield. Lemhi, Marfed, and Idaed had higher viscosity values than Federation but made better cookies. Onas derivatives rated poor for milling score and flour yield (M. A. Barmore, personal communication).

Brevor had marginal milling quality. Of 35 wheats rated by the Western Wheat Quality Laboratory for milling, only two were below Brevor, which ranked lowest in flour yield. Brevor had slightly higher baking absorption, viscosity, and sedimentation than Federation. It ranked the highest of all 35 wheats for cookie diameter (M. A. Barmore, personal communication).

D. Significance of Facultative Wheats

Wheat cultivars that can be sown either in the fall or spring (facultative) have long been considered desirable for some regions of the West. Facultative types have these specific features: they allow diverse management of a single genotype, they assure a ready source of spring wheat seed, they limit the numbers of cultivars grown in a production region, and they offer the option to reseed winter-damaged fall-sown fields with the same wheat. Actually, cultivars such as Pacific Bluestem, Jenkin, Hybrid 123, and Federation have often been managed as facultative types.

During this time period, O. A. Vogel developed two facultative cultivars, Marfed and Orfed. Selected from crosses between spring and winter wheats, these facultative wheats also served as extremely useful parental material for subsequent crosses. These wheats grew during favorable growing conditions in the winter, recovered early and rapidly in the spring, and survived infrequent cold spells of moderate duration. Orfed contributed the traits of rapid spring recovery to later, high-yielding semidwarf types. Marfed has excellent adaptation in Washington and has been grown from 1949 to the present.

The concept of exploiting crosses between winter and spring wheats is receiving renewed emphasis in the West, as in the Oregon State-CIMMYT (International Maize and Wheat Improvement Center) cooperative program. Goals of this venture are increased yield, earliness, winterhardiness, and drought tolerance. The new crosses suggest much new genetic potential has been gained from these recombinations (95).

E. Neutralization of Common Bunt

The bunt problem worsened continuously in the area until about 1956. New resistant cultivars and mercury seed treatments were only partially effective in reducing losses that reached a devastating level when Elmar was smutted by races virulent on the Bt_1 gene. By 1953, the racial dynamics of the pathogen had once again readjusted to host resistance (88). The mercury seed treatments did not protect against soilborne inoculum of the disease, but hexachlorobenzene showed promise shortly after World War II. This chemical had given good results in France (82), so several polychlorobenzenes came under scrutiny as fungicides in the Pacific Northwest. Tests between 1950 and 1955 showed that a 40% formulation of hexachlorobenzene controlled seedborne and soilborne common bunt. On-farm pilot tests verified the practical value of this chemical, and its nearly universal use as a fungicide quickly followed (82).

Almost simultaneous with the discovery of hexachlorobenzene came the release of Omar, which resisted all known races of common bunt. Within four crop seasons the amount of smutty wheat in the Pacific Northwest dropped from 34 to 0.5% (C. S. Holton, personal communication). The incidence of common bunt has remained low in the West because nearly all new cultivars have multiple resistance and seed treatment has become common practice. The threat of future losses to common bunt remains, however. Shortly after the release of Omar, yet another new, highly virulent race arose (88). Apparently seed treatment has kept this race from becoming important for more than 19 years.

F. Changes in Wheat Management

After World War II, use of nitrogen on wheat increased dramatically in the West. During the war, farmers aimed at all-out production and used less summer fallow. Summer fallow allowed available nitrogen to accumulate, but use of this method merely to build up available nutrients rapidly depleted soil organic matter. Therefore, use of commercial nitrogen fertilizer was necessary to achieve the high yields of lodging-tolerant cultivars such as Elgin, Elmar, and Brevor.

Research in this period focused on demonstrating the relationships between available moisture, available nitrogen, and wheat yield. With this information, the amount of nitrogen fertilizer needed for maximal yields could be calculated. An early working formula for the efficient use of nitrogen was 20 kg of wheat produced for each kilogram of added nitrogen (97). Efficiencies varied from 24 to 10 kg of grain produced for each kilogram of nitrogen added per hectare. Low rates of application were most efficient in the high-rainfall areas.

The cooperative wheat breeding program at Pullman, Washington, was one of the first in the nation to routinely test advanced selections, and cultivars at two levels of nitrogen (L. P. Reitz, personal communication). Medium and high levels of nitrogen were used on split plots to identify the genotypes most responsive to high-fertilizer management. Such tests began with the 1956 crop at three eastern Washington sites; by 1959 useful interpretations identified those genotypes capable of maximal N utilization.

Table XI shows average yields of 10 winter wheat cultivars sown at three sites for five crop years, 1955 to 1959. The five most recently developed cultivars—Elgin, Elmar, Marfed, Brevor, and Omar—had relatively short stiff straw, resisted lodging and shattering, and demonstrated high yield potential. The five older cultivars—Triplet, Hybrid 128, Hymar, Red Russian, and Golden—had one or more agronomic faults such as weak straw, tall plant height, or shatter susceptibility and generally had mediocre yield potential. The newer wheats responded more favorably than the five older cultivars to spring-applied nitrogen, with yield increases ranging from 4 to 13% and averaging 10%. In contrast, yields of the older wheats increased 0 to 9% but averaged only 5% (Table XI). The new wheats, more responsive to nitrogen, averaged 3,566 kg/ha at the medium level of nitrogen and 3,906 kg/ha at the high level, versus 3,352 kg/ha and 3,528 kg/ha for the older, less responsive cultivars.

High-yielding types fostered the use of commercial fertilizer on wheat in the West. Semidwarf wheats especially responded to optimal amounts for maximal yield, and they were quickly developed after the high-yielding, standard-height wheats Omar and Brevor were released.

Toward the end of this period, government programs restricting planting decreased the wheat area. Farmers reacted to this change by going back to summer fallow and intensively managing their wheat land in every way possible. Farmers exploited such yield-augmenting practices as commercial fertilizer, early seeding, clean tillage, chemical weed control, irrigation, and fall versus spring planting. The new wheats made available during that time lent themselves to such intensive management.

TABLE XI
Average Yields at Two Levels of Nitrogen Fertilization for 10 Wheat Cultivars (Based on 15 Station-Year Tests in Eastern Washington)

	Yield (kg/ha)		
Cultivar	Moderate Level of N	High Level of N	% Increase Above Moderate N
(responsive)			
Brevor	3770	4250	13
Omar	3970	4380	10
Elgin	3900	4040	4
Elmar	3630	3970	9
Marfed	2560	2890	13
Average	3566	3906	10
(less responsive)			
Triplet	3630	3970	9
Hybrid 128	3570	3770	6
Hymar	3500	3500	0
Red Russian	3100	3300	6
Golden	2960	3100	5
Average	3352	3528	5

G. Relative Changes in Yield

Wheat yields in the 11 western states rose from 1,198 kg/ha to 1,461 kg/ha between 1933-1942 and 1949-1958, an increase of 22%. The release of cultivars with high yield potential accounted for much of this rise. Not all of this increase could be credited to high-yielding cultivars, however; increased use of commercial fertilizer, better management, and irrigation also contributed heavily.

The winter wheats developed during this time did not produce the dramatic increases of the cultivars released toward the end of the 1919-1944 period. For instance, only Omar yielded significantly more than Elgin, based on 25 station-years of tests in eastern Washington. The newer cultivars Omar, Brevor, Elmar, Orfed, and Marfed yielded 34, 29, 27, 12, and 7% more grain, respectively, than Red Russian based on 25 to 61 station-years (C. J. Peterson, personal communication). The main priority during this period was the breeding of common bunt-resistant cultivars: yield potential was somewhat secondary.

Increases in spring wheat yields were less than those of winter wheats. Based on 42 to 132 station-years in eastern Washington, Awned Onas, Onas, Marfed, Lemhi, Idaed, Orfed, and Lemhi 53 yielded 12, 9, 9, 6, 3, 3, and 3% more than Baart, respectively (C. J. Peterson, personal communication).

V. SOFT WHEAT IMPROVEMENT, 1960 TO PRESENT

A. Cultivar Distribution

In 1964, only four soft wheat cultivars were grown on at least 1% of western wheat area. These were Gaines, Omar, Brevor, Ramona, and its prototypes (Table VIII). Soft wheat cultivars made up 29% of the area in the West, only trace amounts of soft red winter were grown, and club wheat was used on 4.0% of the area (Table VII). Only 14 of the 45 soft wheat lines grown in 1964 represented 4,860 ha or more (130).

The most important soft white winter wheats grown during this period were Gaines, Nugaines, Hyslop, Brevor, Golden, Goldcoin, and Druchamp (Tables VIII, IX). Gaines, the first North American semidwarf wheat cultivar, dominated western soft wheat production by 1964 and represented 45% of all of the soft wheat area sown that year. Brevor remained popular until 1964 because it had stable resistance to stripe rust.

By 1969, the popularity of common soft white winter semidwarf wheat was well established in the West. Gaines and its sister line Nugaines comprised 19% of the wheat area in the West and 61% of the region's soft wheat (Table VIII).

In 1974, production of Gaines and Nugaines dropped following the release and acceptance of two new high-yielding soft white common winter wheats, Yamhill and Hyslop. Both cultivars were developed at Corvallis, Oregon. Yamhill was selected from the cross Heines VII × Alba and had excellent adaptation to the coastal regions of Oregon and Washington. It has very strong straw, high yield potential, and moderate resistance to leaf rust and most races of stripe rust (93). Yamhill yields have averaged 19 to 46% more than yields of Nugaines and

Druchamp over a seven-year period in western Oregon. A new stripe rust race that is virulent on Yamhill was found in western Washington in 1974 (101). Hyslop was selected from a cross of Nord Desprez × 2*CI 13438. It is a semidwarf but carries a different gene (Rht_1) for the semidwarf trait than does Gaines (Rht_2). This cultivar is resistant to the prevalent races of stripe rust and has moderate resistance to powdery mildew, leaf blotch, and leaf rust. Until 1978, Hyslop had the highest yield potential among common soft white wheats grown in the Pacific Northwest (94).

Production of soft white spring wheat changed from several old cultivars to new, high-yielding semidwarf types during this period. Ramona backcross derivatives were the most widely grown common soft white spring wheats in the West by 1964 (Table IX); smaller amounts of soft white wheats of Lemhi, Idaed, Federation, and their prototypes as well as Marfed were grown (Table VIII). Marfed became the major soft white spring wheat by 1969, followed by Lemhi and Idaed and their prototypes (Table IX). The area of Federation types declined to their lowest levels since 1924 (Table VIII). By 1974, soft white spring wheat production on the Columbia and Snake river plateaus shifted primarily to the cultivars Twin and Springfield, which were developed at Aberdeen, Idaho. Springfield was the first soft white spring semidwarf wheat released in the United States. Both wheats are Lemhi types selected from a complex cross involving Lee, Chinese Spring, *Aegilops umbelluta*, and two Lemhi backcross derivatives. In tests over several years, Twin has exceeded the yield of Lemhi 66 by 26% (159).

Club wheat area dropped markedly by 1964 mainly because of the change to Gaines, which was high-yielding and resistant to stripe rust. The six most important club wheats from 1960 to 1974 were Moro, Paha, Omar, Elgin, Elmar, and Big Club 60 (Table IX). Between 1959 and 1964, the area sown to Omar was reduced 72%, primarily because of its vulnerability to stripe rust. Moro was the leading club wheat by 1969. Selected from the cross PI 178383 × 2*Omar at Pendleton, Oregon, in 1965, it has a single gene for resistance to stripe rust and some tolerance to strawbreaker foot rot; it emerges better than any other cultivar. The flour of Moro is somewhat stronger than most older clubs. The area of club wheat rose when Paha was released in 1970 (Table VII). Developed at Pullman, Washington, from a cross of Suwon 92 × 4*Omar, Paha is 10 to 20 cm shorter than Omar, more lodging-resistant, and more tolerant to strawbreaker foot rot than either Moro or Omar. Paha generally outyields these two cultivars by 10 to 20% in the 28 to 57 cm rainfall areas. It had specific resistance to stripe rust, and by 1974, a race highly virulent to Paha became prevalent in areas where the cultivar was best adapted.

Areas that produce western soft wheat are more concentrated now than they were in the past. California farmers once grew nearly all soft white wheat but now sow hard red types. The Mexican wheats were introduced to that state in about 1962 to secure stripe rust resistance. The percentage of white wheat with mostly soft wheat properties grown in California at five-year intervals declined as follows: 1959, 96.8%; 1964, 91.4%; 1969, 46.5%; 1974, 6.1%. Soft wheats have declined significantly in recent years in Arizona and Utah. Where once soft wheat was produced in significant quantities in all of the western states, the three Pacific Northwest states now represent the only significant area of soft wheat production. Only nine cultivars accounted for most of the area by 1974.

B. Stripe Rust Epiphytotics and Breeding for Resistance

Stripe rust had not been considered a serious disease of wheat in the West for nearly two decades. Then suddenly, the 1960 crops of the Pacific Northwest and California were hit by a serious outbreak. Losses were even greater in 1961 and reached more than 816,480 tons. The disease has long existed in the United States; preserved grass collections show it was present in the late 1800s. Climate, topography, wind direction, and distribution of grass hosts on the Pacific Slope favor the year-round persistence and rapid spread of the stripe rust fungus (79).

The major reasons for the outbreaks in the 1960s were the occurrence of environmental conditions ideal for the organism, the predominance of highly susceptible cultivars that were often sown early and hence more vulnerable to infection. More than 75% of the Washington wheat area was sown to very susceptible wheats such as Omar, Elgin, and Columbia. Brevor, Burt, Requa, and Orfed gave various levels of resistance. Brevor gave stop-gap resistance between 1959 and 1962, when its production rose from 2 to 31% in the Pacific Northwest. The release of Gaines in 1961 came at an opportune time. It possessed adult plant resistance to the disease. The stripe rust epiphytotics abated almost as quickly as they began because of the effective resistance of Gaines.

Timely research revealed that both large-effect and small-effect genes controlled resistance to the stripe rust pathogen. Older wheats such as Rio, Requa, Ridit, Idaed, Cheyenne, Golden, and Red Russian showed various levels of field resistance (79). Initial screening revealed that field resistance was generally quantitatively inherited, with evidence of transgressive segregation and complementary genetic control (100). It was difficult to find a wheat that did not contribute some resistance to its progeny (149). Genes for adult, or field, resistance were usually recessive and could be readily accumulated to give increased additive resistance (6,100,125). This type of apparently nonspecific field resistance is carried by Gaines, Nugaines, Wanser, McCall, Tendoy, and Luke. Gaines and Nugaines have been extensively grown for more than 15 years and have maintained stable resistance. In contrast, cultivars with exclusively seedling or specific resistance have screened out virulent races at an alarming rate. Shortly after production of Moro, Pitic 62, Yamhill, Paha, and Norco began, each was attacked by a new, but different, virulent race. These cultivars apparently have a single, dominant, large-effect gene for resistance. To date, more than 20 pathogenic races have been isolated (R. F. Line, personal communication).

A desired breeding goal is to incorporate nonspecific and specific forms of resistance to stripe rust into commercial cultivars. Specific resistance would be a valuable addition, because the nonspecific resistance of Gaines still permits significant loss under ideal conditions for stripe rust. Specific resistance can be readily detected by race-testing seedlings, but so far the best proof of nonspecific resistance is the stability of resistance in field tests where different races grow.

According to limited testing, the use of multiline cultivars may be a way to control the highly dynamic stripe rust pathogen. The multilines tested in Washington have comprised 6 to 10 different sources of resistance and have yielded at least equal to the grand mean of their components. In some instances they have yielded as well as the most productive component (R. E. Allan, unpublished data).

C. Identification of the Value of Semidwarf Growth Habit

As stated earlier, a major objective of Pacific Northwest winter wheat improvement has been the breeding of short-strawed, high-yielding, and lodging-resistant cultivars. Vogel et al (170) found that the short-strawed cultivars Elgin, Elmar, and Brevor permitted fall seeding on highly fertile land earlier than was economically feasible with the taller, older cultivars. However, when sown very early to control erosion, these cultivars grew too luxuriantly in the spring, lodged, produced excess straw, and had low grain yields. Erosion could be controlled by early seeding with an intermediate-height wheat that yielded well and resisted lodging. Before 1948, all short types tested at Pullman, Washington, were cytologically unstable, yielded poorly, and had low test weights (170). In 1948, the Japanese wheat Norin 10 (CI12699) received from S. C. Salmon, approached the desired trait combinations of plant height, lodging resistance, and kernel type. Norin 10 grew to about half the height of tall commercial cultivars and had mediocre yield, partially owing to sterility. From a cross of Norin 10 × Brevor, several lines were selected with 13 to 20% increase in yield over productive standard-height cultivars (170). The best performing semidwarf lines were not only notably higher in grain yield but also considerably lower in straw weight when compared to standard-height cultivars. The straw-grain ratios of the semidwarfs averaged 1.61, whereas those of Brevor and Elmar averaged 2.39, indicating that the semidwarfs were therefore more efficient grain producers. Vogel et al (170) recognized that the Norin 10 × Brevor selections held promise as progenitors of winter wheats for highly productive soil and climatic conditions. They emphasized that semidwarfs should be especially adapted to very early fall seedings for controlling erosion. They argued that with semidwarfs the previously accepted wheat production limitations needed to be reevaluated, and they called for more realistic management research on this plant type.

D. Advantages and Disadvantages of Semidwarfs

From the onset of the development of the semidwarfs in the Pacific Northwest, problems associated with this plant type were encountered. Vogel and Craddock (personal communication) found that the early semidwarf selections were very susceptible to powdery mildew and leaf rust and that resistance to these diseases would be needed to achieve their maximal production. They noted that, as a group, the semidwarfs were winter-tender and gave disappointing milling ratings based on a 5-g micromilling test. Seedling emergence of the first semidwarf types was particularly poor; most lines were inferior to Brevor, which itself had mediocre emergence characteristics. These early observations proved that semidwarfs could use much heavier amounts of nitrogen than were economical for other cultivars.

E. Semidwarf Wheats and Their Predominance

The release of Gaines in 1961 began a new era of winter wheat improvement both for the West and for other wheat-growing areas of the world. Gaines was the first semidwarf cultivar to be distributed for commercial production, and its remarkable yields in the Pacific Northwest drew worldwide attention. Verified yields of 8,330 to 11,196 kg/ha were reported by farmers of the region (21). The Pullman, Washington, wheat improvement group led by O. A. Vogel developed Gaines from the cross Norin 10 × Brevor × Orfed × Brevor sib × Burt. The major attributes of Gaines were exceedingly high yield potential under intensive-management systems, efficient use of heavy applications of nitrogen (average, 25% heavier rates), lodging resistance, rapid spring growth recovery, adaptation to early seeding, and improved foliar and common bunt resistance. Within a few years, Nugaines, a sib of Gaines, was released. Nugaines had higher test weight, improved stripe rust resistance, and better milling properties than Gaines. The main problems associated with these semidwarf wheats were lack of seedling vigor, quality inferior to that of club wheats, and less winterhardiness than Turkey type wheats. Despite these shortcomings, semidwarfs comprised 39, 52, and 53% of the wheat area in Idaho, Oregon, and Washington by 1969 (112).

Vogel et al (168) indicated that improvement of semidwarfs from the initial types gave greater yields and combined new plant traits for more economical harvesting. The highest yielding semidwarfs with high resistance to shattering and good threshing qualities also had strikingly similar spike, leaf, culm, awn, and growth traits. These traits consisted of high tiller capacity, early spring recovery, high lodging resistance, coarse awns, and medium height, culm diameter, spike size, leaf length, and leaf width. These traits comprised a general plant pattern, and selections departing from this pattern for one or more traits were notably less efficient in production.

F. Management of Semidwarf Wheat

For the most part, management methods needed for semidwarfs were worked out after their release. Within a few seasons the ways to exploit the semidwarf yield potential were fairly well known. Two essential requirements were adequate moisture and fertility. Areas receiving at least 46 cm of annual precipitation had the potential to produce 6,731 kg/ha (100 bu/acre) of semidwarf wheat (71). In Washington, a wheat needed 10 cm of water to grow before it developed seed, each additional 2.5 cm of water produced 470 kg/ha of wheat, and nitrogen fertilization increased the efficiency of the use of available moisture (71). Other elements such as phosphorus and sulfur could be factors limiting yield. Almost universally, farmers increased fertility levels to fit the new production potentials of semidwarfs. Alternatively, overfertilization created new disease and quality problems. Soil testing was the most reliable way to determine plant nutritional needs.

Early seeding is essential for maximum production of Gaines and Nugaines. The semidwarfs appear to benefit more than other cultivars from a long growing season (71). The main problem with early seeding is achieving adequate stands from deep seedings in warm soils with moisture levels barely adequate for germination. As mentioned earlier, semidwarfs as a group lack seedling vigor. Weed control, seedbed preparation, conservation of seed-zone moisture, and the seeding operation are more important for semidwarfs than for other cultivars. Farmers' ingenuity helped to solve the problem of poor emergence in dry areas. A special chisel-type drill has been developed with split-packer wheels that place the seed in moist soil. The drill rather precisely regulates seeding depth and soil compaction.

Seeding too early courts disease. Losses caused by foot rot, root rot, flag smut, and foliar diseases increase with extremely early seedings. The most successful growers have become experts at judging the best time to seed in order to obtain the highest yields and minimize disease losses.

Proper weed control is essential with semidwarf wheats. When initially released, Gaines was said to compete favorably with weeds, but under commercial conditions, weed problems actually increased. Under poor management, grassy weeds in particular (wild oats and downy brome) overtop semidwarf cultivars, give the wheat more competition, and produce more seed that subsequently infests the seedbed (128). If recommended and timely weed management practices are followed, weeds are no more serious with semidwarfs than with other wheats.

G. Diseases Associated with Semidwarf Culture

The semidwarf growth habit now dominates in several western states. Presently, semidwarf wheats are planted in 75% of the state of Washington and in 98% of northern Idaho (117). In 1974, 91% of the wheat cultivars in California were semidwarf (131). The source of semidwarfing is genetically similar for almost all cultivars. This represents potential genetic vulnerability among our semidwarf wheats (112). This possibility prompted closer examination of the disease loss potential of closely related semidwarf and other wheat lines.

Semidwarfs have suffered proportionately more loss than nonsemidwarfs to stripe rust, flag smut, and dwarf bunt. Three years of tests with two genetic backgrounds showed that stripe rust-susceptible lines of the short semidwarf, medium semidwarf, and normal-height types yielded 34 to 51%, 24 to 40%, and 18 to 22% less than their respective resistant companion sib lines (R. E. Allan, unpublished data). The semidwarf plant conformation and its microenvironment apparently favor stripe rust.

The incidence of flag smut was also related to the semidwarf genes in tests of five isoline populations of four genetic backgrounds with various degrees of flag smut susceptibility. The two-gene semidwarf, one-gene semidwarf, and standard-height phenotypes averaged 63, 56, and 48% smutted tillers, respectively, for all populations (4). The two-gene and one-gene types had more smutted tillers than did standard-height types by factors of 1.32 to 1.00 and 1.18 to 1.00, respectively. Higher incidence of flag smut among semidwarfs may be indirect and could be related to increased opportunity to infect the wheat seedlings. Because infection occurs before emergence and because semidwarfs take longer than other wheats to emerge, they are vulnerable to the flag smut pathogen for longer.

Dwarf bunt tests paralleled those of flag smut. Two- and one-gene semidwarfs averaged 40 to 34%, respectively, more tillers with dwarf bunt than did their normal-height sister isolines in tests conducted at three field sites (R. E. Allan, unpublished data). The implications are that semidwarfs could be severely damaged by dwarf bunt.

The relation between disease damage and semidwarfism is less understood with Cercosporella (strawbreaker) foot rot. The two-gene phenotypes have shown notably more tolerance and less damage than have their standard-height counterparts in some genetic backgrounds, but the reverse has been true in others.

The need to breed cultivars with high, stable forms of disease resistance is even more important with semidwarf than with tall cultivars. Further, with semidwarfs it is even more necessary to exploit alternative methods of disease control such as chemicals and disease-reducing management practices.

H. Breeding for Dwarf Bunt Resistance and Loss of Export Markets

Unlike common bunt, adequate dwarf bunt control measures have not been forthcoming. This disease is similar in many respects to common bunt, but diseased plants are noticeably stunted. There are minor differences in morphology of the teliospores and considerable differences with respect to the environment. Dwarf bunt rarely occurs in regions other than the Pacific Northwest and Intermountain regions (149). Its control is difficult for several reasons. The ecology of the pathogen is such that simple wheat germ plasm screening methods are not possible, present fungicides are ineffective, new virulent races arise at alarming rates, inheritance of resistance is only partially understood, and disease management does not give adequate control.

Recently the dwarf bunt problem has become even more critical. The People's Republic of China represents a potentially important export market for West Coast wheat. In 1973-1974, more than 3.2 million tons of U.S. wheat was shipped to that country (D. W. Stuart, personal communication). Since September 1973, China has quarantined West Coast wheat to prevent the possible introduction of the dwarf bunt fungus.

The quarantine has generated renewed research emphasis to control dwarf bunt. Until now, the breeding approach has been similar to that used against common bunt. In fact, the same sources for resistance have been used in developing cultivars resistant to dwarf bunt. The approach has been only partially effective, since new virulent races rapidly coevolve as resistant cultivars are developed. Several races of the fungus are able to attack each of the 10 genes used as host differentials (J. A. Hoffmann, personal communication). Even the 14 cultivars and selections with resistance to the then-current races of dwarf bunt developed some smut in one or more test sites in 1975. Luke, Raeder, and Barbee are soft wheats with resistance to some races of dwarf bunt, but they are readily attacked by others.

Pope and Dewey (126) have advocated a new breeding approach. They concluded that high levels of resistance to dwarf bunt can be obtained only from interacting gene complexes. Their strategy is to secure high resistance to dwarf bunt via the construction of gene complexes, with emphasis on genes with small effects that can complement the resistance of genes with large effects. Testing with a gradient of disease severity and delayed critical screening of advanced lines are ways to improve the chances of detecting the desired combination of genes with small effects. Research on chemical control, disease management, ecology, and physiology of the pathogen, methods of spore identification, and viability has increased in a multifaceted approach to control this disease.

I. Foot Rot Becomes the Major Disease

After common bunt was controlled in the Pacific Northwest, Cercosporella foot rot (strawbreaker) became the most destructive soilborne disease of winter wheat in the major soft wheat-producing areas (28). The pathogen's range is much greater than was first shown. Its increased geographic extension is believed to be associated with the wheat fallow management system, better moisture conservation, greater use of fertilizer, earlier seeding, and use of high-tillering cultivars (28). Because early seeding in a winter wheat fallow system greatly increases the disease and because medium seedings and relatively late seedings are seldom damaged severely, avoidance of very early seeding is a common control method. Rotations are not economically feasible in dry wheat fallow regions, so they cannot be used for control. Good progress in breeding resistant cultivars has been made and should eventually offer excellent control.

Contrary to popular belief, semidwarf wheats are not more susceptible to Cercosporella foot rot than were the cultivars they replaced. Semidwarfs could be seeded early and they tillered more, which greatly improved the microenvironment of the pathogen. Actually, Gaines and Nugaines suffered less yield loss than Omar and Brevor under tests with controlled levels of disease inoculum (C. J. Peterson, personal communication).

Screening wheat germ plasm for tolerance to Cercosporella foot rot was greatly expedited once Bruehl and Nelson (27) developed a reliable yet simple inoculation procedure. The fungus was cultured on sterilized oat kernels and this inoculum was spread in the fall over plots to be screened for tolerance. With uniform levels of infection, breeders could compare disease symptoms, lodging, yields, and test weight between inoculated and noninoculated plots. Such data have allowed consistent progress in selecting lines for increased tolerance. Several new semidwarf cultivars, including Luke, Paha, and Hyslop, have more tolerance to the disease than do either Gaines and Nugaines. Recently, the line Cerco (CI15922) was released with even higher tolerance to strawbreaker foot rot (124).

J. New Wheats for Specific Production Situations

The number of cultivars with special adaptative or disease-resistant traits has increased, caused undoubtedly by the constantly rising complexities of modern wheat production.

Several soft white winter wheats have been developed for special situations in recent years. Sprague represents the first snow mold-resistant cultivar adapted to the snow mold-infested areas of north central Washington (26). This wheat also possesses excellent seedling vigor and the low level of postharvest dormancy needed for early seeding. Yamhill, a common soft white winter wheat, has been released specifically for the Willamette Valley of Oregon. It produces strong, stiff straw, tolerates several soilborne diseases, and has produced record yields. Another common soft white winter wheat, Peck, was released in northern Idaho where it is particularly adapted to low, poorly drained soils. Peck has good seedling vigor and does well in areas where obtaining suitable stands is a chronic problem. Raeder was recently introduced as a dwarf bunt-resistant common soft white wheat for northern Idaho. Daws, another common soft white winter wheat, was released specifically for its increased level of winterhardiness. Natural and simulated cold tests indicate that Daws has greater winterhardiness than other soft-type semidwarf cultivars. Its release should lessen the genetic vulnerability for this trait in soft wheats of the Pacific Northwest where, in 1973, lack of winterhardiness caused loss of more than 816,480 tons.

Urquie, a common soft white facultative wheat, is to replace Marfed in the 30 to 45 cm rainfall areas of eastern Washington. Urquie has greater winterhardiness, milling quality, and yield potential than does Marfed.

Barbee is a high-yielding soft white club wheat for eastern Washington and northern Idaho. This bearded club lodges less and produces about 5% more grain than Paha in the high rainfall areas of the Palouse.

Several common soft white spring wheats including Springfield, Twin, Fielder, and Fieldwin are well adapted to irrigation. Their main virtues are high yield potential, stiff straw, resistance to certain foliar diseases, and satisfactory soft wheat quality.

K. Impact of Increased Irrigation on Wheat Breeding

Increased irrigation has and will exert strong influence on soft wheat breeding objectives in the western United States. Presently irrigated land is about 7.7 million ha in the West, excluding Colorado, Montana, and Wyoming. Estimates for irrigation potential of the area range from 33 to 47 million ha (61). The potential is particularly great for the traditional soft wheat areas. In Washington, private development alone has added about 12,150 ha annually since 1965 (174). Many wheat farmers in the 20 to 35 cm rainfall areas use 8 to 13 cm of supplemental irrigation water once or twice in the growing season. These quantities of water often increase yields by 1,400 to 2,400 kg/ha.

Irrigation has caused new production hazards. Problems with foliar diseases have intensified. Leaf rust was once considered of minor importance but has become increasingly damaging, particularly on irrigated wheat in the Columbia Basin. Powdery mildew is relatively unimportant on dryland wheat but flourishes on rill-irrigated wheat. Virus diseases, particularly wheat streak mosaic and barley yellow dwarf, have become more serious with expanded irrigation.

Irrigation has intensified problems with soilborne diseases. Take-all disease on wheat rotations often is more serious under irrigation than on dryland, and Cercosporella foot rot attacks wheat grown under supplemental irrigation. In fact, almost all diseases and pests that are serious on early-sown dryland wheat are favored by irrigation as well.

Extensive irrigation probably accounts for the increased problem with green bugs on wheat in the Pacific Northwest. The last two outbreaks of this insect started with early-sown irrigated wheat fields along the Columbia River.

Clearly, wheat improvement priorities must be changed to fit the new problems related to irrigated wheat culture. Some of the problems require expertise and facilities that are presently either limited or not available to western research stations. Easy, rapid solutions to the problems are therefore not expected.

L. Breeding Wheats for Early Seeding and Erosion Control

Erosion is a major problem of the soft wheat-producing regions of the West. It is most threatening in the Palouse region of northeastern Oregon, northern Idaho, and eastern Washington. Soil losses in Whitman County, Washington, have exceeded 224 tons/ha and average 22.4 tons/ha on fall-sown wheat (R. I. Papendick, personal communication). It is disconcerting to realize that the area produces world records for both wheat yield and soil loss. Because of the steep, undulating topography, heavy winter precipitation, and the need for summer fallow, conventional erosion control measures have been at best only partially effective. With the water quality goals of the Federal Water Pollution Control Act of 1972 (PL 92-500) erosion prevention has become even more urgent.

Development of wheats adapted to early seeding for control of erosion has been an important breeding objective in the Northwest for more than two decades. As stated earlier (170), a main impetus for development of semidwarf wheats was the acquisition of types adapted to early seeding for erosion control. The main criteria for early-sown cultivars for erosion prevention are:

DISEASE RESISTANCE

a) Resistance to most foliar diseases, e.g., leaf rust, leaf blotch, powdery mildew, and particularly stripe rust; b) tolerance to foot and root rots, e.g., take-all, dryland foot rot, and particularly Cercosporella foot rot; c)tolerance to viruses, yellow dwarf, and wheat streak mosaic in particular; d) resistance to Cephalosporium stripe.

INSECTS

Resistance to green bugs, other aphids, and wireworms.

ENVIRONMENTAL STRESS

Ability to a) germinate and emerge rapidly from deep seedings in warm, dry, and sometimes surface-crusted soils; b) germinate and grow in seedbeds with suboptimal moisture; c) survive cold spells from emergence to jointing (winterhardiness) yet exploit mild winter periods to achieve maximum yield potential; d) evade spike damage caused by spring frosts; e) use moisture efficiently throughout the life cycle; f) tolerate postmaturity weather damage and specifically resist lodging, shattering, and sprouting; g) form deep, well rooted crowns to resist stand loss to heaving, erosion channeling, and wind.

ADAPTIVE AND ECONOMIC QUALITIES

a) Favorable competition with weeds in early seedings under wheat-fallow management and toleration of herbicides; b) efficient use of soil nutrients; c) high yield potential with optimum grain/straw ratios; d) presence of those plant, leaf, crown, and rooting traits that allow minimum water runoff; e) minimal harvest problems through traits for uniform and moderately early ripening, lodging resistance, good threshability, and manageable straw residue.

QUALITY

a) Low to moderate protein levels not affected by environment; b) stability of those milling and flour traits needed in western soft wheats; c) low α-amylase activity under adverse weather.

No present commercial cultivar comes close to having these traits, yet considerable progress has been made in identifying useful prototypic germ plasm that has one or more of the characteristics needed. For instance, Cerco has the best tolerance level to Cercosporella foot rot. Luke has good tolerance to dryland foot rot and ranks high for water use efficiency. Hyslop probably ranks best for resistance to foliar diseases. Several new selections and Nugaines show tolerance to Cephalosporium stripe disease. Among commercial wheats, Sprague has the highest level of resistance to snow mold and the best overall balance of stand establishment traits for early seedings. Daws exhibits the needed winterhardiness and spring growth recovery characteristics. Raeder and several other new selections have high resistance to flag smut. Peck has those growth traits needed to reduce vulnerability to spike frost damage. Several wheats have strong, lodging-resistant straw; but the two-gene semidwarfs as a group excel in lodging resistance. Cerco probably represents the nearly ideal type for trouble-free mechanical harvesting.

No prototypes have been identified for stable milling and baking properties; tolerance to sprouting and high α-amylase; stable, moderate protein level; ideal plant type for reduced water runoff; tolerance to viral diseases; or the most efficient use of soil nutrients. Plant breeding is only beginning to incorporate these traits into adapted, high-yielding cultivars.

M. Hybrid Wheat Research

Breeding of hybrid wheat with soft wheat properties receives some attention in the West from both the public and private sectors. At least one commercial company has a staffed experiment station in southeastern Washington. The cooperative wheat improvement program at Pullman, Washington, has conducted hybrid wheat research since 1959. Much of the work has involved the *Triticum timopheevi* nucleocytoplasmic system. This program has concentrated on measurement of heterosis levels for yield, cross-pollinated seed production, male restorer improvement, hybrid disease reactions, gametocide studies, and quality of hybrids with normal cytoplasm.

The scope of the yield tests has been rather extensive and has entailed 2,002 comparisons of hybrids vs. conventional wheat and hybrids vs. high parents over five years of testing at four to seven sites in eastern Washington (R. E. Allan, unpublished data). Useful heterosis for grain yield was demonstrated. Hybrids equaled or exceeded the yield of Nugaines in 675 of the 2,002 comparisons and significantly surpassed its yield 15% of the time with increases of 11 to 49% above Nugaines. Hybrids compared favorably to their higher-yielding parents in 384 of the 2,002 comparisons and they significantly out-yielded the high parent 4% of the time.

The main limitation to adaptation of the Washington wheat hybrids has been incomplete and erratic fertility restoration. The best restorers comprise the Rf_1 and Rf_2 genes from *T. timopheevi* and unidentified small-effect genes from CI 13438, Burt, and Itana. Restoration from Primepi has been recently incorporated.

Unwanted heterosis for excessive straw also is a problem. Crosses between normal-height and semidwarf types grow as tall as or taller than their taller parent and lodge in the maximum-yield environments. Often the straw weight of hybrids is proportionately larger than their increase in grain weight. Hybrids from semidwarf types have the most suitable height and straw characteristics (3).

Data on flag smut and common bunt show that susceptibility is almost always inherited as a dominant trait. In fact, some hybrids show transgressive segregation for increased smut even when their parents have moderate or high resistance. Apparently both parents must carry the same resistant factors to obtain the homozygosity required for high resistance. The stripe rust reactions of hybrids frequently are intermediate or like the less resistant parent. Nonspecific adult resistance behaves as recessive or weakly dominant. Only genes for highly specific seedling resistance behave as strong dominants.

Quality evaluations on samples of 233 normal-cytoplasm hybrids and their parents during four years showed that satisfactory soft rather than hard winter wheat hybrids could be readily developed (7). Flour yield was generally inherited as a dominant or partially dominant trait, with 56% of all hybrids similar to or above their high parent. Over 57, 60, and 54% of the hybrids had mixing curve areas, alkaline water retention capacity, and viscosity values equal to or below those of their low parent, respectively, which indicated recessive control of these three criteria. High flour protein content also tended to be recessively inherited and 20% of the hybrids had flour proteins below those of their low parent. Prediction of hybrid wheat quality from parental data was not feasible and quality tests on the hybrids appear to be essential.

N. Quality Research Directly Aiding Plant Breeding

Problems with the milling quality of western wheats reached serious proportions in the mid-1940s when Rex, Rio, and Orfed were grown extensively. With the release of Brevor and the introduction of Norin 10 as a source of semidwarfism, these problems became even more acute. These cultivars had mediocre to poor milling properties but were used extensively as valuable parents. They had superior yield, straw strength, short height, and disease resistance. There was an urgent need to evaluate milling quality on a small amount of seed, preferably on a single plant basis. Seeborg and Barmore (143) reduced the sample size to 5 g by attaching a microfeeder to the first break role of a Bühler mill. Samples were passed through the role and the flour was screened. About 1,000 samples could be milled and rated per day. In this way offspring with inferior milling traits peculiar to Rex, Brevor, and Norin 10 could be readily identified. This screening method was used to select subsequently developed cultivars with satisfactory endosperm-bran separation traits such as Omar, Gaines, Nugaines, Burt, Moro, and Paha. The technique facilitated genetic study of this character. Everson and Seeborg (62) showed that more than four genetic factors control milling quality in a wheat cross. The heritability of the trait was 36%, and good progress was predicted for improved flour yield with recurrent selection for reduced bran weight.

Another member of the Western Wheat Quality Laboratory developed a method for determining crude protein content of wheat and flour by a dye-binding technique (167). This simple, rapid test allows one person to evaluate 30 samples per hour and is used extensively on wheat, other grains, soybean products, and milk. Its use in conjunction with the Kjeldahl nitrogen analysis serves as a rapid estimate of lysine content.

Viscosity is a useful criterion for evaluating soft wheat flour properties, particularly those of club wheats. Determination of this criterion was greatly facilitated when a member of the Western Wheat Quality Laboratory adapted the Brookfield viscometer for measurement of flour samples, replacing the

McMichael viscometer used for more than 50 years by cereal chemists (91). The newer instrument is much faster and simpler to operate, easier to clean, and requires less maintenance.

The Western Wheat Quality Laboratory acquired a Miag pilot mill in 1970 and began a cooperative project involving the laboratory and milling and baking companies in the United States, Japan, and Korea. Samples are made available to nine to 11 collaborative industrial groups in Japan, Korea, Oregon, Montana, Idaho, California, and Washington, which individually evaluate the samples for 14 to 18 different flour traits. These evaluations help wheat researchers make their final judgment of proposed new releases. This procedure identified Norco germ plasm as having outstanding noodle quality and also showed that Paha has noodle and sponge cake properties superior to those of Moro and Nugaines.

O. Export Quality Requirements of Western Soft Wheat

Export markets have become increasingly important for western soft wheat and flour. The first shipload of flour was exported to Japan from the Pacific Northwest in 1857 (173). Generally, the quantity of soft wheat exported from the region has increased, but year to year changes have been erratic in some periods because of fluctuating political decisions and economic situations. Table XII shows exports of all wheat from the Pacific Northwest by decade.

Until 1950, exports varied between 0.1 and 1.2 million tons; since then they have grown rather steadily, increasing by 1.6 to 2.7 million tons every decade. The region exports proportionately more white wheat than any other class. Biennially for 1966, 1968, 1970, 1972, and 1974, the total white wheat exported from the Pacific Northwest represented 89, 95, 75, 99, and 95% of the wheat grown in the region, or an average of 91% per year. Therefore, the needs of the importers of western soft wheat must be given high priority.

The main customers of western soft white wheat during 1965-1975 were Japan, 8.6 million tons; India, 7.2; Korea, 6.5; Pakistan, 2.7; Iran, 2.3; Taiwan, 1.6; and

TABLE XII
Wheat Exported Annually from the Pacific Northwest, Excluding California, by Decades

Period	Metric tons[a]
1875-1880	100,640
1881-1890	209,440
1891-1900	484,160
1901-1910	693,600
1911-1920	549,440
1921-1930	1,215,840
1931-1940	527,680
1941-1950	1,082,560
1951-1960	3,084,480
1961-1970	4,898,720
1971-1976	7,482,720

[a]Tonnage data were obtained from reference 173 for 1875 through 1962. Values for 1963 through 1976 furnished by W. D. Hughes (personal communication).

the Philippines, 1.0 million tons (84). Little is known of the quality requirements and uses for wheat of these customers. Probably the best understood are Japanese requirements for soft wheats used to make Udon noodles, Japanese sponge cake, and certain dry noodles. Udon noodles represent a wet and/or dry process noodle, and noodles of the Kishimen, Hiya-mugi, and Somen type are dry-process noodles. Generally western soft white wheat types are not suited for Soba and Chinese (Ramen) noodles because they usually are too low in protein and lack the flour strength needed for these products. Nevertheless, soft white wheats originating in the United States are regularly included in flour blends used to make these two noodles, as well as in bread and several confectionary products consumed by the Japanese (G. L. Rubenthaler, personal communication). Udon noodles require a protein level of 8.3 to 10.5% and ash contents of 0.38 to 0.50%, with flour strength properties similar to those of soft white and white club wheats. The three types of dry noodles require even lower protein, such as 8.0 to 10.0%, and ash of 0.35 to 0.45%. Udon noodles and these three dry noodles account for about 15% of Japanese wheat usage. Sponge cakes are made from about 6% of the imported wheat milled in Japan. They require protein of 7.1 to 7.8% and ash levels at 0.33 to 0.38, and they should have weak flour quality traits like those of the club wheats Paha, Elgin, and Omar (G. L. Rubenthaler, personal communication).

The Western Wheat Quality Laboratory makes preliminary noodle quality tests of advanced soft white lines. Measurements and ratings are assigned for noodle volume, texture, color, and solubilities lost during cooking. Noodle taste is an important criterion for Japanese food technologists, but because their standards are unknown to us, more empirical evaluation properties are presently stressed.

P. The High-Alpha-Amylase Problem

Markets in Japan, Korea, Philippines, and Taiwan also require wheats with low diastatic activity for noodles and sponge cakes. Preharvest rains on wheat can initiate germination, increase the α-amylase activity, and cause poor quality in these products.

Untimely rains have caused considerable sprout-damage, high α-amylase proportions of wheat production in the West at various times. In 1968, this problem cost the United States about $18 million in business and another $0.5 million in redress.

High α-amylase wheat is a complex problem involving seed dormancy, seedling vigor, diverse environments, and disease incidence; it is aggravated by certain management practices. Farmers prefer wheats that have low postharvest dormancy or types that can be planted very soon after harvest. In some western areas harvest and planting are done almost simultaneously. If wheat cultivars have a long dormancy requirement, poor stands result from early seeding in warm soils. Use of aged seed alleviates this problem, but most growers have been indifferent to this practice. Further, because of the diverse environments of the region, rain damage usually happens somewhere in any year, and exactly where the damage will occur is difficult to predict. Overriding these considerations is the need for cultivars that emerge rapidly. Those cultivars with low postharvest dormancy usually are among the fastest to emerge, particularly when soil moisture is limited.

Breeding tests now focus on identifying lines that resist sprouting and maintain a high falling number value (low α-amylase activity) after exposure to harvest rains (123). These tests are done in conjunction with postharvest dormancy evaluations. Some wheat selections have low postharvest dormancy but resist sprouting and maintain a high falling number. Others show little or no sprouting but have a low falling number. Lines that have a low falling number after moderate rain damage are discarded. Certain sprout-resistant lines apparently possess germination inhibitors in the glumes and once thrashed they germinate readily. This would be a most useful trait to incorporate into western club and common white wheats. Wheat prone to lodging, whether owing to disease or weak straw, also is apt to be higher in α-amylase activity; development of cultivars with foot rot resistance and stiff straw can help alleviate this problem.

Q. Relative Changes in Yield

Cultivars released during this period set new yield records throughout the West and led the way to an expanded wheat industry. Gaines and Nugaines were estimated to have added $1 billion to the farm economy of the Pacific Northwest for the period 1964-1975 (60). The common soft white semidwarf winter wheats have been particularly productive. Nugaines and Hyslop exceeded the yield of Kharkof by 48 and 57%, respectively, based on 95 station-years of tests conducted in five western states from 1971 to 1975 (F. H. McNeal, personal communication). They exceeded Elgin by 41 and 49%, respectively, for the same period. In similar tests the club wheat Paha surpassed the yield of Kharkof by 39% and Elgin by 32%.

Excellent advancement in yield potential was made by the new semidwarf common soft white spring wheats as well. In 102 station-years of tests from 1970 to 1975, Twin and Fielder exceeded the yields of Federation by 27 and 31%, respectively (F. H. McNeal, personal communication).

APPENDIX

Common and Scientific Names of Diseases, Insects, and Plants

DISEASES

Cephalosporium stripe—*Cephalosporium gramineum* Nis. & Ika.
Cercosporella foot rot—*Pseudocercosporella herpotrichoides* (Fron) Dei.
Common bunt—*Tilletia caries* (DC) Tul. and *T. foetida* (Wallr.) Liro
Dryland foot rot—*Fusarium culmorum* W. G. Smith and *F. graminearum* Schwabe
Dwarf bunt—*Tilletia contraversa* Kühn
Flag smut—*Urocystis agropyri* (Preuss) Schroet.
Glume blotch—*Septoria nodorum* Berk.
Leaf rust—*Puccinia recondita* Rob. ex Desm. f. sp. *tritici*
Loose smut—*Ustilago tritici* (Pers.) Rostr.
Powdery mildew—*Erysiphe graminis* DC ex Merat. f. sp. *tritici*
Septoria leaf blotch—*Septoria tritici* Rob. em. Desm.

Snow mold—*Typhula idahoensis* Remsb., *Fusarium nivale* (Fr.) Ces., *Sclerotinia borealis* Bub & Vleug.
Stem rust—*Puccinia graminis* Pers. f. sp. *tritici* Eriks & Henn.
Stripe rust—*Puccinia striiformis* West
Take-all—*Gaeumannomyces graminis* (Sacc.) Arx & Oliv. var. *tritici*

INSECTS
Cereal leaf beetle—*Oulema melanopus* L.
Greenbug—*Toxoptera graminum* (Rond)
Hessian fly—*Mayetiola destructor* (Say)
Wireworms—*Ctenicera pruinina* (Horn)

PLANTS
Downy brome—*Bromus tectorum* L.
Wild garlic—*Allium oleraceum* L.
Wild oats—*Avena fatua* L.

LITERATURE CITED

1. Agricultural Experiment Stations of the Western States, Land Grant Universities and Colleges. 1964. Soils of the Western United States. A joint regional publication, pub. at Washington State Univ., Pullman, WA. 69 pp.
2. Alexander, G. L. 1933. The results of bleaching Michigan soft winter wheat cake flour by the Brabender electric bleaching apparatus. Cereal Chem. 10:623-626.
3. Allan, R. E. 1973. Yields of wheat hybrids of *Triticum timopheevi* nucleo-cytoplasmic system. In: Proc. 4th Int. Wheat Genetics Symp., Aug. 6-11, Columbia, MO. pp. 311-317.
4. Allan, R. E. 1975. Relation of flag smut incidence to semidwarf growth habit in wheat isolines. Crop Sci. 15:427-429.
5. Allan, R. E., Heyne, E. G., Jones, E. M., and Johnson, C. O. 1959. Genetic analysis of 10 sources of Hessian fly resistance, their interrelationships and association with leaf rust reaction in wheat. Kans. Agric. Exp. Stn. Tech. Bull. 104. 51 pp.
6. Allan, R. E., Purdy, L. H., and Vogel, O. A. 1966. Inheritance of seedling and adult reaction of wheat to stripe rust. Crop Sci. 6:242-245.
7. Allan, R. E., and Rubenthaler, G. L. 1972. Genetic aspects of hybrid wheat quality. Agron. Abstr. p. 2.
8. American Association of Cereal Chemists. 1962. AACC Approved Methods, 7th ed., Am. Assoc. Cereal Chem., Inc., St. Paul, MN, Method 10-50, Baking test for cookie flour, Method 10-90, Baking test for cake flour.
9. Barmore, M. A., and Bequette, R. K. 1965. Weight per bushel and flour yield of Pacific Northwest white wheat. Cereal Sci. Today 10:72-77, 87.
10. Bayfield, E. G. 1932. The soft wheat improvement program for Ohio. Cereal Chem. 9:322-324.
11. Bayfield, E. G. 1935. Observations on the whole wheat meal fermentation time test. J. Am. Soc. Agron. 27:241-250.
12. Bayfield, E. G. 1936. The influence of climate, soil, and fertilizers upon quality of soft winter wheat. Ohio Agric. Exp. Stn. Bull. 563. 77 pp.
13. Bayfield, E. G., and Shiple, V. 1937. Soft winter wheat studies. V. Evaluating the quality and the strength of some varieties. Cereal Chem. 14:551-577.
14. Bayles, B. B., and Clark, J. A. 1954. Classification of wheat varieties grown in the United States in 1949. U.S. Dept. Agric. Tech. Bull. 1083. 173 pp.
15. Berg, S. O. 1947. Is the degree of grittiness of wheat flour mainly a varietal characteristic? Cereal Chem. 24:274-283.
16. Bever, W. M. 1947. Physiologic races of *Ustilago tritici* in the eastern soft wheat region of the United States. Phytopathology 37:889-895.
17. Bitzer, M. J., and Patterson, F. L. 1973. Pollen dispersal and cross-pollination of soft red winter wheat. (*Triticum aestivum* L.). Crop Sci. 13:92-95.

18. Bitzer, M. J., Patterson, F. L., and Nyquist, W. E. 1971. Hybrid vigor and gene action in a six-parent diallel cross of soft winter wheat. Can. J. Genet. Cytol. 13:131-137.
19. Blair, B. O., and Patterson, F. L. 1962. Controlled climate tests for evaluating the adaptation of winter wheats. Crop Sci. 2:374-377.
20. Bode, C. E. 1959. The Soft Wheat Quality Laboratory. Cereal Sci. Today 4:259-264.
21. Briggle, L. W., and Vogel, O. A. 1968. Breeding short-stature, disease resistant wheats in the United States. Euphytica 17:107-130. Supp. 1.
22. Briggle, L. W., Yamazaki, W. T., and Hanson, W. D. 1968. Heritability of three soft wheat quality characters in the F_2 and F_3 generations. Crop Sci. 8:283-285.
23. Briggs, F. N., and Holton, C. S. 1950. Reaction of wheat varieties with known genes for resistance to races of bunt, *Tilletia caries* and *T. foetida*. Agron. J. 42:483-486.
24. Brown, C. M., Weibel, R. O., and Seif, R. D. 1966. Heterosis and combining ability in common winter wheat. Crop Sci. 6:382-383.
25. Bruehl, G. W. 1967. Diseases other than rust, smut and virus. In: K. S. Quisenberry and L. P. Reitz, eds. Wheat and Wheat Improvement, Monograph 13. Am. Soc. Agron., Madison, WI. pp. 375-410.
26. Bruehl, G. W., Kiyomoto, R., Peterson, C., and Nagamitsu, M. 1975. Testing winter wheats for snowmold resistance in Washington. Plant Dis. Rep. 59:566-570.
27. Bruehl, G. W., and Nelson, W. L. 1964. Technique for mass inoculations of winter wheat in the field with *Cercosporella herpotrichoides*. Plant Dis. Rep. 48:863-865.
28. Bruehl, G. W., Nelson, W. L., Koehler, F., and Vogel, O. A. 1968. Experiments with *Cercosporella* foot rot (strawbreaker) disease of winter wheat. Wash. Agric. Exp. Stn. Bull. 694. 14 pp.
29. Bryan, W. E., and Pressley, E. H. 1929. Milling and baking qualities of pure lines of Arizona-grown wheat. Univ. Ariz., Agric. Exp. Stn. Tech. Bull. 27. pp. 67-100.
30. Caldwell, R. M. 1966. Advances and challenges in control of plant diseases through breeding. U.S. Dep. Agric. Bull. 33-110. pp. 117-126.
31. Caldwell, R. M. 1968. Breeding for general and/or specific plant disease resistance. In: Proc. 3rd Int. Wheat Genetics Symp. Aust. Acad. Sci., Canberra, Australia. 3:263-272.
32. Caldwell, R. M., Cartwright, W. B., and Compton, L. E. 1946. Inheritance of Hessian fly resistance derived from W38 and Durum P.I. 94587. J. Am. Soc. Agron. 38:398-409.
33. Caldwell, R. M., Cartwright, W. B., Compton, L. E., Patterson, F. L., and Schafer, J. F. 1960. Redcoat soft red winter wheat. Purdue Univ. Agric. Exp. Stn. Mimeo ID-44.
34. Caldwell, R. M., Cartwright, W. B., Compton, L. E., Schafer, J. F., and Patterson, F. L. 1959. Dual, a Hessian fly resistant soft red winter wheat. Purdue Univ. Agric. Exp. Stn. Res. Bull. 681. 20 pp.
35. Caldwell, R. M., Compton, L. E., Cartwright, W. B., Patterson, F. L., and Schafer, J. F. 1959. Monon wheat. Purdue Univ. Agric. Exp. Stn. Mimeo ID-36.
36. Caldwell, R. M., Compton, L. E., Schafer, J. F., and Patterson, F. L. 1959. Knox and Vermillion, early maturing soft red winter wheats. Purdue Univ. Agric. Exp. Stn. Res. Bull. 681. 20 pp.
37. Caldwell, R. M., Roberts, J. J., and Eyal, Z. 1970. General resistance "slow rusting" to *Puccinia recondita* f. sp. *tritici* in winter and spring wheats (Abstr.) Phytopathology 60:1287.
38. Caldwell, R. M. Schafer, J. F., Compton, L. E., and Patterson, F. L. 1957. A mature-plant type wheat leaf rust resistance of composite origin. Phytopathology 47:691-692.
39. Carleton, M. A. 1900. The basis for the improvement of American wheats. U.S. Dep. Agric. Div. Veg. Physiol. Pathol. Bull. 24. 87 pp.
40. Cartwright, W. B., and LaHue, D. W. 1944. Testing wheats in the greenhouse for Hessian fly resistance. J. Econ. Ent. 37:385-387.
41. Clark, J. A. 1936. Improvement in wheat. In: Yearbook of Agriculture. U.S. Dep. Agric. pp. 207-302.
42. Clark, J. A., and Bayles, B. B. 1933. Varieties of common white wheat. USDA Farmers Bull. 1707. 22 pp.
43. Clark, J. A., and Bayles, B. B. 1935. Classification of wheat varieties grown in the United States. U.S. Dep. Agric. Tech. Bull. 459. 164 pp.

44. Clark, J. A., and Bayles, B. B. 1951. Distribution of the varieties and classes of wheat in the United States in 1949. U.S. Dep. Agric. Circ. 861. 71 pp.
45. Clark, J. A., Martin, J. H., and Ball, C. R. 1922. Classification of American wheat varieties. U.S. Dep. Agric. Dep. Bull. 1074. 238 pp.
46. Clark, J. A., Martin, J. H., Quisenberry, K. S., Hooker, J. R., Leighty, C. E., and Dubois, C. N. 1929. Distribution of the classes and varieties of wheat in the United States. U.S. Dep. Agric. Bull. 1498. 68 pp.
47. Clark, J. A., and Quisenberry, K. S. 1929. Varieties of hard red winter wheat. USDA Farmers Bull. 1585. 13 pp.
48. Clark, J. A., and Quisenberry, K. S. 1933. Distribution of the varieties and classes of wheat in the United States in 1929. U.S. Dep. Agric. Circ. 283. 76 pp.
49. Clark, J. A., and Quisenberry, K. S. 1937. Distribution of the varieties and classes of wheat in the United States in 1934. U.S. Dep. Agric. Circ. 424. 68 pp.
50. Clark, J. A., and Quisenberry, K. S. 1942. Distribution of the varieties and classes of wheat in the United States in 1939. U.S. Dep. Agric. Circ. 634. 75 pp.
51. Clark, J. A., and Quisenberry, K. S. 1948. Distribution of the varieties and classes of wheat in the United States in 1944. U.S. Dep. Agric. Circ. 761. 80 pp.
52. Clark, J. A., Stephens, D. E., and Florell, V. H. 1920. Australian wheat varieties in the Pacific coast area. U.S. Dep. Agric. Bull. 877. 25 pp.
53. Collins, W. E. 1975. The wild garlic problem. What to do about it in 1975 and subsequent years. Commodity Activities, American Farm Bureau Federation, Park Ridge, IL. Mimeo, 11 pp.
54. Committee of Genetic Vulnerability of Major Crops. 1972. Genetic vulnerability of major crops. National Academy of Sciences, Washington, DC. 307 pp.
55. Cutler, G. H., and Brinson, G. A. 1935. The granulation of whole wheat meal and a method of expressing it numerically. Cereal Chem. 12:120-129.
56. Cutler, G. H., and Mulvey, R. R. 1953. Winter wheat production in Indiana. Purdue Univ. Agric. Ext. Bull. 390.
57. Cutler, G. H., and Worzella, W. W. 1931. A modification of the Saunders Test for measuring "quality" of wheats for different purposes. J. Am. Soc. Agron. 23:1000-1009.
58. Cutler, G. H., and Worzella, W. W. 1933. The wheat-meal fermentation time test of "quality" in wheat as adapted for small plant breeding samples. Cereal Chem. 10:250-262.
59. Economic Research Service. 1975. Wheat situation. U.S. Dep. of Agric. p. 11 (May).
60. Engibous, J. C. 1976. Research means livelihood. Wheat Life No. 2, Feb. p. 9.
61. Engibous, J. C., and Starr, W. A. 1975. Additional lands available for food and fiber production in the western states. Paper presented ann. meeting of Soil Cons. Soc. Am. San Antonio, TX. Aug. 10-13.
62. Everson, E. H., and Seeborg, E. F. 1958. The heritability of milling quality in wheat as measured by the separation of the bran and endosperm. Agron. J. 50:511-513.
63. Fifield, C. C., Bode, C. E., Fellows, H. C., Weaver, R., Hayes, J. F., Christie, A., Rothgeb, B. E., Hartsing, T. F., and Hoffecker, E. 1950. Milling and baking experiments with wheat varieties grown in the western United States 1936-45. U.S. Dep. Agric. Tech. Bull. 1014. 35 pp.
64. Finney, K. F., Morris, V. H., and Yamazaki, W. T. 1950. Micro versus macro cookie baking procedures for evaluating the cookie quality of wheat varieties. Cereal Chem. 27:42-49.
65. Finney, K. F., Yamazaki, W. T., and Morris, V. H. 1950. Effects of varying quantities of sugar, shortening, and ammonium bicarbonate on the spreading and top grain of sugar-snap cookies. Cereal Chem. 27:30-41.
66. Fonseca, S., and Patterson, F. L. 1968. Hybrid vigor in a seven-parent diallel cross in common winter wheat (*Triticum aestivum* L.). Crop Sci. 8:85-88.
67. Gaines, E. F. 1923. Genetics of bunt resistance in wheat. J. Agric. Res. 6:445-480.
68. Gaines, E. F. 1928. New physiologic forms of *Tilletia levis* and *T. tritici.* Phytopathology 18:579-588.
69. Gallun, R. L., and Reitz, L. P. 1971. Wheat cultivars resistant to races of Hessian fly. U.S. Dep. Agric. Agric. Res. Serv., Production Res. Rep. 134. 16 pp.
70. Gallun, R. L., Roberts, J. J., Finney, R. E., and Patterson, F. L. 1973. Leaf pubescence of field grown wheat: A deterrent to oviposition of the cereal leaf beetle. J. Environ. Qual. 2:333-334.
71. Guettinger, D. L. 1966. 100-bushel wheat. Crops & Soils, Aug.-Sept. pp.

14-15.

72. Gyawali, K. K., Qualset, C. O., and Yamazaki, W. T. 1968. Estimates of heterosis and combining ability in winter wheat. Crop Sci. 8:322-324.
73. Hanson, W. D. 1959. The theoretical distribution of parental gene blocks in gametes of an F_1 individual. Genetics 44:197-209.
74. Hanson, W. D. 1959. The breakup of initial linkage blocks under selective mating systems. Genetics 44:857-868.
75. Hanson, W. H. 1943. Report of the 1941-42 committee on biscuit and cracker flours. Cereal Chem. 20:314-318.
76. Hayward, C., and Poehlman, J. M. 1967. Registration of Lewis wheat. Crop Sci. 7:169.
77. Hayward, C., and Poehlman, J. M. 1967. Registration of Stadler wheat. Crop Sci. 7:169.
78. Heald, F. D., and Woolman, H. M. 1915. Bunt or stinking smut of wheat. State Col. Wash. Agric. Exp. Stn. Bull. 126. 24 pp.
79. Hendrix, J. W. 1964. Stripe rust, what it is and what to do about it. Wash. Agric. Exp. Stn. Circ. 424.
80. Heyne, E. G. 1960. Registration of improved wheat cultivars, XXIV. Agron. J. 52:655-658.
81. Hoffmann, J. A., and Metzger, R. J. 1976. Current status of virulence genes and pathogenic races of the wheat bunt fungi in the northwestern USA. Phytopathology 66:657-660.
82. Holton, C. S. 1957. Wheat smut control in the Pacific Northwest. Agric. Chem. July-Aug.
83. Holton, C. S. 1967. Smuts. In: Wheat and Wheat Improvement. K. S. Quisenberry and L. P. Reitz, eds. Monograph 13. Am. Soc. Agron., Madison, WI. pp. 337-353.
84. Hughes, W. D. 1975. Grain Market News, Pacific Northwest Weekly Ed. 59:No. 29. U.S. Dep. Agric. Mkt. Serv., Portland, OR.
85. Jackson, A. O., Bracker, C. E., Huber, D. M., Scott, D. H., and Shaner, G. 1975. The occurrence and transmission of a disease in Indiana with properties of wheat spindle streak mosaic virus. Plant Dis. Rep. 59:790-794.
86. Jensen, N. F. 1968. Registration of Yorkstar wheat. Crop Sci. 8:641-642.
87. Johnson, J. W., and Patterson, F. L. 1973. Pollen production of fertility restored lines of soft red winter wheats. Crop Sci. 13:92-95.
88. Kendrick, E. L., and Holton, C. S. 1961. Racial population dynamics in *Tilletia caries* and *T. foetida* as influenced by wheat varietal populations in the Pacific Northwest. Plant Dis. Rep. 45:5-9.
89. Kissell, L. T. 1959. A lean-formula cake method for varietal evaluation and research. Cereal Chem. 36:168-175.
90. Kissell, L. T., and Marshall, B. D. 1962. Multi-factor responses of cake quality to basic ingredient ratios. Cereal Chem. 39:16-30.
91. Kitterman, J. S., and Rubenthaler, G. L. 1971. Application of the Brookfield viscometer for measuring the apparent viscosity of acidulated flour-water suspension. Cereal Sci. Today 16:275-276.
92. Knake, E. L., and McGlamery, M. D. 1975. Wild garlic. Univ. Ill. Coop. Ext. Circ. 1109.
93. Kronstad, W. E., Foote, W. H., and Kolding, M. F. 1972. Registration of Yamhill wheat. Crop Sci. 12:397-398.
94. Kronstad, W. E., Foote, W. H., Kolding, M. F., and Rohde, C. R. 1972. Registration of Hyslop Wheat. Crop Sci. 12:398.
95. Kronstad, W., Klatt, A., Bayraktar, A., and Rajaram, S. 1975. Spring × winter crosses for winter and spring wheat improvement. In: 2nd Int. Winter Wheat Conf. Zagreb, Yugoslavia, June 9-19, pp. 105-107.
96. Latta, W. C. 1900. Field tests of varieties of wheat covering 19 years. 12th Annu. rep. Indiana Agric. Exp. Stn. pp. 54-58.
97. Leggett, G. E. 1959. Relationships between wheat yield, available moisture, and available nitrogen in eastern Washington dryland areas. Wash. Agric. Exp. Stn. Bull. 609. 16 pp.
98. Leggett, G. E., Reisenauer, H. M., and Nelson, W. L. 1959. Fertilization of dryland wheat in eastern Washington. Wash. Agric. Exp. Stn. Bull. 602. 21 pp.
99. Leighty, C. E., and Martin, J. H. 1922. The soft red winter wheats. USDA Farmers Bull. 1305, 54 pp.
100. Lewellen, R. T., Sharp, E. L., and Hehn, E. R. 1967. Major and minor genes in wheat for resistance to *Puccinia striiformis* and their responses to temperature changes. Can. J. Bot. 45:2155-2172.
101. Line, R. F. 1976. Stripe rust is still a problem. Wheat Life No. 7, July. p. 6.
102. Livers, R. W. 1964. Fertility restoration and its inheritance in cytoplasmic male

sterile wheat. Science 144:420.

103. Mains, E. B., and Jackson, H. S. 1926. Physiologic specialization in the leaf rust of wheat, *Puccinia triticina*. Phytopathology 16:89-120.

104. Mains, E. B., Leighty, C. E., and Johnston, C. O. 1926. Inheritance of resistance to leaf rust, *Puccinia triticina* Eriks. in crosses of common wheat, *Triticum vulgare* Vill. J. Agric. Res. 32:931-972.

105. McFadden, E. S. 1930. A successful transfer of emmer characters to vulgare wheat. J. Am. Soc. Agron. 22:1020-1034.

106. McKinney, H. H. 1967. Virus diseases. In: Wheat and Wheat Improvement, K. S. Quisenberry and L. P. Reitz, eds. Monograph 13. Am. Soc. of Agron., Madison, WI. pp. 355-374.

107. Middleton, G. K., Bennett, J. R., and Newton, H. C., Jr. 1973. Registration of McNair 4823 wheat. Crop Sci. 13:585.

108. Miyamoto, T., and Everson, E. H. 1958. Biochemical and physiological studies of wheat seed pigmentation. Agron. J. 50:733-734.

109. Miyamoto, T., Tolbert, N. E., and Everson, E. H. 1961. Germination inhibitors related to dormancy in wheat seeds. Plant Physiol. 36:739-746.

110. Morgan, R. W. 1924. A practical application of the viscosimeter to the mill. Cereal Chem. 1:288-292.

111. Murphy, C. F. 1967. Registration of Blueboy wheat. Crop Sci. 7:82.

112. National Research Council. 1972. Wheat. In: Genetic vulnerability of major crops. National Academy of Sciences, Washington, DC. pp. 119-154.

113. Nelson, C. A., and Loving, H. J. 1963. Mill stream analysis; its importance in milling special flours. Cereal Sci. Today 8:301-302,304,326.

114. Nelson, L. R., Morey, D. D., and Brown, A. R. 1974. Wheat cultivar responses to severe glume blotch in Georgia. Plant Dis. Rep. 58:21-23.

115. Ohm, H. W., and Shaner, G. E. 1976. Three components of slow leaf-rusting at different growth stages in wheat. Phytopathology 66:1356-1360.

116. Pacific Northwest Crop Improvement Association. 1943. Description and geographical distribution of the major wheat varieties of the Pacific Northwest wheat area in 1942. Bull. No. 1.

117. Pacific Northwest Crop Improvement Association. 1975. Wheat production estimates by varieties in certain Pacific Northwest counties, 30 pp.

118. Park, E. L., Cole, R. W., Sands, J. K., and Lynch, B. T. 1974. Fertilizer use on selected crops in selected states, 1973. Field crops. Indiana Crop & Livestock Rep. Serv., Lafayette, IN Mimeo.

119. Patterson, F. L., and Gallun, R. L. 1973. Inheritance of resistance of Seneca wheat to race E of Hessian fly. In: Proc. 4th Int. Wheat Genetics Symp., Missouri Agric. Exp. Stn., Columbia, MO. pp. 445-449.

120. Patterson, F. L., Gallun, R. L., and Roberts, J. J. 1974. Registration of Arthur wheat. Crop Sci. 14:910.

121. Patterson, F. L., Schafer, J. F., and Caldwell, R. M. 1968. Effects of selected linkage blocks on yield and yield components in wheat. In: Proc. 3rd Int. Wheat Genetics Symp. Aust. Acad. Sci., Canberra, Australia. pp. 350-354.

122. Pelshenke, P. 1933. A short method for the determination of gluten quality of wheat. Cereal Chem. 10:90-96.

123. Peterson, C. J., Jr., George, D. W., and Rubenthaler, G. L. 1975. Relationship of sprouting, dormancy, and alpha-amylase activity in winter wheat. Agron. Abstr. p. 60.

124. Peterson, C. J., Jr., Vogel, O. A., Bruehl, G. W., and Allan, R. E. 1974. Registration of Cerco, *Cercosporella* foot rot resistant wheat germplasm. Crop Sci. 14:612.

125. Pope, W. K. 1968. Interaction of minor genes for resistance to stripe rust in wheat. In: Proc. 3rd Int. Wheat Genetics Symp. Aust. Acad. Sci., Canberra, Australia. pp. 251-257.

126. Pope, W. K., and Dewey, W. G. 1975. Quantitative resistance to dwarf smut (*Tilletia controversa* Kuhn.) of wheat as influenced by gene interactions and non-specific genes. In: Proc. 2nd. Int. Winter Wheat Conf. June 9-19. Zagreb, Yugoslavia. pp. 486-496.

127. Quisenberry, K. S., and Reitz, L. P. 1967. Wheat and Wheat Improvement. Monograph 13. Am. Soc. Agron., Madison, WI. 560 pp.

128. Reitz, L. P. 1976. Wheat in the United States. U.S. Dep. Agric. Agric. Inf. Bull. 386. 57 pp.

129. Reitz, L. P., and Briggle, L. W. 1960. Distribution of the varieties and classes of wheat in the United States in 1959. U.S. Dep. Agric. Stn. Bull. 272. 83 pp.

130. Reitz, L. P., and Briggle, L. W. 1966. Distribution of the varieties and classes of wheat in the United States in 1964.

U.S. Dep. Agric. Stn. Bull. 369. 68 pp.

131. Reitz, L. P., and Hamlin, W. G. 1978. Distribution of the varieties and classes of wheat in the United States in 1974. USDA-ARS Stn. Bull. 475.
132. Reitz, L. P., Lebsock, K. L., and Hasenmyer, G. D. 1972. Distribution of the varieties and classes of wheat in the United States in 1969. U.S. Dep. Agric. Stn. Bull. 475. 70 pp.
133. Roberts, J. J., and Caldwell, R. M. 1970. General resistance (slow mildewing) to *Erysiphe graminis* f. sp. *tritici* in Knox wheat. (Abstr.) Phytopathology 60:1310.
134. Rodenhiser, H. A., and Holton, C. S. 1945. Distribution of races of *Tilletia caries* and *Tilletia foetida* and their relative virulence on certain varieties and selections of wheat. Phytopathology 35:955-969.
135. Salmon, S. C., Mathews, O. R., and Leukel, R. W. 1953. A half century of wheat improvement in the United States. In: Advances in Agronomy. A. G. Norman, ed. Academic Press, New York. pp. 1-151.
136. Salmon, S. C., and Reitz, L. P. 1957. Distribution of the varieties and classes of wheat in the United States in 1954. U.S. Dep. Agric., Agric. Handb. 108. 86 pp.
137. Salmon, S. C., and Taylor, J. W. 1939. Growing wheat in the eastern United States. U.S. Dep. Agric. Farmers Bull. 1817. 59 pp.
138. Saunders, H. A., and Humphries, S. 1928. The Saunders Test. In: Humphries, A. E., Clover, J. P., and Hutchinson, R. Report on the quality of wheats from the Institute's 1925-26 trials. J. Nat. Inst. Agric. Bot. 2(1):28-37.
139. Schafer, E. G., and Gaines, E. F. 1915. Washington wheats. State Col. Wash. Agric. Exp. Stn. Bull. 121. 16 pp.
140. Schafer, E. G., and Gaines, E. F. 1928. A wheat variety survey in Washington. J. Am. Soc. Agron. 20:171-181.
141. Scott, J. H. 1951. Flour milling processes, Second ed. Chapman and Hall, Ltd., London. 670 pp.
142. Sears, E. R. 1956. The transfer of leaf rust resistance from *Aegilops umbellulata* to wheat. In: Genetics and Plant Breeding. Brookhaven Symposia in Biology No. 9. Brookhaven Nat. Lab., Upton, NY. pp. 1-22.
143. Seeborg, E. F., and Barmore, M. A. 1957. A new five-gram milling-quality test and its use in wheat breeding. Cereal Chem. 34:299-303.
144. Shands, R. G. 1941. Disease resistance of *Triticum timopheevi* transferred to common wheat. J. Am. Soc. Agron. 33:709-712.
145. Shands, R. G. 1974. Registration of Timwin wheat. Crop Sci. 14:908.
146. Shands, R. G., and Cartwright, W. B. 1953. A fifth gene conditioning Hessian fly response in common wheat. Agron. J. 45:302-307.
147. Shaner, G. 1973. Evaluation of slow mildewing resistance of Knox wheat in the field. Phytopathology 63:867-872.
148. Shaner, G., Finney, R. E., and Patterson, F. L. 1975. Expression and effectiveness of resistance in wheat to Septoria leaf blotch. Phytopathology 65:761-766.
149. Sharp, E. L. 1973. Wheat. In: Breeding Plants for Disease Resistance—Concepts and Applications. R. R. Nelson, ed. Penn. State Univ. Press, University Park. pp. 110-131.
150. Sharp, P. A., and Gortner, R. A. 1923. Viscosity as a measure of hydration capacity of wheat flour and its relation to baking strength. Minn. Tech. Bull. 9.
151. Shellenberger, J. A., and Ward, A. B. 1967. Experimental milling. In: Wheat and Wheat Improvement, K. S. Quisenberry and L. P. Reitz, eds. Monograph 13. Am. Soc. Agron., Madison, WI. pp. 445-469.
152. Shipton, W. A., Boyd, W. R. J., Rosielle, A. A., and Shearer, B. I. 1971. The common Septoria diseases of wheat. Bot. Rev. 37:231-262.
153. Slykhuis, J. T. 1970. Factors determining the development of wheat spindle streak mosaic caused by a soil-borne virus in Ontario. Phytopathology 60:319-331.
154. Spillman, W. J. 1909. The hybrid wheats. State Col. Wash. Agric. Exp. Stn. Bull. 89. 27 pp.
155. Sopher, R. F. 1939. Soft wheat milling. Cereal Chem. 16:194-197.
156. State College of Washington Extension Service. 1951. Recommended Pacific Northwest wheat varieties. Pacific Northwest Coop. Ext. Publ. 4. May.
157. Steen, H. 1963. Flour milling in America. T. S. Denison & Co., Inc., Minneapolis. 455 pp.
158. Stroike, J. E., Johnson, V. A., Schmidt, J. W., and Mattern, P. J. 1972. Results of the second international winter wheat performance nursery. Neb. Agric. Exp. Stn. Bull. 248. 158 pp.
159. Sunderman, D. W., and Wise, M. 1972.

Registration of Twin wheat. Crop Sci. 12:259.

160. Suneson, C. A. 1947. An evaluation of nine backcross-derived wheats. Hilgardia 17:501-510.
161. Swanson, C. O. 1936. Physical test to determine quality in wheat varieties. Cereal Chem. 13:179-201.
162. Swanson, C. O., and Working, E. B. 1933. Testing the quality of flour by the recording dough mixer. Cereal Chem. 10:1-29.
163. Taylor, J. W., Bayles, B. B., and Fifield, C. C. 1939. A simple measure of kernel hardness in wheat. J. Am. Soc. Agron. 31:775-784.
164. Thatcher, R. W. 1907. Wheat and flour investigations. State Col. Wash. Agric. Exp. Stn. Bull. 84. 48 pp.
165. Thatcher, R. W., Olson, G. A., and Hadlock, W. L. 1911. Wheat and flour investigations. State Col. Wash. Agric. Exp. Stn. Bull. 100. 52 pp.
166. Tisdale, W. H., Martin, J. H., Briggs, F. N., Mackie, W. W., Woolman, H. M., Stephens, D. E., Gaines, E. F., and Stevenson, F. J. 1925. Relative resistance of wheat to bunt in the Pacific coast states. U.S. Dep. Agric. Dep. Bull. 1299. 30 pp.
167. Udy, D. C. 1956. Estimation of protein in wheat and flour by ion binding. Cereal Chem. 33:190-197.
168. Vogel, O. A., Allan, R. E., and Peterson, C. J. 1963. Plant and performance characteristics of semidwarf winter wheats producing most efficiently in eastern Washington. Agron. J. 55:397-398.
169. Vogel, O. A., and Barbee, O. E. 1944. Comparative performance of wheat varieties in eastern Washington. State Col. Wash. Agric. Exp. Stn. Bull. 450. 28 pp.
170. Vogel, O. A., Craddock, J. C., Jr., Muir, C. E., Everson, E. H., and Rohde, C. R. 1956. Semidwarf growth habit in winter wheat improvement for the Pacific Northwest. Agron. J. 48:76-78.
171. Vogel, O. A., Swenson, S. P., and Holton, C. S. 1944. Orfed Wheat. State Col. Wash. Agric. Exp. Stn. Bull. 451. 10 pp.
172. Vogel, O. A., Swenson, S. P., and Holton, C. S. 1951. Brevor and Elmar—Two new winter wheats for Washington. State Col. Wash. Agric. Exp. Stn. Bull. 525. 8 pp.
173. Western Wheat Associates, U.S.A., Inc. 1963. Wheat supply, distribution, and value in the Pacific Northwest. State Bull. No. 2. Western Wheat Associates, U.S.A. Inc. coop. with U.S. Dep. Agric. Stn. Rep. Serv. 78 pp.
174. Whittlesey, N. K., and Butcher, W. R. 1975. Irrigation development potential in Washington. Wash. State Univ. Col. Agric. Res. Center Circ. 579. 34 pp.
175. Wiancko, A. T., and Fisher, M. L. 1906. Winter wheat. I. Results of variety tests. Purdue Univ. Agric. Bull. 114. pp. 289-308.
176. Wiese, M. V., Saari, E. E., Clayton, J., and Ellingboe, A. H. 1970. Occurrence of wheat streak mosaic and a new variegation mosaic disorder, wheat spindle streak mosaic, in Michigan wheat. Plant Dis. Rep. 54:635-637.
177. Wilson, J. A. 1968. Problems in hybrid wheat breeding. Euphytica 17 (Supp. 1):13-33.
178. Wilson, J. A., and Ross, W. M. 1962. Cross breeding in wheat, *Triticum aestivum* L. II. Hybrid seed set on a cytoplasmic male sterile winter wheat composite subjected to cross-pollination. Crop Sci. 2:415-417.
179. Wilson, J. T., Donelson, D. H., and Sipes, C. R. 1964. Mechanism of improver action in cake flours. I. The relation between flour specific surface and chlorine distribution. Cereal Chem. 41:260-274.
180. Yamazaki, W. T. 1953. An alkaline water retention capacity test for the evaluation of cookie baking potentialities of soft winter wheat flours. Cereal Chem. 30:242-246.
181. Yamazaki, W. T. 1959. The application of heat in the testing of flours for cookie quality. Cereal Chem. 36:59-69.
182. Yamazaki, W. T. 1972. A modified particle size index test for kernel texture in soft wheat. Crop Sci. 12:116.
183. Yamazaki, W. T., and Briggle, L. W. 1969. Components of test weight in soft wheat. Crop Sci. 9:457-459.
184. Yamazaki, W. T., and Donelson, D. H. 1972. The relationship between flour particle size and cake-volume potential among Eastern soft wheats. Cereal Chem. 49:649-653.
185. Yamazaki, W. T., and Donelson, J. R. 1972. Evaluating soft wheat quality of early generation progenies. Crop Sci. 12:374-375.

CHAPTER 3

SOFT WHEAT BREEDING IN EUROPE AND AUSTRALIA

F. G. H. LUPTON
Plant Breeding Institute
Cambridge University, Cambridge, England

N. F. DERERA
Plant Breeding Institute
University of Sidney, Narrabri, New South Wales, Australia

PART ONE: BREEDING IN EUROPE

I. HISTORY OF SOFT WHEAT CULTIVATION

Bread wheat (*Triticum aestivum* L.) is a hexaploid species with 42 chromosomes, unknown as an uncultivated form. It arose as a result of natural chromosome doubling of a cross between a primitive emmer wheat *T. dicoccum* (with 28 chromosomes) and a diploid 14-chromosome goat grass (*Aegilops squarrosa*). The emmer wheats are themselves tetraploids, arising from chromosome doubling of a cross between a diploid einkorn (*T. monococcum*) and another species of *Aegilops*, possibly *Aegilops speltoides* or one of its progenitors (Fig. 1).

A hexaploid species morphologically and cytologically similar to bread wheat has been synthesized from *A. squarrosa* and *T. dicoccum* (41) and an amphiploid resembling *T. dicoccum* has been obtained from *A. speltoides* and *T. monococcum*. The distribution of these species suggests that bread wheat or a closely allied spelt wheat (*T. spelta*) probably first appeared during the last 6,000 to 8,000 years, somewhere in the Fertile Crescent of the Jordan Valley or Mesopotamia.

Wheat has been cultivated since neolithic times, the earliest cultivation probably taking place in the Near East in a region extending from the Caucasus mountains south to Turkey and the Levant and east to Iran and India. The earliest cultivated crops were the diploid einkorns and tetraploid emmers, but hexaploids, probably spelts or club wheats (*T. compactum*), have been found in the Balkans and at Knossos in Crete from the fifth millenium B.C. From the

Figure 1. Origin of cultivated wheat.

Balkans, wheat cultivation spread across Europe, following the Mediterranean coastline and the valleys of the Danube and the Rhine, reaching the North Sea coast and the British Isles about 3500 B.C. (53).

Cultivation of wheat by the ancient Greeks in the fifth century B.C. was outlined by Theophrastus, who gave details of the morphological and physiologic characters of the grain and described how winter and spring varieties were grown and how it was sometimes necessary to cut or graze the young crop to prevent lush growth and formation of long, weak straw.

Virgil in the "Georgics" described wheat cultivation in early Roman times and referred to a "vile blight" on the stalks of the corn. Columella, writing in the first century A.D. referred to several varieties of wheat, recommending siligo or winter wheat for its bread-making capacity. In his "Natural History," Pliny preferred Italian wheat for its whiteness but considered the best foreign wheats to be those from Spain.

Wheat was an important item of trade during the Roman period, when considerable quantities were imported from Spain and from Carthage in North Africa. There was also substantial traffic in grain along the inland waterways, although Britain did not establish trade routes until later in the Roman era.

Agricultural methods changed little during the Dark Ages following the end of the Roman era, although the introduction of the moldboard and colter by the Germans or Belgians led to the replacement of the Roman square or rectangular fields by the open-strip system characteristic of the medieval period, which is still seen in many parts of Europe (18).

The replacement of the two-field cropping and bare-fallow rotation by the three-field system during the latter part of the Middle Ages led to a general increase in productivity. Bennett (3) estimated that yields in Britain during the 13th and 14th centuries were between 450 and 500 kg/ha and increased to about 650 kg/ha by the 17th century as rotations and farming techniques improved. But the most spectacular increase in productivity, leading to yields of more than 1,500 kg/ha, occurred in the 18th and early 19th centuries following the establishment of four-course rotations with legumes and root crops and the introduction of machinery permitting interrow cultivation and row crop planting. Improved mechanization and better land drainage resulted in a slow increase in productivity throughout the 19th and early 20th centuries, the average wheat yields in Britain in the first decade of the 20th century being 1,850 kg/ha. The expanding agriculture in the New World and elsewhere led, however, to an acute depression in Britain and some other countries in western Europe during this period.

The subsequent improvement in agricultural productivity, following the introduction of the gasoline-driven tractor and later the combine harvester, with corresponding mechanization of other phases of crop production, has led to a revolution in intensive agricultural systems. These changes, combined with the introduction of high-yielding short-strawed varieties that benefit from high applications of nitrogenous fertilizers and the development of highly specific chemical fungicides and herbicides have led to increases in productivity on a scale unthinkable in the past. As a result, the mean average yield of wheat in Britain during the period 1976-1980 approached 5 te per hectare, and individual yields of 11 te per hectare and more have been obtained.

A. Early Efforts at Wheat Breeding

The first and perhaps most significant advance in crop production took place in neolithic times, when primitive people selected solid rachis mutants from the wild wheats that they gathered for food. Such mutants had probably occurred at intervals during geological time but would have been at a disadvantage in a natural environment owing to their less efficient seed dispersal. Their selection by early man was, however, of fundamental importance in the establishment of a settled agriculture and hence for the development of modern civilization.

No further evidence of conscious selection for improved wheats is available until the time of the Greek and Roman writers, and even then selection was mostly concerned with the maintenance of good stocks of seed. Theophrastus, for example, stressed that seed corn should only be taken from plants harvested in their prime, and Virgil advised that grain from the largest heads should be retained for sowing. At the same time, however, "land races" adapted to specific areas were being developed, and although no conscious selection was involved, the merits of locally adapted varieties came to be recognized.

The evolution of these land races was an important stage in wheat improvement and led over the centuries to the proliferation of varieties with names like Taunton White and Burwell Red, indicating their ear type and place of origin. These varieties consisted essentially of populations of morphologically similar varieties with specific local adaptation and provided the raw material for the early plant breeders—and also a reserve of genetic variability that is still called upon by breeders and should be preserved for future needs.

Scientific plant breeding is largely founded on the work of Linnaeus (1707-1778) who first recognized the importance of hybridization in crop improvement. Knight (1759-1838) was, however, the first to use hybridization in the improvement of cereals, though his "crosses" were obtained by growing parental varieties close together and probably consisted mainly of mixtures of the parental forms.

The greatest British wheat breeder of the 19th century was undoubtedly Patrick Sherriff, who used both selection and hybridization to produce new varieties. He recognized the mixed nature of the so-called varieties cultivated in his day but realized at the same time that characters can only be changed by hybridization followed by selection in the mixed population so created. His selection work led to a range of varieties such as Mungoswells (1819) and Sherriff's Bearded White (1857), and his work on hybridization to the variety King Richard (Sherriff's Bearded White × Talavera).

The improved British varieties were widely grown in many countries of western Europe, where they were used by breeders whose work was greatly stimulated by the publication of Charles Darwin's "Origin of Species" in 1857. Prominent among these were Vilmorin in France, Rimpau in Germany, and later, Nilsson Ehle in Sweden and Broekema in the Netherlands.

B. Development of Modern Plant Breeding

The development of modern plant breeding dates from the rediscovery of Mendel's laws at the beginning of the present century, although by this time schools of plant breeding were thriving in many parts of the world. As has been discussed, much of the progress up to this time came from selection within the mixed varieties then cultivated, though the limitations to this approach became clear in 1903 when Johannsen defined the concept of the pure line in relation to the breeding of self-pollinated crops.

Application of Mendel's laws to characters of economic importance was first demonstrated by Biffen in 1906 and was followed by a great expansion of interest in plant breeding, which was thought likely to solve all problems of crop improvement. The next major advance was given by Nilsson-Ehle's demonstration of the inheritance of complex factors (47), thus giving a scientific basis to transgressive segregation.

Practical experience soon tempered enthusiasm for the application of Mendelian principles to plant breeding problems, but the scene had been set for an era of steady progress that has expanded over the years into the complex structure of modern plant breeding. In Britain, for example, Biffen's work on the inheritance of yellow rust resistance led to the release of the variety Little Joss in 1912. Biffen was also interested in the possibility of combining high yield with good bread-making quality, partly because of complaints that the high yielding Squarehead wheats were inferior in quality to the old land varieties that they had replaced. A full account of this work, which led to the release of the variety Yeoman in 1916, is given by Biffen and Engledow (5), who drew attention to the problems of genotype and environment interactions, and called for a study of the physiology of the crop throughout its growth. Yeoman was followed in 1935 by the variety Holdfast, which was not effectively challenged until the release of Maris Widgeon in 1964, although higher yielding varieties of inferior quality were introduced from France and elsewhere. These include the varieties Wilma (1936) from Holland and Hybrid 46 (1946), bred by a private seed firm in Britain and widely grown in the years after their release. British varieties were, however, unable to compete with the French wheats until a major breakthrough was achieved following the introduction of the wheats Maris Ranger (1968), Maris Huntsman (1972), Hobbit (1975), Mardler (1977), Brigand (1979), and Avalon (1980) bred by the government-sponsored Plant Breeding Institute at Cambridge.

Scientific plant breeding in France began with the work of Henri de Vilmorin who crossed Sherriff's Squarehead with the variety Noé, itself imported from Odessa by the Marquis de Noé. This led to the variety Dattel, released in 1883, followed by a series of varieties bred by succeeding generations of the Vilmorin family, which were to dominate French agriculture for the next 50 years. These included Bon Fermier (1905), Hatif Inversable (1908), and the well-known varieties Vilmorin 23, 27, and 29, released in 1923, 1927, and 1929. Other private plant breeding firms were established in France during the years between the wars; government-sponsored plant breeding was also started during this period, following the establishment of the Institut National de la Recherche Agronomique in 1921. Some of the better known varieties bred by these organizations include Bersée (1936), Cappelle Desprez (1946), and the better quality varieties Yga Blondeau (1938) and Flanders (1975).

In Germany, Wilhelm Rimpau first undertook scientific breeding work. His earliest crosses, to combine the yield of the Squarehead wheats with winterhardiness and grain quality from German and North American varieties, were made in 1875. Although Mendel's laws were not then known, Rimpau noted the uniformity of the F_1 and the great variability and recombination of characters in the F_2 generation. His first variety, Rimpau Früh, was released in 1889 and was still widely cultivated 40 years later. Squarehead wheats continued to dominate winter wheat breeding in North Germany for many years; winterhardiness, disease resistance, and grain quality were introduced with varieties from other countries. German spring wheats have a somewhat wider genetic base and are much influenced by the French variety Noé and good quality wheats from North America.

Plant breeding in Sweden started with the establishment of the well-known research station at Svalöf in 1886. Earlier work there was of limited value as it was based on mass selection within existing mixed populations. The work was put on a more scientific basis by Nilsson-Ehle who developed the pedigree system of selection, while at the same time conducting growing trials of residual seed in early generations. Crosses of the British Squarehead wheats with Swedish land races formed the basis of his and subsequent Svalöf winter wheat breeding programs, leading to the varieties Anchor (1928), Iron (1920), and more recently Odin (1960). A parallel winter wheat program was developed at the private Weibull's breeding station, established at Landskrona in 1904. Their work also concerned crosses of Swedish varieties with Squarehead, but grain quality was introduced as well, using Hungarian wheats and leading to a series of varieties culminating in Starke (1959) and Holme (1973). In contrast to the somewhat narrowly based winter wheat program, a very successful and widely based spring wheat program has been conducted in Sweden, exploiting the high quality Kolben wheats and giving rise to a sequence of good quality varieties including Atle (1936), Svenno (1957), Sappo (1972), and Timmo (1975).

Major wheat breeding programs were established in several other European countries at the turn of the century. Special mention should be given to the station established at Wageningen in the Netherlands in 1886 and to that at Gembloux in Belgium, the work at which developed in close liaison with that in France. A remarkable feature of all these programs is their widespread use of the British Squarehead wheats, which have had a major influence on wheat breeding throughout northwestern and central Europe.

II. GENETIC RELATIONSHIPS AND BREEDING SYSTEMS

Bread wheat is a naturally self-pollinating species and is therefore adapted to inbreeding, although it sometimes shows heterotic vigor when suitable lines are crossed, and breeders have recently attempted to utilize this vigor in the production of "hybrid wheat," in which F_1 grain is sold for crop production. Because it is normally self-pollinated, controlled pollinations need only be made at the beginning of a breeding program, self-pollination taking place in all subsequent generations. The scale of F_1 seed production is limited as only one seed is obtained from each flower pollinated, but this is of little importance in most breeding programs because 300 to 400 seeds can readily be obtained from well-grown F_1 plants.

All varieties of bread wheat are fully compatible, though plants of some crosses display a lethal or sublethal hybrid necrosis determined by two independently inherited but mutually incompatible genes ne_1 and ne_2 (20). Many plant characters such as straw height and disease resistance are determined by groups of three genes, one on each of the genomes. This gives the crop a better capacity for adaptation to varying environments than that of a diploid organism such as barley.

A. Construction of the Ideotype

The main objectives of a wheat breeding program may be defined as improvement in yielding capacity, reliability of performance, and the quality of the end-product. In selecting for improved yield, the breeder must consider the range of soil type in which the variety will be grown, bearing in mind that many crops will be grown under suboptimal conditions in terms of fertility and contamination with weeds and soilborne diseases. There is considerable evidence that varieties differ in response to these conditions; selection and yield assessment in particular should therefore include tests done under suboptimal conditions.

Selection for reliability of performance includes also tests for tolerance of adverse conditions that may occur in some seasons but not in others. It includes, for example, resistance to cold and drought, tolerance of selective herbicides that may be needed to control graminaceous weeds such as wild oats (*Avena fatua* and *A. ludoviciana*) or black grass (*Alopecurus agrestis*), and the most important problems of breeding for resistance to diseases. These are considered in more detail later, but mention should be made that breeding strategy has changed because of the continual breadkown of resistance to many diseases, especially the rusts (*Puccinia* spp.), as a result of the continually changing spectrum of physiologic races of these pathogens.

The grain quality characters to be considered by the breeder may be divided into three main groups, concerned first with milling quality, which is inherited independently of yield and is relatively easy to select. Secondly, the breeder is concerned with the quality and quantity of grain protein, his work being complicated by the negative association of protein content and yield. Finally, he must avoid varieties that are unduly susceptible to premature germination when harvesting conditions are wet, and those with a high endogenous content of α-amylase even when harvested under dry conditions.

The progress that the breeder can make by avoiding the defects that may affect his crop are essentially negative. If he is to make real progress, he must also consider how best to alter the structure of his crop so that it may be better adapted to its environment. As long ago as 1923, Engledow and Wadham (15) suggested that plant breeders should find a "touchstone" on which yielding capacity depended. The problem was put on a quantitative basis by Watson (60), who introduced the concepts of net assimilation rate and relative growth rate. It was further clarified by Donald (12) who suggested the term "ideotype" for the model that the plant breeder should define and seek to create. Donald emphasized that a successful variety should exploit its environment fully by competing strongly with plants of other species but weakly with plants of its own species. It should therefore be considered as a member of a crop community and not as a single plant grown in isolation.

The first essential in a wheat ideotype is that the stems should be strong enough to support the ears without lodging. This normally involves selection for short straw but not to the extent that the crop is insufficiently aggressive to control weeds adequately. Its leaves should have high photosynthetic capacity and should allow maximal penetration of light through the crop. They should therefore be erect and should remain photosynthetically active as long as possible. The ears should also be efficient photosynthetically and should remain green as long as possible. There is some argument about the value of awns, which Donald considers necessary for high yield, though this opinion is not supported by experience in western Europe. Bingham (6) suggested that the high yielding capacity of the semidwarf wheats may be partly caused by a reduced demand for carbohydrate by their stems at the time of rapid ear growth shortly before anthesis, thus increasing supplies available to the developing ear. The wheat ideotype should therefore make provision for adequate ear development. It should also provide for a minimal production of nonearing tillers, which would otherwise compete with productive tillers for photosynthates.

Varietal differences in root development have rarely been considered in selection because of the obvious difficulties in making the necessary observations. Such differences may, however, be of considerable importance although criteria for selecting a variety for a maritime climate, where rapid exploitation of the soil environment is required, may differ from those for a dry climate, where limited water supplies must not be exhausted too early in the growing season.

In addition to considerations of crop morphology, the relative lengths of the growth phases must be adapted to the conditions in which the crop will be grown. In maritime temperate conditions where water is unlikely to be scarce, the most important limiting factor is likely to be light incidence, and the crop should be adjusted so that grain filling takes place when the light intensity is greatest, that is, during the long days of June and July. Under drier conditions, however, development may have to be concentrated a shorter period while adequate moisture is available, to ensure in particular that enough moisture is available during grain filling.

The yielding capacity of a crop is determined partly by its photosynthetic ability and partly by its capacity to store carbohydrates in the developing grain. These two aspects, which have been described as the "source" and "sink" capacities of the crop, must also be considered in determining the ideotype. In particular, it is essential that the period of ear development be as long as possible, so that the sink capacity will be adequate, but this period should not be extended by shortening the period of grain ripening. Some form of compromise is clearly required, possibly associated with a high growth rate of the ear primordium (31). These aspects are essentially under hormonal control, and the hormonal interactions must be considered in constructing the crop ideotype. For example, the semidwarf wheats now widely cultivated have a very high content of gibberellic acid (52), but one action of this hormone, causing stem elongation, is blocked by an antigibberellin. Other activities of the hormone, such as reduced apical dominance resulting in increased tillering, are not blocked, however, and may be exploited by the breeder.

B. Choice of Parental Material

The choice of suitable parental material is of the greatest importance in planning a wheat breeding program. Because the crop is self-pollinated, the material available comprises a wide range of homozygous lines, each with a precisely defined genetic constitution. Many breeders, therefore, make a large number of crosses each year in order to sample a wide range of genetic variability. Many of the crosses are made between segregating lines from previous crosses, choices based largely on subjective judgment. This method has been used with considerable success, notably by those associated with the Centro Internacional Mejoramiento Maiz y Trigo (CIMMYT) in Mexico. Other breeders, however, choose parental combinations with more precision, preferring to make smaller numbers of crosses and handle each more extensively.

The choice of parental material when handling a simply inherited character such as disease resistance is relatively straightforward, provided an adequate collection is available within which to look for a genotype with the desired attribute. This genotype can then be crossed to a variety lacking the desired attribute and then handled by a backcrossing technique. In most cases, however, the breeder needs to consider more complex character combinations, most of which are not simply inherited, and needs to use more sophisticated techniques to select the best parental combinations. These comprise first, those in which the potential value of possible crosses is assessed using techniques of biometrical genetics, and second those in which the physiologic attributes of the parental varieties are studied in detail and parents are chosen with complementary advantages.

The biometrical techniques used in cross prediction generally follow the principles defined by Mather (37) and Mather and Jinks (38). These involve the analysis of diallel sets of crosses, in which potential parental varieties are either crossed in all combinations or crossed onto a group of tester varieties chosen to represent the genotypes in general cultivation (28,29). The latter method, known as the incomplete diallel analysis, lacks some of the precision of the complete diallel but is useful in assessing the potential of large numbers of potential parents. Analysis of the data obtained from such crosses identifies parental combinations showing nonallelic interaction and therefore likely to exhibit transgressive segregation, in a positive or negative direction, for the character being considered. A second analysis measures the general combining ability and specific combining ability of the parents. The general combining ability of a variety assesses its average performance as a parent in hybrid combinations, and the specific combining ability the extent to which the performance of a single parental combination differs from expectation as estimated from the general combining abilities of the parents concerned (56).

A more direct estimate of the potential value of parental combinations can be obtained by measuring the mean and variance of yield of randomly selected F_2 plant progenies. This method can be used to compare a range of crosses without the limitations imposed by the diallel analysis. It was used by Lupton (28), who found that the results agreed reasonably well with those obtained from a diallel analysis of the same group of crosses, although there were rather wide differences between two years of trial. The labor to obtain the necessary data was very considerable, however, and the results were not available until the crosses had been handled for a number of years.

The most useful parents for a breeding program may also be assessed by multivariate or canonical analysis. This method, for which no crosses need be made, requires the breeder to define a model variety by a range of quantitative characters. These characters are then measured on a number of plants of each potential parent and are used to define positions in multidimensional space appropriate to each variety. Although difficult to appreciate, the concepts of multidimensional space can readily be analyzed by a computer, which can allow for correlations between the characters measured and evaluate which parental combinations approach the model variety most closely. The use of the method was described by Whitehouse (61), who compared six morphological or physiological components of yield of eight barley varieties, checked his results from a diallel set of crosses between the varieties, and predicted a cross that might provide useful transgressive segregation.

Comparison of the merits of methods of cross prediction is difficult, because of the diverse circumstances in which the various tests were done, although reasonable agreement between methods was reported by Lupton (29). An objective comparison was, however, reported by Bhatt (4), who compared cross predictions in wheat made in four ways—a conventional cross using one adapted parent, a choice based on multivariate analysis, a choice based on geographical diversity, and a random choice. His results suggested that the greatest promise could be expected from crosses based on multivariate analysis and that conventional crosses were less efficient than those based on geographical diversity. His results were based on a rather limited number of parental combinations but indicate the potential value of the more sophisticated methods of cross prediction.

C. Mutation Breeding

As an alternative to crop improvement by genetic recombination, desirable new genotypes may be obtained by exposing existing varieties to mutagenic chemicals or irradiation. The mutant genotypes so obtained may be useful as varieties or may be used as parents in subsequent breeding programs.

Artifical mutations in wheat were first obtained by Stadler in 1929, who produced a range of chlorophyl mutants in seedling plants using x-rays. Workers in the Soviet Union (11,54) later demonstrated the induction of speltoids, awned mutations, and other morphological abnormalities. But the first evidence of mutants with possible agricultural value was not obtained until 1941 when Froier and co-workers in Sweden produced wheat mutants showing differences in ear density, leaf width, tillering capacity, and earliness of maturity (17). The frequency of occurrence of major morphological mutants in wheat, however, was found to be less than that in many other crops, possibly as a result of buffering action resulting from the polyploid nature of the crop (36). Yet economically useful gene mutations in bread wheat have been reported, and MacKey reported mutants causing reduced plant height, improved straw stiffness, and increased adult plant resistance to black rust (*Puccinia graminis*). There have also been reports of improvement in grain protein, some associated with reduced 1,000-grain weight (14), but in other cases, total protein production per hectare is increased (33).

Despite the valuable results obtained by a number of breeders, the variability available in wheat is so great that there seems to be little need to extend it through artificially induced mutations. There may, however, be merit in using mutagens to increase the variability in early segregating generations of hybrid material, and they have been effectively used in causing translocation of segments of alien chromosomes when introducing such characters as disease resistance by the use of intergeneric crosses (13,55).

D. Selection Methods in Segregating Populations

A wide range of selection methods has been developed for handling segregating populations, and it has been claimed with some justification that no two breeders use the same procedure and even that no two crosses are handled in precisely the same way.

The backcrossing technique may be used when introduction of a simply inherited character in an otherwise acceptable variety is desired. This technique was at one time widely used in breeding for disease resistance, as it is particularly suited to handling simply inherited major genes determining hypersensitive resistance. When using this method, the donor parent carrying the desired character is crossed to the recipient variety. If the character being transferred is dominant, half the population derived from this first backcross will show the desired character and will be crossed again to the recipient variety. This process would be repeated until the plants obtained were equivalent in all respects to the recipient parent, apart from expression of the character being transferred; six backcrosses normally are considered sufficient for this purpose (9). If the gene being transferred is recessive, selection after each backcross would not be possible, and a large number of plants would therefore have to be backcrossed in each generation to ensure with reasonable statistical certainty that some would carry the desired character. Alternatively, selections could be made in the second generation after each backcross, but this inevitably delays completion of the program.

The backcrossing technique is only practicable when the character to be transferred can easily be recognized in a segregating population. This may not be possible when breeding for a quantitative character, in which case the chromosome of the donor variety on which the character is determined can sometimes be exchanged for the corresponding chromosome of the recipient variety, by using cytological techniques involving monosomic plants (27). Such plants lack one chromosome of the normal complement, and hence have 20 pairs of chromosomes and one single unpaired chromosome. Sets of monosomic plants, deficient in turn for one member of each of the 21 pairs of chromosomes of a normal (euploid) wheat, have been derived from a number of leading varieties and have been used to identify the chromosomes determining important quantitative characters such as milling and baking quality (39), winterhardiness (27), and nonrace specific disease resistance (23). This technique may have special attractions because it enables the genes for disease resistance, or for any other character, to be handled with purely cytological methods and without need to test for their expression during the process. An additional source of disease resistance can therefore be added to the genotype of a resistant variety without the need to test for one form of resistance in the presence of another (23).

In most breeding programs, the breeder wants to combine several independently inherited characters in a new variety and must devise a more empirical selection system, suited to the characters and the selection pressure he wishes to impose. At one end of the range of possibilities he may follow the "bulk selection" technique, as proposed by Åkerman and MacKey (1), in which no selection is undertaken for at least six generations after the initial cross. At the end of this time, the population will consist of more or less true breeding lines that can be selected and multiplied as new varieties. The procedure has the disadvantage of being slow and of exposing the population to several years of natural competition during which an agriculturally desirable genotype with poor competitive value, such as one with short straw, would be liable to be eliminated. Another serious drawback is that selection, when eventually attempted, must be carried out on a much larger population than is required using the pedigree system if all desirable recombinants are to be identified. But it has considerable attractions in conducting an extensive program with limited facilities, especially if the successive bulk populations can be exposed to selection pressure for such characters as winterhardiness or disease resistance or if the grain can be exposed to mechanical screening for weight and density during the segregating years.

At the other end of the range of possibilities, detailed pedigree selection may be attempted on a single-plant basis until a fully homozygous progeny is obtained. The selected plants are then multiplied until sufficient seed is available for yield tests and estimates of other quantitative characters. Pedigree systems of this type are used by most breeders, although most of them estimate yield and grain quality before their material is completely homozygous. Such tests are frequently based on the progeny of single F_3 or F_4 lines, as it is generally agreed that these are the most critical stages of breeding programs, when the merits of a selection can be considered in terms of the performance of the progeny of a single F_2 or F_3 plant.

Lupton and Whitehouse (32) describe two methods of selection that have been used at the Plant Breeding Institute, Cambridge, and that attempt to combine the advantages of detailed selection using the pedigree system with yield assessment in early generations. In the first of these systems, known as the F_2-progeny method, single plants are selected by eye judgment in F_2 and their F_3 progeny are grown in small plots. These plots are again assessed by eye, and the more promising plants harvested as sheaves to provide grain for yield trials. They are grown without further selection in F_4, F_5, and F_6, after which single plants are selected within the more promising families. Selections from several crosses may be compared in these trials and the better parental combinations identified. This method has, however, been largely displaced at Cambridge by a procedure similar to the other system described by Lupton and Whitehouse. In this, known as the progeny-trial system, (Fig. 2), detailed pedigree selection is performed in F_2, F_3, and F_4. After the better lines have been selected in F_4, however, the remaining F_3 plant progenies in each of the families to be retained are harvested as bulks to provide seed for yield estimation in F_5. The same procedure is applied in F_5 and F_6 to provide seed for trials to be grown alongside the pedigree plots in F_6 and F_7, respectively. The material entering trial each year thus differs from

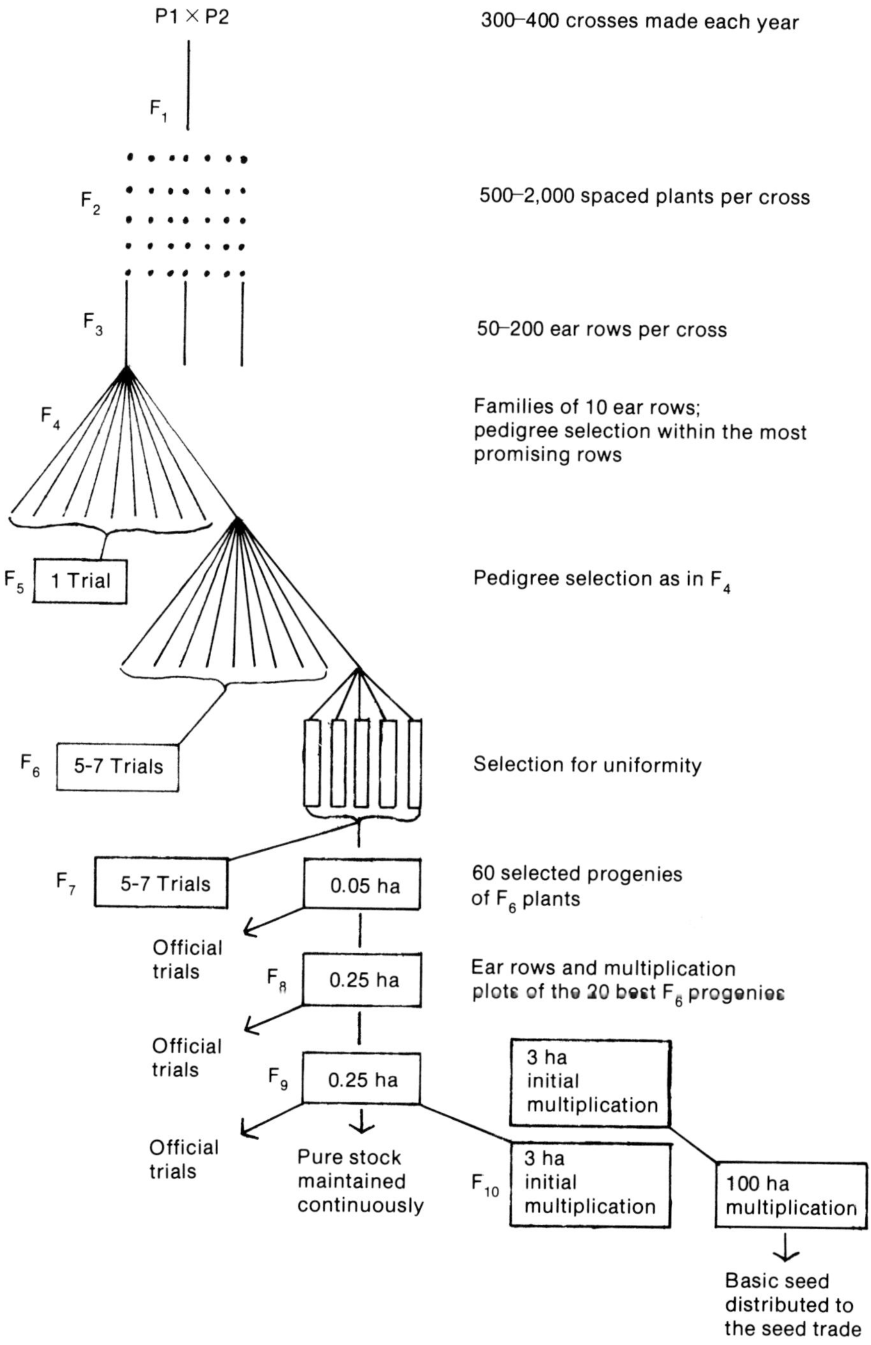

Figure 2. Winter wheat pedigree: trail selection and multiplication system[a] (after Bingham, 1975, [7]).

that in trial the previous year by the elimination of the less desirable genotypes. The trial stocks in each year also become more uniform, homozygous, and genetically similar to the corresponding ear rows. This system is essentially similar to those used by breeders of many crops for many years (21,47). It has been widely used in the development of the high yielding varieties of winter wheat recently developed at Cambridge (Fig. 3). It enables breeders to produce varieties much more quickly than would have been possible with the F_2 progeny method, as there is no need for reselection at the end of the trial period. The continued screening during and immediately after the small-scale trials is also important to meet the stringent standards of genetic purity and stability required in western Europe. The program is, however, limited to a single generation per year, apart from the very earliest generations, which may be grown in a greenhouse from artificially vernalized seed, when necessary. Problems of vernalization seriously restrict the extent to which a winter wheat program can be further accelerated, but spring wheat breeding can be speeded by growing alternate generations at different localities where growing conditions are similar. Such systems are relatively easy to operate when breeding for low latitude countries but become more difficult when breeding for more northerly or southerly environments; European wheat breeders need to grow their alternate generation crops in the southern hemisphere.

Thus, a scheme was recently introduced in which alternate generations of the Cambridge spring wheat breeding program are grown in New Zealand, under conditions very similar to those in Britain, although midsummer days are somewhat shorter. The New Zealand generation can therefore be used for selection as well as for grain multiplication, and this method has considerably shortened the time required to produce a new variety.

The early generations of a spring wheat program may also be accelerated using the single seed descent system (19,46). In this, plants are grown under lights in sand or vermiculite in seed boxes irrigated with a minimal nutrient medium. The plants mature very rapidly, producing small ears each with two or three grains. A single grain is selected from each plant for growing in a subsequent generation of single seed descent. Little selection is possible from such generations, but the plants become progressively more homozygous. Four or more generations of single seed descent may be completed in a year, after which the seed is sown under normal growing conditions and selected for agronomic characters.

E. Breeding for Disease Resistance

Losses to disease are one of the principal factors limiting crop production in western Europe. Although total crop failures are rare, losses of 20 to 30% are quite common, and the estimated annual crop loss to leaf diseases in Britain averages about 10% of the national crop (22).

The principal diseases attacking the wheat crop in western Europe are: the leaf diseases—yellow, brown and black rusts (*Puccinia striiformis, P. recondita,* and *P. graminis*), powdery mildew (*Erysiphe graminis*), and leaf and glume blotch (*Septoria tritici* and *S. nodorum*); root and stem rot diseases—eyespot (*Pseudocercosporella herpotrichoides*), take-all (*Gauemannomyces graminis*), and Fusarium root rot (*Fusarium culmorum*); the seedborne diseases—loose smut

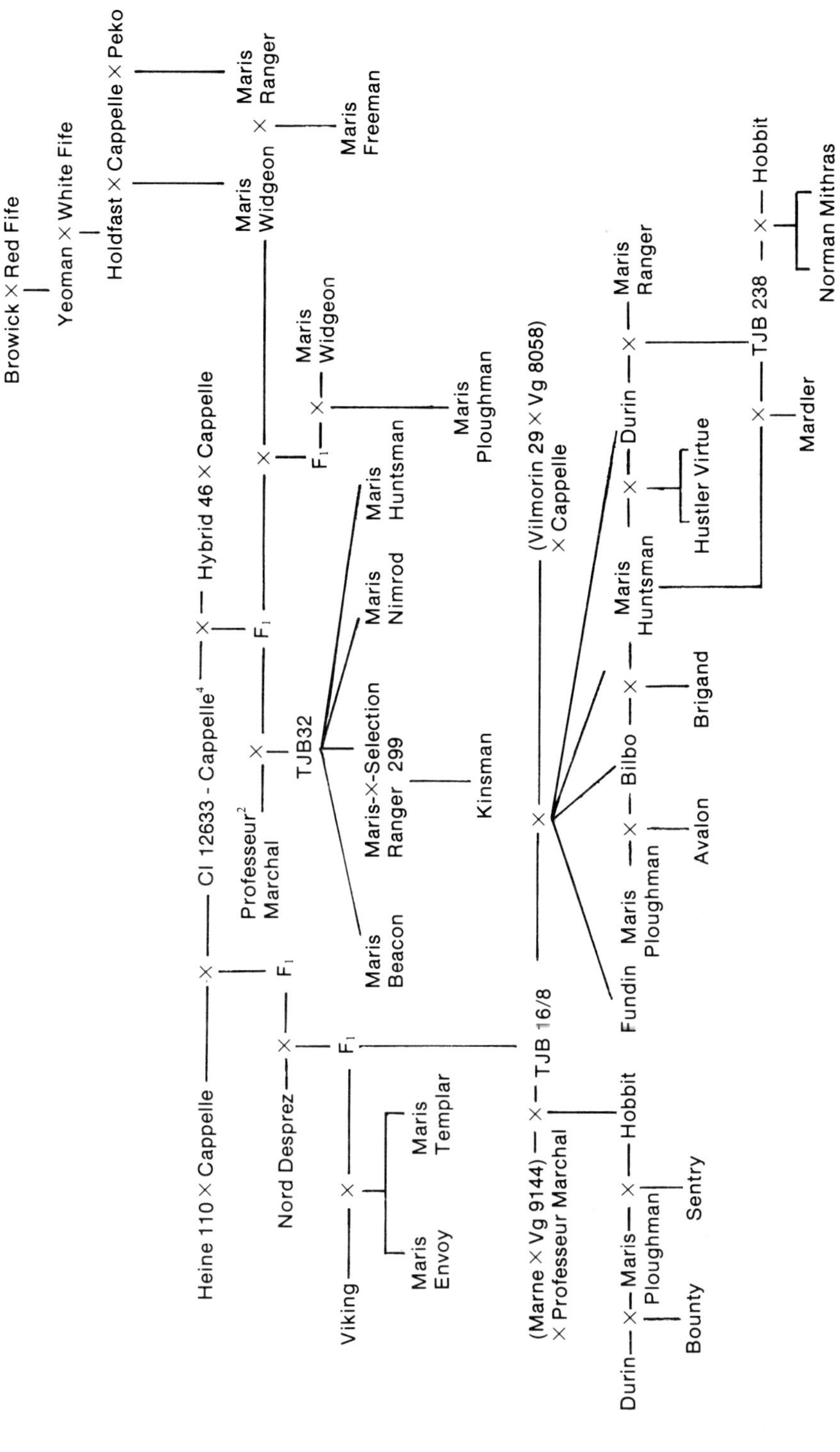

Figure 3. Pedigrees of winter wheat varieties and lines bred at the Plant Breeding Institute.

(*Ustilago tritici*), bunt (*Tilletia caries*), and leaf stripe (*Pyrenophora sativa*).

The rusts are probably the most important of these diseases. Until recently, resistance was obtained by using a backcrossing technique to introduce simply inherited genes for high expression of resistance obtained from exotic varieties. Varieties obtained in this way lost their resistance, however, because the rust fungus was able to produce new strains, or physiologic races, capable of attacking varieties into which each new gene for resistance had been introduced. Such new physiologic races frequently arose while a new variety was being multiplied for release or shortly after it had been released and become widely grown, often causing serious economic loss. A new policy had to be adopted in breeding for resistance, and breeders attempted to select for varieties showing durable or nonrace-specific resistance. It was thought that such varieties would show some susceptibility to a wide range of physiologic races of the pathogen, particularly at the juvenile stages when many varieties are more susceptible, but would never be severely attacked under field conditions. In order to select for such resistance, selections were first tested as seedlings to ensure that they did not carry major genes for resistance, which might break down to a newly developing physiologic race of the pathogen. They were then exposed as mature plants in the field to a high inoculum pressure of rust spores and selected rigorously for low expression of disease. This procedure, which is currently followed in breeding for yellow rust resistance at the Plant Breeding Institute, Cambridge, has met with some success but must be considered with caution, because new physiologic races have arisen capable of attacking varieties such as Joss Cambier, which previously had resistance. As an alternative, the breeder may use as parents long established varieties that, though not fully resistant, are rarely attacked severely. A few such varieties are available. They include the yellow rust resistant varieties Little Joss and Browick (30), but they tend to have undesirable agronomic characters, and it is difficult to select for their type of resistance in the segregating generations. This problem may, however, be avoided by transferring the relevant chromosomes using monosomic lines, as described earlier (23).

The risk of serious losses to leaf diseases has been greatly increased by the widespread and intensive cultivation of a limited number of varieties. This near-monoculture of some varieties provides an ideal environment for spread of a new physiologic race of a pathogen and exposes the agricultural industry to severe loss if this occurs. Farmers in the U.K. are therefore advised to diversify their cropping as much as possible and not to sow varieties with the same factors for disease resistance. Alternatively, they may sow each field with a mixture of varieties each with a different factor for resistance. Such a mixture clearly needs to comprise varieties of similar agronomic character and grain quality. Experience with crown rust in oats in North America (16), with powdery mildew in barley (62), and with brown rust in wheat (M. V. Rao, personal communication) has, however, shown that the rate of spread of disease through such mixtures is much slower than that in pure stands of a single variety. The method shows considerable promise as a new means of reducing disease losses, and work is being started to produce morphologically similar "isolines" of leading varieties, each with different forms of resistance that can be combined to produce agriculturally acceptable "multiline" varieties.

Powdery mildew is not normally as serious in wheat as the rust fungi, though it is frequently present and may be serious in some seasons. It also occurs in physiologic races, which evolved as new sources of resistance were exploited by the breeders, but a number of agriculturally useful varieties show a high level of field resistance that can be used in breeding.

Leaf and glume blotch have increased greatly in importance since the introduction of the high-yielding semidwarf wheats. This is partly because the short straw and high tillering capacity of these varieties create a microclimate that is particularly favorable to these diseases, which are spread by spores splashed from leaf to leaf within the crop canopy. There is, however, considerable variability in susceptibility to leaf and glume blotch within both semidwarf and taller varieties, so the breeder has adequate scope to select for resistance in both plant types.

Stem and root rots are of considerable importance where wheat is grown repeatedly on the same land. No reliable varietal resistance is known to take-all, but this disease can be readily controlled by a short break in cereal cultivation. Resistance to eyespot is shown by the variety Cappelle Desprez, and this resistance has been widely exploited by breeders in western Europe in the past 20 years. It is, however, only partial, and a new and more effective resistance, derived from the wild grass *Aegilops ventricosa* is now being used with some success.

Little is known of the genetics of resistance to Fusarium root rot, though some varietal differences are known. But the effects of this disease can be greatly reduced by treating the seed with a mercurial seed dressing. In the same way, convenient seed treatments are now available for controlling the seedborne diseases, which are no longer of importance to the plant breeder.

F. Breeding for Processing Quality

Breeding for processing quality involves consideration of a range of independently inherited characters, some of which are negatively correlated with yielding capacity. Broadly, these characters can be divided into those concerned with milling and those concerned with use of the flour.

Milling texture is a strongly inherited character, the expression of which is little affected by environmental conditions. Although a full milling procedure is required to determine the finer distinctions between varieties, an approximate classification can be made with very small samples such as may be available at the early stages of a breeding program. At the Cambridge Plant Breeding Institute, for example, preliminary selection for hardness of milling is done in the earliest segregating generations, by using a modified Brabender stone mill. Half the grain from a single plant can be milled by using only the upper two stones of this mill. The bran obtained is then sieved clean of flour and inspected; the harder milling plants give a cleaner bran sample than do softer millers.

Plants can be readily graded into flour of five categories according to the cleanness of the bran, and when breeding for bread-making quality, seed from plants with harder milling characters is retained for further selection. Softer milling plants are, however, retained in crosses designed for breeding for biscuit manufacture. Milling quality is subsequently tested with Quadrumat and Bühler experimental mills, the use of which is essential to obtain flour of suitable extraction rate for estimation of bread-baking quality.

The most important characters pertaining to use of flour are those determining protein quantity and quality and α-amylase activity. Protein content is a weakly inherited character, the expression of which is greatly affected by the environment. Differences in grain protein content are almost entirely associated with a negative link of this character with yielding capacity so that, when grown under similar conditions, modern varieties differ little in total yield of protein per hectare. This is hardly surprising, as increased protein yield could only be obtained by more efficient uptake of nitrogen or by redistribution of protein within the plant. Although there is little evidence of varietal differences in either of these characters, a few genotypes that deviate from the negative regression of protein content on yield indicate the possibility of breeding for improvement in protein yield (51), and active programs with this objective are being conducted at Cambridge and elsewhere. The most promising parents being used in this work are Atlas 66, a tall weak-strawed variety from the U.S.A., and the Indian variety Nap Hal. There are suggestions that the high protein contents of the latter two varieties are independently inherited, that of Atlas 66 being found in the endosperm and that of Nap Hal in the endosperm and aleurone layers (24).

Protein quality is a much more strongly inherited character than protein quantity but is difficult to assess, partly because of its complex heredity and partly because it can only be done on the basis of grain samples grown under strictly comparable conditions. Although tests have been devised for making loaves from very small quantities of flour (42), most breeders make preliminary screenings of protein quality with empirical methods such as the Zeleny sedimentation test (63,64) used widely in Austria, Germany, and Sweden, the Pelshenke dough ball test (48,49), or more recently, the sodium dodecyl sulfate (SDS) sedimentation test. More elaborate tests, such as the Chopin alveograph, the Brabender farinograph, and the Simon extensometer have been developed for use when larger quantities of flour are available, but no fully effective artificial test has been devised, and the experimental bakery provides the final assessment of quality in any breeding program.

The first tests of protein quality at Cambridge are done using the SDS sedimentation test on F_4 grain. This test has the advantage of being largely independent of the environment in which the sample is grown. It is rapidly performed and thus a large number of selections can be screened to provide valuable information to the breeder in determining which selections should be grown in preliminary yield trials. Grain from these trials is used for more detailed tests in later generations using specially designed small-scale equipment for the Chorleywood bread process. Similar selection procedures are used by most other European breeders, though many of them give more emphasis to empirical tests in early generations and follow this with the "rapid mix test," similar to a small-scale Chorleywood bread process, which produces batches of small bread rolls from each flour sample.

α-Amylase activity is a strongly inherited character that must be considered in relation to both resistance to premature germination (sprouting) in the ear and to activity in the absence of sprouting. Resistance to sprouting is strongly associated with grain color; all white grained varieties show a greater tendency to sprout during a wet harvest than do those with red grains. But there is a considerable range of variability within both white and red grained varieties, and even the most resistant varieties show some sprouting when harvesting conditions are difficult. The character is, however, fairly strongly inherited within each grain color group, giving considerable scope for selection by the breeder.

Low α-amylase in the absence of sprouting is a strongly inherited character even though many of the highest yielding varieties currently grown in western Europe have a high endogenous α-amylase content. These varieties are mostly derived from the variety Professeur Marchal (8), which was widely used as a parent before this deleterious character was identified. There is, however, no reason why future varieties should have high levels of endogenous α-amylase. It may also be possible in the long term to breed varieties with a gene for very low α-amylase activity that has been identified in the agriculturally poorly adapted varieties Tom Thumb and Minister Dwarf.

In addition to these characters that are of direct importance in relation to grain quality, the breeder can also select for anatomical characters such as grain size and shape and depth of crease, though the environmental effect is greater. Grain shape and depth of crease may be important in relation to milling quality as they determine to some extent the ease with which the endosperm can be separated from the bran. They are also important in determining test weight, but this latter character, though easy to measure, is a poor criterion of varietal differences in grain quality and appears to be given exaggerated importance in current regulations of the European Economic Community (50).

III. MULTIPLICATION AND MAINTENANCE OF NEW VARIETIES

The production and maintenance of stocks of seed of improved varieties is an essential part of the responsibility of the plant breeder. Multiplication methods must ensure that the seedstocks released for commercial use do not differ in performance or diagnostic characters from those used in trial and that they satisfy the agricultural, legal, and certification standards of the countries where the variety is to be grown.

In many countries, particularly those where royalties are paid to plant breeders, strict regulations are in force to assure the uniformity of seedstocks sold to the farmer. These ensure that new varieties meet required standards of distinctness, uniformity, and genetic stability, and stocks are subject to field and laboratory inspection to make certain that the required standards are maintained. Further legislation also limits the number of generations of multiplication that a seed merchant may grow without renewing his seedstocks from the breeder.

When devising a system of varietal multiplication, the breeder is anxious to start building up stocks as soon as he can identify promising selections in his program. This can normally be done in F_4 or F_5, at a stage when the material is unlikely to be homozygous for the morphological characters used in varietal identification. It is therefore necessary to fashion a system in which selection for genetic uniformity can be achieved while multiplication is in progress.

In the system used at the Cambridge Plant Breeding Institute (Fig. 2), multiplication is normally based on a single F_4 plant progeny selected in F_5 as showing special promise and normally represented by about 100 plants. Sixty progenies are grown from these plants at wide spacing, so as to occupy an initial multiplication plot of about 1/20 ha.

This plot provides seed for official trials and also for further multiplication of the variety. The 20 most uniform progenies are used for further multiplication and for official tests of distinctness, uniformity, and stability; the remainder for trial purposes. These 20 plots are examined in detail for uniformity in morphological and physiologic characters and if no deviating plants are found, 100 ears are taken from each and the remainder of the plots harvested separately, each plot yielding approximately 3 kg of grain. From the 2,000 ears and grain harvested, 500 ears and 6 kg of grain are required for official tests of distinctness, uniformity, and stability.

Further multiplication is based on ¼-ha plots comprising drilled plots derived from each of the 20 separately harvested progenies sown so as to surround 200 ear rows, comprising 10 rows from each of the 20 progenies, sown so that the relationship between ear row and progeny can be maintained. The ear rows and progenies are examined critically throughout their development, and if any deviants are found, the ear rows and the corresponding progeny are eliminated. The surviving progenies are harvested in bulk to provide seed for trials and for large-scale multiplication as "breeders' seed." Single ears are then taken from each row to provide seed for ear rows in the following year, and the remainder of each ear row is harvested and threshed separately to provide seed for the corresponding ear row progenies. This procedure can be repeated indefinitely, providing "breeders' seed" for large-scale multiplication so long as demand for the variety continues.

IV. FUTURE PROSPECTS

The national average yield of wheat in Britain has increased from 2.5 tonnes (te)/ha in 1950 to 5.2 te/ha in 1979. Increases in yielding capacity have been similar in other countries of Western Europe. About half of this advance in yielding capacity can be attributed to new varieties and half to improved husbandry techniques and better control of weeds and pathogens. There is no reason to suppose that this upward trend in yield levels will not continue and no evidence that a plateau of yield has been reached.

Much of the advance in yield in Britain and elsewhere has come about through shorter strawed wheats. Many of these varieties carry the Norin 10 genes that have been widely exploited in the semidwarf wheats developed by CIMMYT in Mexico and are now widely grown in lower latitude countries. It is likely that wheats of this type will be increasingly cultivated in Western Europe in the next decade.

The more intensive cultivation of wheat in Europe and elsewhere has led to an increased incidence of many pathogens. Losses to the rusts and powdery mildew have given considerable cause for alarm, especially as the resistance of widely cultivated varieties has broken down. It is likely, therefore, that breeders will pay increased attention to the selection of varieties with durable resistance to these pathogens, though much work is required to devise suitable selection techniques; increased use of mixtures of closely related varieties to limit the spread of disease within or between crops is also likely. The increased cultivation of short-strawed varieties may be associated with wider incidence of the splashborne diseases caused by *Septoria* spp., though adequate genetic variability seems to be available.

The production of higher yielding varieties is likely to be associated with a reduction in the bread-making quality and in particular in the protein content of European wheats. This tendency may, however, be compensated by increased economic and sociological pressure for the production of varieties suitable for bread-making. It is therefore likely that breeders will pay increased attention to component characters of bread-making, such as low α-amylase content and hard milling, which are not negatively associated with yield. Techniques probably will be developed also to enable breeders to select for protein quality more efficiently and to exploit the limited available variation in total protein as fully as possible.

PART TWO: BREEDING IN AUSTRALIA

I. HISTORY OF CULTIVATION

The early history of soft wheat culture in Australia cannot be separated from that of wheat in general. Although wheat growing was initiated in the colony by Governor Phillip in 1788, the real development of the wheat industry traces back only about 80 years (34,35). Little is known about the cultivars grown during the early days, but it is assumed that they were of English origin. During the first half of the 19th century, Red Lammas and White Lammas were popular wheat cultivars, together with Talavera, Red Tuscan, and White Tuscan. About 1860, the cultivar Purple Straw was introduced (probably from England), and it played an important role in the breeding of many Australian wheat cultivars. Another popular cultivar was Du Toit, introduced from South Africa in 1881, which matured early and had partial resistance to stem rust. Du Toit was part of the parentage of many Australian cultivars. A spectacular increase in wheat growing took place between 1860 and 1880; the area under cultivation increased from 400,000 to 1,760,000 ha during this period.

At the turn of the 19th century, the Australian wheat industry changed dramatically. The wheat area increased from 2.03 million ha in 1900 to 4.86 million ha by 1915 because of the entry of Australian wheats into world markets. From then until the 1930s, the Australian product was accepted as a soft white wheat with good milling quality and low moisture content. It was sought for blending with higher quality wheats produced elsewhere. According to A. T. Pugsley (personal communication), the term "soft" was not used until the late 1950s; previously wheats were termed either "weak" or "strong."

In 1932, the Australian wheat production was largely comprised of "weak" or unspecified quality types, but by 1965 the production of medium-strong and strong wheats had increased to more than 30% of the total. For example, in 1965 the area sown to weak wheats amounted to more than 90% of the crop in Victoria, 50 to 60% in New South Wales and South Australia, and 30 to 40% in Western Australia.

Until quite recently, no segregation was practiced because most wheats grown were the weak wheats, particularly in the southern wheat belt comprised of southern New South Wales, Victoria, parts of South Australia, and most of Western Australia. This region has a relatively cool climate, predominantly winter rainfall, and a relatively long growing season. The northern wheat belt (northern New South Wales, Queensland, and some parts of Western Australia) has a shorter, warmer growing season and mostly summer rainfall and produces largely hard and prime hard types of wheats. Although most of the weak wheats were intended for bread production, millers had no difficulty in obtaining low-protein, extensible soft wheats for biscuit making. In fact, so much soft, weak wheat of low protein was available in Victoria until the 1960s that wheat suitable for bread making was difficult to obtain. With the improvement in cultural practices, the use of fertilizer, the introduction of clover ley farming, and the release of so-called soft wheats with low extensibility, it became more difficult to obtain low protein (8%) biscuit wheats. This led to the formation of the Soft Wheat Millers Group in the early 1960s. The growing of soft wheats in southern New South Wales was, however, limited to the Riverina district (G. McCorquodale, personal communication).

Even though Australia was originally a soft white wheat producer, the actual segregation of soft wheats did not take place until 1963 in Western Australia and 1972 in Victoria. The Australian Wheat Board Annual Reports mentioned separate soft (biscuit) wheat receipts only during 1972-1973 (81,000 te) and 1973-1974 (22,400 te); the corresponding export figures for these two years were 37,000 and 12,000 te, respectively.

II. USE OF GENETIC VARIABILITY

As mentioned previously, early wheat cultivars in Australia were introductions mainly from England and South Africa. Cultivar purity was not a concern and therefore no selection work was done until the late 1800s. In 1884, Steinwedel, a South Australian, selected a number of earlier maturing, tall plants from a crop of Champlain Hybrid. The cultivar Steinwedel was drought-tolerant and high yielding, but it was highly susceptible to stem rust and had very weak flour. Between 1890 and 1920, Steinwedel was widely grown, particularly in drier areas. James Ward, also from South Australia, selected from DuToit a very popular wheat called Ward's Prolific which subsequently featured in the parentage of many important cultivars. Baroota Wonder and Gluyas Early were direct selections from Ward's Prolific. The cross of Gluyas Early with Clubhead gave rise to the popular soft wheat Gluclub, which is still grown in Western Australia.

The most important factor that permitted a vastly increased wheat production in Australia was the pioneer breeding work of William James Farrer. The most significant among the numerous cultivars produced by Farrer was Federation (Purple Straw 14A × Yandilla A); it was popular in Australia from 1901 to 1938 and also widely grown on the Pacific coast of the United States during the 1920s and was introduced into India. Federation and some other Farrer wheats became the basis of many modern Australian wheat cultivars (35). That most of the soft and/or weak wheats take their ancestry from Federation or Gluyas Early has been aptly demonstrated by Moss and Wrigley (45).

Although Australia produced predominantly weak wheats until the late 1960s, no breeding programs were directed solely toward production of soft wheat cultivars suitable for the cake, confectionery, and biscuit industries until 1928. A number of cultivars existed, however, which, under low protein conditions, were suitable for the manufacture of cakes and biscuits. Most of the soft or weak wheats were not sufficiently extensible or contained excessively high protein levels for this purpose. Most cultivars used as soft wheats were actually dual-purpose wheats used both for the bread and confectionery, depending on protein content.

Between 1928 and 1939, some attempt was made to breed wheat cultivars specifically for the biscuit trade (2). Apollo (Clarendon × Hurst's 14), a late-maturing wheat with plump soft grain was crossed with promising rust-resistant, crossbred lines. A South American wheat, San Martin, was also used. During World War II, the project was abandoned in favor of more important challenges associated with bread wheats. Although the requirements of the soft wheat trade could be satisfied by selecting low protein samples from weak cultivars, more recently the Agricultural Research Centre at Wagga Wagga, New South Wales, and the Victorian Department of Agriculture at Werribee and Horsham, have initiated successful programs for the production of cultivars particularly for the soft wheat trade. The classification "Soft" has recently been used to denote cultivars acceptable to the premium soft grade, which is strictly a biscuit type. The other soft wheats used for cakes, confectionery, etc., are either weak bread wheats or selected general-purpose soft wheats.

III. QUALITY ASPECTS OF SOFT WHEATS

Before 1961, classification of hard and soft wheat in Australia was based mainly on grain appearance, and it was assumed that vitreous grain was hard and chalky grain was soft. This situation often led to confusion, particularly when the grain was slightly bleached. In 1961, however, Symes (57) applied a simple and reliable "particle size index" test, in which 10 g of clean wheat are ground in a mill set as finely as possible and fitted with a gravity feed designed to ensure a uniform rate of grinding. The meal is sieved on a number 15 silk (aperture 75 μ), the proportion passing through being defined as the particule size index.

Symes (57) tested 41 Australian cultivars and three crossbreeds, and classified them into two groups according to this index. He also assessed the influence of 1,000-grain weight and moisture and grain protein on particle size index within a cultivar, but the correlation was not significant. The influence of moisture content on particle size index was studied on one soft (Heron) and one hard (Gabo) wheat, and significant correlation coefficients of 0.829 and 0.925, respectively, were obtained, demonstrating the importance of storing samples in uniform conditions before testing for particle size index. Protein level of the grain was found to influence particle size index, and genotypic differences were apparent. However, Symes stated:

> Of even greater importance than the overriding varietal character of particle size index is the fact that this index divides Australian wheats into two clearly defined classes. These classes are so clearly defined that not even the variation due to protein content is likely to cause confusion.

Symes (58) also studied the inheritance of grain hardness as measured by particle size index in Australian wheats and found that a single gene controls the difference in particle size index between the hard wheat cultivar Falcon and the soft wheat Heron. The results from several other crosses also pointed to monogenic inheritance, possibly modified by minor genes. He further indicated the possibility of a number of different major genes or alleles. His work suggests that, by the use of backcrossing and particle size index tests, it would be simple to convert a hard wheat to a soft wheat and vice versa.

The influence of the gene causing hardness on milling and baking qualities was investigated by Symes (59), using near-isogenic lines of Falcon (hard) and Heron (soft), differing only in the alleles that produce hardness and softness as determined by particle size index. He found that the milling extraction, loaf texture, and dough handling characteristics were strongly associated with the gene controlling hardness.

While the particle size index became a very useful laboratory measure of hardness of the wheat grain, Australian breeders were looking for a simpler and perhaps quicker test. Kramer and Albrecht (25) had demonstrated that a laboratory pearler can be used to measure grain hardness of wheat, but the original laboratory barley pearler has disadvantages; for example, the slotted silicon carbide wheel did not allow ready cleaning between samples. Chesterfield (10) modified the barley pearler principally by replacing the slotted silicon carbide wheel with a smooth-edged wheel. His equipment was further improved by B. E. Miskle and N. F. Derera (unpublished data), who applied a timer that semiautomatically feeds and automatically empties the instrument every 60 sec. The correlation between particle size index and pearling resistance was high ($r = 0.94$), indicating that the modified barley pearler can be used with confidence to determine grain hardness in a breeding program.

While Symes and Chesterfield were more concerned with differentiation of soft and hard Australian wheats, Moss et al (43,44), were seeking a method to select cultivars for soft wheat quality. They reported the results of an investigation of the factors contributing to softness and the role of genotype-environment interactions on soft wheats in Australia. They found that flour yield, water absorption, and cookie spread showed slight genotype-season interaction, but no genotype-site interaction was observed. They emphasized the importance of the identification of factors limiting soft wheat quality in order to select appropriate genotypes for soft wheat breeding. Water absorption or retention, starch damage, flour color, paste viscosity, and cookie spread are all influenced by the hydrophilic components. The farinograph water absorption gave the best instrumental measurement of the combined effects of these characteristics.

Moss et al (44) attempted to find a *single* small-scale component of soft wheat quality that plant breeders could use for screening. They measured 10 such components: 1,000-grain weight, fiber content, protein percentage, ash content, pearling resistance, wheat meal granularity and fermentation time, milling yield (using a Brabender Quadrumat mill), sedimentation value, and alkaline water retention capacity (AWRC) of the flour. These workers concluded that no single criterion could be used to classify the milling yield and flour water absorption of cultivars from a single trial. The high degree of interaction between genotype, site, and season rendered the small-scale test of questionable value, though pearling and AWRC resistance measurements were relatively unaffected by environmental factors.

IV. CURRENT STATUS AND OUTLOOK

The world market demand for high quality soft wheats is more limited than that for hard bread wheats and rarely exceeds 20% of the total. The quality requirements of soft wheats for the Australian domestic market are very strict. Besides the softness of the grain, the protein level must be less than 10%, and the water absorption should be low. Emphasis is also placed on resistance to extension and extensibility of the dough. The quality specifications for the classes of Australian soft wheat are shown in Table I.

In New South Wales, a grade of "soft varieties" has been established in which grain of very soft cultivars is received and stored separately. This grading method is to satisfy the requirements of buyers who need a soft wheat with stronger gluten than is usual in biscuit flour. In Victoria, the "Australian Soft" (soft wheat of the cultivars Pinnacle, Summit, and Zenith) is handled as a separate grade for biscuit manufacture. The "soft varieties" segregation from the Australian Standard White (ASW) is similar to that in New South Wales.

For sweet biscuits, which require low protein, extensible dough with minimal resistance to extension and low water absorption, the cultivars Pinnacle and perhaps Gluclub are most suitable. Egret and Zenith have been bred specifically as biscuit and general-purpose soft wheats, but neither of them has the desirable softness and dough extensibility of Pinnacle. The other so-called soft wheats may be suitable for the variable requirements of the cake, confectionery, and pasta industries. Gluclub is also suitable for the selective requirements of the cake industry. The cultivar Jabiru, released in 1975, seemingly approaches Pinnacle in quality.

The use of wheat to manufacture starch and gluten is widespread in Australia. Indeed, the quantity of wheat used for these purposes (approximately 327,000 te per year) is double that consumed by the biscuit, cake, and pasta industries. The starch and gluten industries have attempted to purchase soft wheats with higher protein levels from the southern wheat belt of Australia. However, freight cost was so prohibitive for factories in the northern wheat belt that they were forced to use some of the northern hard wheats. Whereas gluten quality and yield (approximately 13.5%) from the northern hard wheats were satisfactory, the

Table I.
Quality Specification for the Australian Soft Wheat Class[a]

	General Soft Varieties	Soft Biscuit
Protein range (%)	9.0-12	8-9.5
Approximate hardness (particle size index)	> 20	Soft-> 25
Mixing time (min) (farinograph)	<3	<2
Extensograph, maximum height (BU)	200-300	100-200
Extensibility (cm)	> 17	> 17
Water absorption (%)	59-63	<57

[a]E. E. Bond, personal communication.

starch yield was 4% less than anticipated (G. K. Adkins, personal communication). The starch and gluten industries are best served by soft wheat cultivars with more than 12% protein. Such wheats should have free milling characteristics with a minimum of starch damage. The flour must have a relatively high protein content, contain minimal water-soluble carbohydrates and nitrogen, and have very low enzyme activity, particularly of proteases and amylases. The extracted gluten should be of a good baking quality with adequate cohesive and extensibility properties, and starch yields should also be as high as possible. Adkins (personal communication) has commented that the industry would welcome a genotype that not only complies with these requirements but also has an increased starch content at the expense of the nongluten components and/or an increased ratio of large lenticular starch granules to small spherical ones.

Breeding for both stem and leaf rust resistance in soft wheat improvement programs was somewhat neglected until recently. Both diseases caused considerable losses, particularly in southern New South Wales, Victoria, and South Australia, which are the traditional soft wheat-producing regions. The losses to stem rust in 1973 were estimated to be 681,000 te and 600,000 te in Victoria and South Australia, respectively, (26,40). After these major losses, a more concentrated effort was initiated to control this disease throughout Australia. A National Rust Control Programme has been established, and the University of Sydney's Plant Breeding Institute at Castle Hill has been appointed to act as a national rust control laboratory. The various wheat breeding institutions will submit breeding materials to the Plant Breeding Institute for resistance screening as seedlings and adult plants. It is anticipated that approximately 20,000 lines can be handled by the institute each year. As a result of this program, significant improvement in rust resistance in future soft wheat cultivars can be foreseen.

The concept of breeding soft wheats on the basis of industrial requirements is relatively new in Australia. This change was induced by the recent tendency to breed harder wheats for the southern, traditionally low-protein wheat regions. But this process could, in the long term, be detrimental to the wheat industry by producing large quantities of hard, low-protein samples. The present situation of wheat breeding in relation to market requirements was summarized by E. E. Bond (personal communication):

> Looked at critically, wheat breeding programmes in Australia should be coordinated to a greater extent than they are at the present time. The effort that is put into this work should be more consciously directed into defined hard and soft wheat breeding programmes. Both are needed and both are of equal importance. Standards of quality need to be lifted in relation to: (a) imparting greater strength to the hard wheat class while retaining the desirable general Gabo-type qualities; (b) the development of suitable soft wheat varieties to overcome deficiencies that are becoming increasingly apparent in the two main categories of this class, the Olympic type or better and the soft biscuit.

ACKNOWLEDGMENTS

I wish to express my appreciation to Professor I. A. Watson, Messrs. E. E. Bond, G. McCorquodale, J. T. Reeves, R. W. Fitzsimmons, R. M. Young, and Drs. A. Pugsley and

G. K. Adkins for their helpful and ready assistance.
Special thanks are due to H. J. Moss for his helpful suggestions.

LITERATURE CITED

1. Åkerman, A., and MacKey, J. 1948. The breeding of self-fertilized plants by crossing. In: Svalof 1886-1946. A. Åkerman, O. Tedin, and F. Froier, eds. Lund. pp. 46-71.
2. Anonymous. 1950. Crop, pasture and fruit breeding in New South Wales 1930-1950. Sci. Bull. 72, N.S.W. Dept. of Agric.
3. Bennett, M. K. 1968. British wheat yields per acre for seven centuries. In: Essays in Agrarian History, Vol. I, pp. 55-72. W. E. Minchington, ed.
4. Bhatt, G. M. 1973. Comparison of various methods of selecting parents for hybridization in common bread wheat (*Triticum aestivum* L.) Austral. J. Agric. Res. 24:457-464.
5. Biffen, R. H., and Engledow, F. L. 1926. Wheat breeding investigations at the Plant Breeding Institute, Cambridge. Stationery Office, London. p. 114.
6. Bingham, J. 1972. Physiological objectives in breeding for grain yield in wheat. In: The Way Ahead in Plant Breeding, Proc. 6th Cong. of Eucarpia, Cambridge 1971, F. G. H. Lupton, G. Jenkins, and R. Johnson, eds. pp. 15-19.
7. Bingham, J. 1975. Winter wheat breeding methods and prospects. J. Royal Agric. Soc. England 136:65-77.
8. Bingham, J., and Whitmore, E. T. 1966. Varietal differences in wheat, in resistance to germination in the ear and α-amylase content of the grain. J. Agric. Sci. Cambridge, 66:197-201.
9. Briggs, F. N., and Allard, R. W. 1953. The current status of the backcross method of plant breeding. Agron. J. 45:137-138.
10. Chesterfield, R. S. 1971. A modified barley pearler for measuring hardness of Australian wheats. J. Aust. Inst. Agric. Sci. 37:148-151.
11. Delaunay, L. 1931. Resultate eines dreijährigen Röntgenversuches mit Weizen. Züchter 3:129-137.
12. Donald, C. M. 1968. The breeding of crop ideotypes. Euphytica 17:385-403.
13. Driscoll, C. J. 1968. Alien transfer by irradiation and meiotic control. Proc. 3rd Int. Wheat Genetics Symp., Canberra, 1967.
14. Dumanovic, J., Ehrenberg, L., and Denic, M. 1970. Induced variation of protein content and composition in hexaploid wheat. In: Improving Plant Protein by Nuclear Techniques, IAEA, Vienna. p. 107.
15. Engledow, F. L., and Wadham, S. M. 1923. Investigations on yield in cereals I. J. Agric. Sci., Cambridge. 13:390-439.
16. Frey, K. J., Browning, J. A., and Simons, M. D. 1975. Multiline cultivars of autogamous crop plants. SABRAO J. 7:113-123.
17. Froier, K., Gelin, O., and Gustafsson, A. 1941. The cytological response of polyploidy to x-ray dosage. Botanisker Notiser (Lund), pp. 199-216.
18. Fussell, G. E., and McGregor, O. R. 1961. In: English Farming Past and Present, introduction by Lord Ernlie, 6th ed. William Heine, London.
19. Goulden, C. H. 1939. Problems of plant selection. In: Proc. 7th Int. Genetics Cong. Cambridge University Press. pp. 132-133.
20. Hermsen, J. G. TH. 1963. Sources and distribution of the complementary genes for hybrid necrosis in wheat. Euphytica 12:147-160.
21. Hutchinson, J. B., Panse, V. G., Apte, N. S., and Pugh, B. M. 1938. Studies in plant breeding techniques. III. Crop analysis and varietal improvement in Malvi Iowar. (*Andropogon sorghum*). Ind. J. Agric. Sci. 8:131-152.
22. Jenkins, J. E. E. 1973. An evaluation of the control of cereal leaf diseases by fungicides in England and Wales. Proc. 7th Brit. Insecticide and Fungicide Conf., pp. 781-790.
23. Johnson, R., and Law, C. N. 1975. Genetic control of durable resistance to yellow rust (*Puccinia striiformis*) in the wheat cultivar Hybride de Bersee. Ann. Appl. Biol. 81:385-391.
24. Johnson, V. A., Mattern, P. J., and Vogel, K. P. 1975. Cultural, genetic and other factors affecting quality of wheat. In: Bread. A. Spicer, ed. Allied Science Publishers Ltd.: London. pp. 127-140.
25. Kramer, H. H., and Albrecht, H. R. 1948. The adaptation to small samples of the pearling test for kernel hardness in wheat. J. Am. Soc. Agron. 40:422-431.
26. Krause, W. R., and Heard, T. G. 1975. Cereal variety recommendations for 1975.

Dep. Agric. S. Aust. Bull. 7/74:4.
27. Law, C. N., and Worland, A. J. 1974. Chromosome substitutions and their use in the analysis and prediction of wheat varietal performance. Proc. 4th Int. Wheat Genetics Symp., 1973.
28. Lupton, F. G. H. 1961. Studies in the breeding of self-pollinating cereals. III. Further studies in cross prediction. Euphytica 10:209-224.
29. Lupton, F. G. H. 1965. Studies in the breeding of self-pollinating cereals. V. Use of the incomplete diallel in wheat breeding. Euphytica 14:331-352.
30. Lupton, F. G. H., and Johnson, R. 1970. Breeding for mature plant resistance to yellow rust in wheat. Ann. Appl. Biol. 66:137-143.
31. Lupton, F. G. H., Oliver, R. H., and Ruckenbauer, P. 1974. An analysis of the factors determining yield in crosses between semidwarf and taller wheat varieties. J. Agric. Sci., Cambridge, 82:483-496.
32. Lupton, F. G. H., and Whitehouse, R. N. H. 1957. Studies on the breeding of self-pollinating cereals. I. Selection methods in breeding for yield. Euphytica 6:169-184.
33. Lukyanenko, P. P., Zhogin, A. F., and Ryadchikov, V. G. 1974. Obtaining new genetic sources of high protein and lysine content in the grain of bread wheat. Selskok hozgaistvennaya Biologiya 9:661-666.
34. MacIndoe, S. L. 1974. History of Production of Wheat in Australia. Australian Field Crops. Vol. 1. Wheat and Other Temperate Cereals. A. Lazenby and E. M. Matheson, eds. pp. 99-121. Angus and Robertson, publishers.
35. MacIndoe, S. L., and Brown, C. M. 1968. wheat breeding and varieties in Australia. Sci. Bull. 76. N.S.W. Dep. Agric.
36. MacKey, J. 1959. Mutagenic response in *Triticum* at different levels of ploidy. Proc. 1st Int. Wheat Genetics Symp., Winnipeg 1958. pp. 88-111.
37. Mather, K. 1949. Biometrical Genetics. Methuen, London.
38. Mather, K., and Jinks, J. L. 1971. Biometrical Genetics. Cornell University Press, Ithaca, NY.
39. Mattern, P. J., Morris, R., Schmidt, J. W., and Johnson, V. A. 1974. Locations of genes for kernel properties in the wheat variety Cheyenne using chromosome substitution lines. Proc. 4th Int. Wheat Genetics Symp. pp. 703-707.
40. McCann, J. M. 1974. A look back at wheat rust. J. Agric. Victoria, 72:193-196.
41. McFadden, E. S., and Sears, E. R. 1946. The origin of *Triticum spelta* and its free threshing hexaploid relatives. J. Hered. 37:107-116.
42. Meppelink, E. K., Belderok, B., and Ruiter, D. 1959. Ergebnisse einer Vergleichsuntersuchung von Methoden zur Bestimmung des Backwertes von kleinen Weizenmustern. Getreide und Mehle 9:125-130.
43. Moss, H. J., Goodchild, N. A., and Edwards, C. S. 1973. Selection of cultivars for soft wheat quality. I. Tests for softness in flour. Aust. J. Exp. Agric. Anim. Husb. 13:292-298.
44. Moss, H. J., Edwards, C. S., and Goodchild, N. A. 1973. Selection of cultivars for soft wheat quality. 2. Small scale tests of soft wheat quality. Aust. J. Exp. Agric. Anim. Husb. 13:299-305.
45. Moss, H. J., and Wrigley, C. W. 1974. Interrelationships between the pedigrees of Australian wheats. J. Aust. Inst. Agric. Sci. 40:207-211.
46. Mukade, K., Kamio, M., and Hosoda, K. 1973. The acceleration of generation advancement in breeding rust resistant wheat. Proc. 4th Int. Wheat Genetics Symp., Columbia, MO. pp. 439-444.
47. Nilsson-Ehle, H. 1910. Arbetena med hvete och havre vid Svalof under ar 1909. Sv. Utsadesf. Tidskr. 20:332-353.
48. Pelshenke, P. 1933. A short method for the determination of gluten quality in wheat. Cereal Chem. 10:90-96.
49. Pushman, F. M., and Bingham, J. 1975. A reappraisal of the Pelshenke test as a screening technique in the breeding of bread-making wheats for the United Kingdom. J. Agric. Sci., Cambridge. 85:221-231.
50. Pushman, F. M., and Bingham, J. 1975. Components of test weight of ten varieties of winter wheat grown with two rates of nitrogen fertilizer application. J. Agric. Sci., Cambridge. 85:559-563.
51. Pushman, F. M., and Bingham, J. 1976. The effects of granular nitrogen fetilizer and a foliar spray of urea on the yield and bread-making quality of ten winter wheats. J. Agric. Sci., Cambridge. 87:281-292.
52. Radley, M. 1970. Comparison of endogenous gibberellins and response to applied gibberellins in some dwarf and tall wheat cultivars. Planta 92:292-300.
53. Renfrew, J. A. 1973. Paleoethnobotany. The Prehistoric Food Plants of the Near

East and Europe. Methuen, London.

54. Sapehin, A. A. 1934. X-ray mutations as a source of new varieties of agricultural plants. Priroda 9:28-31.
55. Sears, E. R. 1956. The transfer of leaf rust resistance from *Aegilops umbellulata* to wheat. In: Genetics in Plant Breeding. Brookhaven Symposia in Biology 9, 1.
56. Sprague, G. F., and Tatum, L. A. 1942. General vs. specific combining ability in single crosses of corn. J. Am. Soc. Agron. 34:923-932.
57. Symes, K. J. 1961. Classification of Australian wheat varieties, based on the granularity of their wholemeal. Aust. J. Exp. Agric. Anim. Husb. 1:18-23.
58. Symes, K. J. 1965. The inheritance of grain hardness in wheat as measured by the particle size index. Aust. J. Agric. Res. 16:113-123.
59. Symes, K. J. 1969. Influence of a gene causing hardness on the milling and baking quality of two wheats. Aust. J. Agric. Res. 20:971-979.
60. Watson, D. J. 1952. The physiological basis for variation in yield. Adv. Agron. 4:101-145.
61. Whitehouse, R. N. H. 1970. An application of canonical analysis to plant breeding. Genet. Agric. 23:61-96.
62. Wolfe, M. S., and Barrett, J. A. 1976. Population genetics of powdery mildew epidemics. Ann. NY Acad. Sci.287:151-163.
63. Zeleny, L. 1947. A simple sedimentation test for estimating the bread baking and gluten qualities of wheat flour. Cereal Chem. 24:465-475.
64. Zeleny, L., Greenaway, W. T., Gurney, G. M., Fifield, C. C., and Lebsock, K. L. 1960. Sedimentation value as an index of dough mixing characteristics in early generation wheat selections. Cereal Chem. 37:673-678.

CHAPTER 4

MILLING IN THE EUROPEAN ECONOMIC COMMUNITY

M. FORD
RHM Research Ltd.
The Lord Rank Research Centre, High Wycombe, England

K. KINGSWOOD
RHM Research Ltd.
The Lord Rank Research Centre, High Wycombe, England

I. INTRODUCTION

For centuries, water and wind were almost the only means of powering mills until the 19th century. A water mill was first described by the Roman engineer Vitruvius in the first century B.C. and this design was widely used wherever the flow of water was suitable. The Romans are thought to have introduced water mills into Britain and the rest of Europe and by 1086, at the time of the Domesday survey, there were 5,624 water mills in Britain (44).

The Crusaders are said to have introduced the windmill into Europe and over the centuries developments in design were gradual. The mill became a common feature of the landscape. Because throughput was limited by the grinding action, a large number of mills was required. Also, as the wheat was locally grown and transport was by means of horse and cart, the mill tended to serve the local area.

With the coming of the Industrial Revolution and the use of steam power, it was logical that the use of steam to drive millstones would soon occupy milling engineers. However, although steam-driven, the milling process remained unchanged; wheat still passed in a horizontal direction between two stones, which were dressed by hand to make furrows so that the meal produced would be removed from the grinding faces (or land) of the stones and pass to the circumference, where it was collected as meal. White flour was produced by using a cylindrical sifter or bolter to remove the coarsest bran. It is believed that even with this system, some millers added water before milling to toughen the bran and improve flour color.

The first roller mill in Britain for use on grain was patented in 1753 (27), but it had no effect on the use of millstones, which were not displaced by rollers until the second half of the 19th century. Before then, rolls were applied to grain milling by Collier of Paris (1823), Bollinger at Venice (1824), and Helfenberger in Switzerland. Innovations were made and Sulzberger introduced a roller mill with an improved system and built mills at Milan, Venice, Stettin, Leipzig, Mannheim, and Pest. Only one was a success; this was the Pesther Walzmühle founded by Count Szechenyi, a Hungarian nobleman, using Sulzberger's chilled iron roller mills.

Further developments were introduced by Naeff and Friedrich Wegmann. By 1875, rollers were beginning to be used extensively in Germany.

British millers, alarmed at the way cheap, imported, white flour was affecting their trade, visited Hungary in 1877 to see these mills in action. From 1881, when a demonstration of the new roller milling system was shown at the Agricultural Hall in London, the new system was rapidly adopted in the United Kingdom.

Today the process is still essentially the same but with various modifications brought about by improved technology, such as the use of plan-sifters instead of centrifugals.

II. SOFT WHEAT—MILLING IMPLICATIONS

Although this chapter is devoted to the milling of soft wheat, i.e., *Triticum aestivum*, this wheat can be classified by millers as having hard and soft characteristics, depending on milling performance.

A. The Wheat Kernel

Grain size of wheat grown in the European Economic Community, now the European Community (EC) is somewhat larger than the Canadian or U.S. spring type wheats and smaller than durum. European spring wheats are generally smaller than winter types.

Milling everywhere requires the separation of bran and grinding of endosperm to produce flour.

The endosperm contributes 83 to 84% of the weight of the grain and consists of starch granules embedded in a protein matrix within large thin-walled cells. The size and shape of these cells differ according to their position within the grain.

There are two types of starch granule, different in size and shape. The larger A-type is lenticular; the smaller B-type is spherical. Their mean diameters are approximately 25 and 10 μm, respectively. The B-type constitutes 30% of the total weight of starch (17). Brocklehurst and Evers (11) found different sizes of grain within the one variety grown in the United Kingdom (cv. Maris Huntsman) and that the large, plump grains contained a significantly larger proportion of B-type granules than did smaller or shriveled kernels. This was related to the kernels having undergone growth under optimal conditions whereby a maximal number of A-type granules, which are the first to be formed during cell division, are laid down at a fast rate and can accumulate a maximal number of B granules during cell development leading to the well-filled appearance. Weather conditions could therefore affect this ratio during these two stages. Bingham (8) indicated that the U.K. climate, which is cooler and damper, is more favorable for high grain yields than hotter, drier climates where grain filling and ripening are more rapid.

Fisher and Jones (19) noted that a protein network was formed in wheat on drying from 40% moisture content, and stresses were set up within the grain as it shrank. Certain types could resist this stress, giving a coherent, vitreous, or hard endosperm. Other types that could not resist this stress produced an endosperm containing minute cracks, giving the grain an opaque or mealy appearance owing to internal light reflection.

B. Kernel Texture

Scott (36) suggested that grain size and shape, protein content, and quality could be involved in producing hard or soft wheats. Parish and Halse (30) showed that rate of drying and temperature were related to development of opacity, or translucency, in grain.

Others have discussed the genetic basis of hardness (21,42), the cementing effect between starch and protein (6), and influence of protein content and site (41).

Stenvert (40), using a simple laboratory hammer mill, related differences in hardness to a grinding resistance value. The results obtained on varieties grown in England are shown in Table I.

A good correlation between grinding resistance and starch damage value was established. A grinding time greater than 30 sec enabled hard varieties of wheat samples to be identified.

The introduction of hard North American wheats into Europe in the 19th century had an immediate effect on the stone grinding process. A reduction in flour yield occurred that could not be corrected unless excessive heat buildup was produced. This led to the development of the high-grinding process to produce middlings for a low-grinding process. Here was the beginning of the break and reduction stages of modern flour milling. At this stage of the development of milling, the effect of the crease on extraction was not important, although the use of this two-stage process led to greater yield of flour owing to the effectively larger surface used for grinding.

As the roller mill process developed, the presence of the crease as a source of bran contamination became increasingly important. It became necessary to open

TABLE I
Wheat Hardness (Grinding Time) for Some Cultivars Grown in England

Cultivar	Time (sec)
Kleiber	42
Cama	40
Hardi	37
Kolibri	36
Huntsman	34
Champlein	24
Cappelle Desprez	22
Atou	20
Val	18
Joss Cambier	16

the grain from the crease, leaving the bran intact and using the successive fluted rolls to scrape the endosperm away from the bran. At the end of the break passages, the bran was removed from the process. It became obvious that the amount of flour removed from the bran depended on the type of wheat being milled. Wheats that left a minimal quantity of flour on the bran and at the same time produced gritty, free-flowing particles that sieved easily and did not produce chokes were identified as hard wheats and were desirable for good milling. The other type of wheat left much flour adhering to the bran and produced woolly stocks that blinded sieves and stuck to spouts and were referred to as soft wheats. The soft types produced higher break release and the balance of the mill was shifted. It was found that the imported wheats from North America were generally hard and of good milling quality, whereas the home-grown wheats were soft and lower in milling quality.

The problem of milling soft wheats was partly solved by milling them together with the hard North American wheats, but poor cleanup of soft U.K. wheat depressed extraction rates.

In the 1950s and '60s, with the need to rebuild or reequip mills and the cost effectiveness of newer machines, shorter-surface milling was widely adopted. With the use of softer grists in the EC, excessive flake production was occurring on the head reduction rolls in an attempt to maintain flour extraction. The centrifugal sieving machines broke up the flakes to a certain extent, but with the introduction of higher capacity plan-sifters and even heavier roll pressures, the production of flakes remained a problem. This was finally overcome by the use of flake disrupters after A, B, C, D rolls. These disrupters were entoleter units originally included for disinfestation, but they were very effective for breaking down flakes. Also, the introduction of pneumatic conveying after 1945 allowed higher capacity milling owing to the cooling effect of the fluidized system.

C. Water Absorption and Starch Damage

Another serious milling problem created by the use of softer grists was the failure of the process to meet the water absorption requirements expected of flour milled from harder wheats. Before the introduction of the Chorleywood bread process in the United Kingdom, for instance, water absorption levels were guaranteed by the use of high-protein North American wheats. Following the reduction in protein that the CBP allowed, the use of softer wheat was increased. The water absorption fell because of the reduction in starch damage levels. This need to maintain water absorption levels led to the introduction of greater roll pressures on the reduction side, increasing roll wear and stock temperatures.

The phenomenon of starch damage has been investigated for some years. Jones (23) microsopically examined roller-milled flours and found that certain granules were flattened and appeared faint in outline and were given the name of "ghosts," the number of ghosts within a sample being proportional to its maltose figure (a measure of starch damage at constant α-amylase level). Starch granules have elastic properties and can be changed into ghosts by taking them beyond their elastic limit. Evers (16) published scanning electron microscopy photographs showing flattened, damaged granules that he termed "relaxed granules." He maintains that this relaxation of the internal tension of starch granules allows them to take up water easily.

This relaxing effect is created during the milling process. As Evers and Jones showed, enzyme attack leading to microscopically visible pitting does not relax the granule or cause it to absorb water more easily. Also, radial cracking induced by compression does not contribute to water absorption. Four types of damage can be identified (46): 1) radial cracks, 2) sheared or split granules, 3) abraded, and 4) flattened granules. The latter two items only contribute to water absorption and are fortuitously produced during milling.

Because of the importance of water absorption of bread flours, methods of maintaining the level when greater quantities of soft wheat are being used are important. In addition, the reduced protein level of EC wheats works against maintaining water absorption levels, so the miller is forced to examine ways to modify the properties of starch.

First of all, the miller can choose cultivars that are known to be hard milling in order to raise starch damage level, and he may pay premiums to obtain them. However, the bulk of research into starch damage has been aimed at the milling process and especially the reduction operation. Break flour contribution to starch damage has been reported as only 20% of the total (47).

With regard to the effect of the reduction system on starch damage the following factors must be considered: feed particle size, feed rate, roll gaps, roll speed, roll differential, roll surface, roll pressure, and roll temperature.

The influence of these variables has been the subject of much research. It is difficult to isolate the effect of any one owing to the interaction between them.

FEED PARTICLE SIZE

The distribution of starch damage within a sample of break stock can be altered by light rolling using smooth rolls. The initial material has low starch damage in the semolina and dunst fractions but higher starch damage in the flour. After rolling, the level in dunst and flour is unchanged, but the coarse fraction has high starch damage. This level of starch damage in semolina is referred to as the internal factor and is caused by stresses set up in the unreduced endosperm by rolling. The effect of abrasion on the surface of the coarse particles is referred to as the surface factor. Jones states that starch damage is greater in coarse particles than in finer particles, owing to this internal factor, assuming that all particles are equally resistant and that pressures are equal. Hard wheat tends to produce a greater proportion of larger particles and higher resistance to reduction than soft wheat does. Hence their internal factor will be higher.

FEED RATE

Jones et al (25) showed that if the feed rate is decreased, flour release and starch damage increase.

When the feed rate is high, rolls apparently are at a disadvantage for starch damage production, unless coupled with higher roll speeds.

ROLL CHARACTERISTICS

Willm showed that decreasing the roll gap increases starch damage (47).

Roll speeds vary widely between mills. A rolls can vary from 200 to 600 rpm, between long and short surface mills.

Roll speed, differential, and roll peripheral velocity difference are related by the following formula for a 10-in. roll:

$$\Delta V = \frac{2.618 \text{ RF (D-1) ft/sec}}{60D}$$

where Δ V = peripheral velocity difference between fast and slow roll, RF = fast roll speed rpm, and D = differential.

Any change in one variable therefore affects the others. The effect of solely increasing roll speed at constant D without altering Δ V cannot be assessed as it is directly proportional.

Few data are available for judging the effect of increasing roll speed on starch damage at constant differential. Willm (47) shows that flour release increases with roll speed, and between 250 and 400 rpm at 1.25:1 differential, starch damage falls and then remains constant. It appears that roll speed or Δ V alone is insufficient for increased starch damage production and that roll differentials must be altered.

Jones et al (25) reported that increasing differentials on both smooth and fluted rolls (80/in. at 5°) increased starch damage level in A flour, with fluted rolls more so than smooth. This test was done using 6-in. rolls and a constant fast roll speed of 420 rpm; hence differential was produced by altering slow roll speed. Flour release was maintained by adjusting roll pressures. The maltose figure of coarse fraction from the grind (greater than No. 1 silk) using smooth rolls was higher than that from fluted rolls.

Ziegler (50) showed that starch damage increases with differential with any type of roll surface, although different types of surface affected the degree of increase. Wanzenreid (45) also reported increase of starch damage with differential.

Work on the effect of roll surface on starch damage production was reported by Jones et al (25) and Ziegler (50).

Jones described experiments with different types of roll surfaces on a laboratory mill. On A stock, a coarse matt surface gave the highest increase in the maltose figure. The effect of this surface was detectable in later reduction flours. Thus, using a reduction system of A, B2, C, D, and E rolls, when smooth A rolls were substituted by fine matt, starch damage of the final flour increased over that contributed by A flour. The use of fine matt rolls down the system, however, led to some flours, e.g., B_2, D, and E flours, having reduced starch damage, and Jones recommended their use on A rolls only.

The effect of increasing differential was enhanced by using coarse matt rolls. This was confirmed by Ziegler who reported that starch damage production by differential was increased further by increased roughness of roll surface. Also, roll surface and pressure interact, leading to still higher starch damage with high pressure and rough surfaces.

Ziegler's work on coarse middlings indicates that higher maltose figures could be achieved using scratched rolls rather than frosted or polished rolls. Crosscuts of 1 to 3 mm applied at 45 and 135° were used to provide scratched rolls; this type of roll produced the highest flour release but with highest ash value.

The use of mixed pair, i.e., fluted (80/in.) and smooth roll, has been described

(25). Coarse particle reduction was facilitated, although no increase in starch damage occurred.

Effect of roll pressure depends on the feed particle size. With increasing roll pressure, coarse semolina (A) reduction shows little increase in the maltose figure of the flour produced, but flour release is increased. The flour contains more coarse material, i.e., the particle size distribution is shifted to the coarse end. This coarse material has a low starch damage and hence the overall starch damage level of A flour tends to remain constant. Maltose figures of the fractions greater in particle size than flour, i.e., B2, C feeds, show that maltose figures have been increased substantially over the same fractions in the A feed. This internal factor appears later in the flour streams. Hence, the importance of pressure on A rolls is to increase starch damage levels later in the reduction system. Flake disrupters have made the use of higher roll pressures more practical in reducing the loss of flour tailing over the sieves, and to some extent any flakes broken down will increase starch damage levels. No evidence of the effect of roll pressure on flour color is available.

The effect of increased roll pressure on finer material, such as C stock, is to increase both flour release and starch damage level. Levels of starch damage in dunst produced are also increased.

Ziegler (50) reported that increasing roll temperatures were associated with increased maltose figures and that smooth rolls heated more than rough or fluted rolls during milling.

D. Extraction Rate

As mentioned previously, the increase in the use of softer EC wheats, apart from the water absorption effect, also has other milling quality implications, especially for flour extraction rates.

When discussing extraction rates, it is important to state the basis of the calculation because the yield of flour can be expressed in many different ways (28), e.g., as a percentage of: 1) wheat entering the silo from intake and including screenings, 2) wheat entering silo from intake but excluding screenings, 3) wheat to first break roll, or 4) total production basis.

Whatever the extraction rate, it can only be considered in relation to the flour quality in terms of bran content.

Various methods are used in the EC to express this quality.

ASH CONTENT

This method is favored by most EC countries. There is a general relationship between extraction rate and ash content. Figure 1, constructed by Russell-Eggitt and Hartley (34), shows this relationship. It also shows the relationship between ash content and flour color grade for EC flour types 1–5.

The ash method is only an approximate guide to extraction rate. It depends on the quantity of aleurone present, rather than on the outer layers. The laboratory test is lengthy and difficult to perform routinely. In addition, evidence suggests that North American wheats are inherently higher in ash content and this is not related to final flour color.

FLOUR COLOR GRADE

The ash test is not applicable to flours to which materials such as calcium carbonate are added. Because this material is compulsory in the United Kingdom, other tests must be used. The original subjective assessment of bran in the Pekar test has been superseded by the use of the Flour Color Grader developed by Kent Jones and Martin. The advantage of this method is its rapidity, although the relationship between ash and color grade is somewhat tenuous. A further advantage is that the method indicates darkening of endosperm resulting from wet harvest, whether caused by mold (cladosporium) growth or increase in dark endosperm pigments caused by metabolic activity (18).

III. EC WHEAT MARKETING

(with particular reference to France, West Germany, and the United Kingdom)

A. France

The organization of the wheat market in France is:

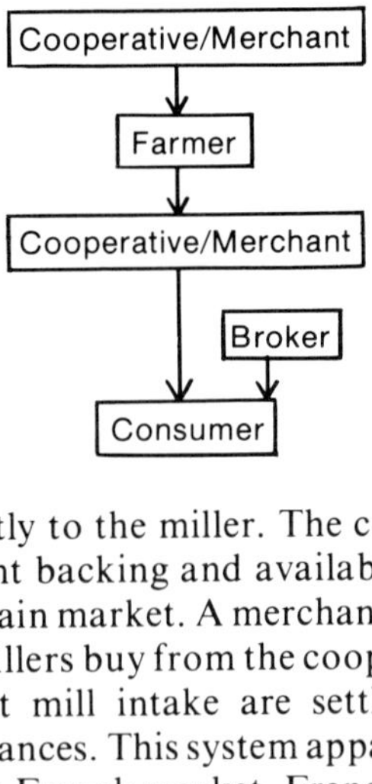

The farmer may sell directly to the miller. The cooperative system has many advantages, with government backing and availability of financing, but is not allowed to speculate in the grain market. A merchant has this right but must have an official license. French millers buy from the cooperative on contract, and any quality differences found at mill intake are settled in accordance with the contract by a system of allowances. This system apparently works well because of the competitive nature of the French market. French millers normally buy their wheat on the following specifications: test weight, moisture and impurities, less frequently on alveograph, Hagberg, or protein, after establishing a region or supplier as a reliable source early in the season.

The miller often buys on a maximum moisture content and minimum test weight. Then, if the weight of the wheat is better than the contract guarantees, he is obliged to pay an allowance to the supplier. As an example, if the contract is for a test weight of 75 to 76 kg, the miller pays the seller allowances per tonne along the following lines: up to 77kg/hl, 2.60 franc per tonne (te); 77 to 78 kg/hl, 5.10; 78 to 79 kg/hl, 7.60; 79 kg/hl, 9.30 franc/te.

There is also a system for weight below guarantee, normally on the basis of no allowance unless below 74 kg. (73 to 74 kg/hl, 6.40 franc/te; 73 to 72 kg/hl, 10.60 franc/te). Below 72 kg the miller has the right of refusal. There is a scale of premiums payable by the miller for moisture lower than the contract's 15.5%.

Wheat is bought according to the price per 100 kg based on August shipment in each crop year. To this August price are added the majorations, or monthly premiums, which are set yearly by the EC Commission. For the 1978-1979 crop year the majoration was approximately 146 francs per half month. As in any market economy, millers may buy their wheat when they choose. Large millers have the right to trade wheat in the paper market as a form of futures market trading, but the mills have to have a registered and licensed merchanting company (C. J. Haycroft, Soufflet [U.K.] Ltd., personal communication).

B. West Germany

Millers buy their wheat on a regional basis from local merchants on intervention terms (E. Amme, ABK Hamburg, personal communication). There are special allowances for moisture content and test weight. For example, up to

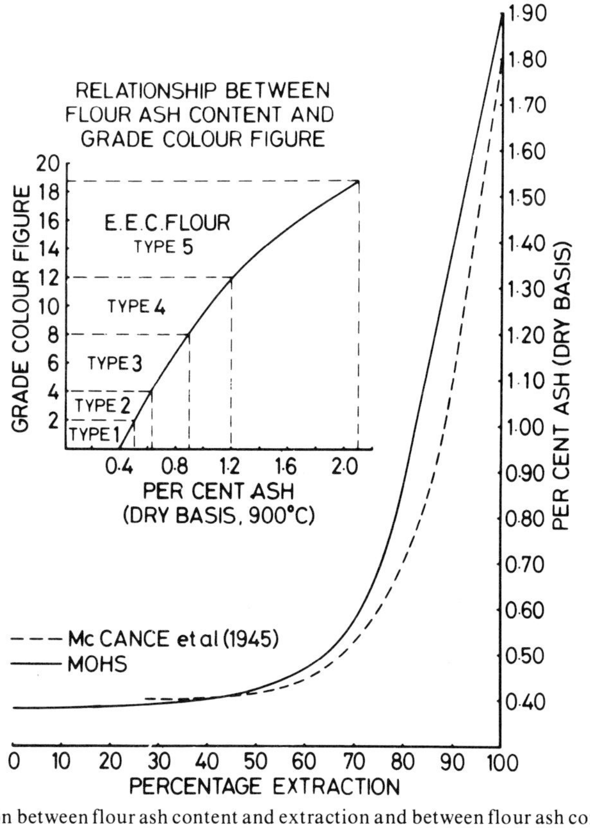

Figure 1. Relation between flour ash content and extraction and between flour ash content and grade color value (34).

16.2% moisture, 7 DM/te and above this level 8 DM/te is deducted. No premium is paid for protein level, although premiums can be paid for wheat of certain sedimentation values. Table II shows wheat usage by region. (Note that the quality of wheat is inferior from North Rhine-Westphalia; this region uses a high proportion of EC wheat.) West Germany currently imports about 1 million te of French wheat and takes 30% of all EC exports including significant quantities of wheat from the Netherlands and Denmark. The Federal Research Centre of Cereal and Potato Processing at Detmold monitors the quality of French imports (10). West Germany has reduced third-country imports from 25 to 13% by switching to French wheat.

C. United Kingdom

The importance of U.K. wheat to U.K. millers was increased by two factors in the early 1970s. The first was the entry of the United Kingdom into the EC, whereby a levy was payable to the EC Commission on imported wheat; the second was that world wheat prices rose sharply following the purchase of U.S. wheat by the Soviet Union.

The ability of the U.K. miller to use home-grown wheat depends on its quality in terms of protein level and α-amylase, two factors important for U.K. bread production. In order to ensure a supply of high protein, good quality wheats, which have generally lower yields, the miller has to pay a premium of up to 10% for certain varieties. Generally, the U.K. miller spot buys on protein and Hagberg and excludes feed type wheat or contracts with farmers via a merchant to supply wheat as stipulated.

The total U.K. use of wheat for flour milling is 5 million te and because the miller can only use half of that produced in the United Kingdom, he needs to import from the EC and other countries the wheat of required quality for bread. The remaining U.K. wheat is used in feed milling and for industrial purposes.

Two thirds of U.K. imports now come from other EC countries, mainly France, Germany, Denmark, and the Netherlands, although the amounts may vary widely from year to year, depending on quality. Nearly 1 million te is imported from North America, mainly from Canada. A small amount comes from Sweden (Table III).

TABLE II
Proportions of Wheats Used by West German Millers in Various Regions in 1974-1975

	Total (1,000 te)	Homegrown (%)	EC (%)	Third Country (%)
Baden-Württemberg	778	78.1	10.5	11.4
Bavaria	597	93	0.6	6.4
Hesse	156	87.6	0.0	12.4
Lower Saxony	581	84.3	8.2	7.5
North Rhine-Westphalia	1,032	57.4	33.5	9.1
Rhine-Pfalz	144	88.7	2.3	9
Schleswig-Holstein	292	87.3	2.9	9.8
Total	3,580	76.3	14.3	9.4

[a] Data from APRIA (5).

IV. EC GRAIN MARKET REGULATIONS

The Treaty of Rome (1957), which established the European Economic Community also laid down the principles of the EC's Common Agricultural Policy, which were: 1) to increase the productivity of farming, 2) to increase farmers' incomes, 3) to stabilize markets, 4) to ensure security of food supplies. and 5) to provide food at reasonable prices.

In July 1967, grain market regulations came into force for the EC, which at that time consisted of six members: France, West Germany, Belgium, Luxembourg, the Netherlands, and Italy. The objectives of these regulations were to: 1) abolish all government controls, e.g., on grist compositions; 2) control price by external and home trade; 3) reintroduce free competition; and 4) adjust prices to offset freight charges from surplus to deficit areas.

A new grain marketing policy was introduced in 1976 (Silo system), which aims to introduce a) an identical intervention price for soft wheat, barley, and maize; b) reference prices for bread wheat; c) development of prices for different kinds of grains; and d) the extension of the difference between intervention and threshold prices.

Under the Common Agricultural Policy, farmers are guaranteed certain minimum prices for wheat irrespective of the level of production. Each year, the EC commission sets the internal market price structure for the crop year starting on 1 August and the following prices are set:

A. Grain Prices

REFERENCE PRICE

Introduced in 1976, this price is for wheat of bread-making quality and is set at a level above the feed wheat intervention price. In 1977-1978 it was 15.53 units of account (u.a.). This price is set to encourage farmers to grow bread-making type wheats rather than poorer quality, high yielding types. The difference in price is to compensate for the lack of yield of the baking-quality type.

THRESHOLD PRICE

External trade is regulated by the application of a threshold price acting as a

TABLE III
Origin of Wheat Imported by the United Kingdom, 1970-1975 (in 1,000 te)

Country of Origin	1970-1971	1971-1972	1972-1973	1973-1974	1974-1975
France	98	621	1,255	774	600
Netherlands	224	170	240	299	584
Belgium/Luxembourg	126	74	35	75	284
Denmark	...	...	84	109	74
West Germany	...	...	2	4	15
EC Total	448	865	1,616	1,267	1,557
Canada	1,604	1,182	1,293	1,310	1,231
Australia	1,769	491	515	...	...
U.S.A.	1,177	605	748	394	229
Others	182	...	160	61	13
Total third country	4,732	2,278	2,716	1,765	1,473

trade wall against wheat on the world market that is generally lower in price than EC wheat. Table IV shows the threshold prices for 1977-1978.

TARGET PRICE

Threshold prices apply at Rotterdam but are calculated on the basis of the target price at Duisburg because that inland port is the intervention center with the greatest deficit of grain, hence the highest wheat price. It therefore represents the threshold price plus the cost of transshipment and transporting wheat to Duisburg from Rotterdam. Target prices are shown in Table V.

IMPORT LEVY

Any wheat entering the EC is subject to a levy that amounts to the difference between the threshold price and the world market price. In this way, cheap wheat from outside the EC is made to cost at least as much as EC wheat. The levy is based on a sliding scale and can change daily as the value of wheat on the world market fluctuates according to its price and relative quality. The differences in quality of external wheat are made comparable by applying coefficients of equivalence with reference to a European Standard. The higher the quality difference, the greater the coefficient of equivalence. Table VI shows coefficients for 1977-1978. The levy, therefore, would be based on the cheapest price CIF Rotterdam on the basis of the adjusted price using coefficients of equivalence.

TABLE IV
EC Threshold Prices for 1977-1978 (EC regulation 1360/77 [43])

	1977-1978		1976-1977	
	u.a.[a]/te	**DM/te**	**u.a./te**	**DM/te**
Soft wheat Mixed grain	155.15	529.46	149.30	519.69
Durum	221.30	755.20	216.10	752.21
Rye	152.15	519.22	146.45	509.77
Barley Maize	142.00	484.59	135.10	470.26
Oats	136.60	466.16	130.00	452.51
Buckwheat Millet Sorghum Canary seed	139.80	477.08	133.00	462.95

[a] u.a. = unit of account.

TABLE V
EC Target Prices for 1976-1978 (EC Regulation 1152/77 [43])

	1977-1978		1976-1977	
	u.a.[a]/te	**DM/te**	**u.a./te**	**DM/te**
Soft wheat	158.08	539.46	152.00	529.09
Durum	224.27	765.34	218.80	761.61
Barley	144.97	494.74	137.80	479.66
Rye	155.12	529.36	149.15	519.17
Maize	144.97	494.72	137.80	479.66

[a] u.a. = unit of account.

For example, if No. 1 CWRS 14% protein was priced 90 u.a./100 kg, 14%, Hard Winters at 85 u.a./100 kg, and Argentinian Plate wheat at 80 u.a./100 kg, then, using coefficients of equivalence, their prices are adjusted to 77, 73, and 71 u.a., respectively. If the threshold price was 90 u.a. a levy of 19 u.a. would have to be paid on all wheats based on the cheapest (Argentinian) price. Hence, the price of No. 1 CWRS 14% for use inside the EC would be 109 u.a.; Hard Winters and Plate would be 102 u.a. and 99 u.a., respectively.

INTERVENTION PRICE

This is the lowest price at which EC wheat is likely to be sold and is the support price set so the EC authorities will buy wheat from the farmer when the market price falls below this level. Table VII shows 1977-1978 Intervention prices.

In order to be bought by the various intervention boards, minimum quality specifications have to be set and are given in Table VIII.

These minimum quality specifications can vary slightly between countries.

B. Testing of Wheat

With the introduction of the reference price for bread-making quality wheat and the so-called two-tier intervention system, a test had to be established to distinguish between bread-making wheat and feed quality wheat. Some baking test was needed, but the main problem in setting such a test is that bread-making processes and wheat requirements vary widely within the EC.

STICKINESS TEST

One approach to the problem was the use of a stickiness test suggested by Bolling (9), devised at Detmold.

Doughs that broke down on mixing became sticky, and the Bolling method relied on subjective hand determination of this degree of stickiness. Another suggested alternative was the use of variety identification using gliadin electrophoresis.

To study these methods, the technical representatives within the National Milling Associations of the EC countries, known as the "Groupement," discussed the Bolling test and an alternative method based on a baking test for intervention purposes. They rejected the Bolling test, owing to its lack of reproducibility, and decided to develop an EC baking test. The first baking test was based on the work of the ICC Baking Test Study Group. Basically, the test involved mixing flour to a predetermined energy level with salt, sugar, reconstituted dried yeast, and no fat. The results of a collaborative test within the Groupement showed that, generally, laboratories had placed the flours in the same order of quality, but wide variations in results on individual samples were apparent between laboratories. In performing the test, laboratories were asked to comment on physical characteristics of the dough at various stages, and good agreement was found on recordings of the physical state of the dough up to the molding stage and on its machinability. It was, therefore, recommended, and accepted, that dough stickiness and observations on stickiness and machinability be applied as intervention criteria for the crop year 1976-1977.

A modified test using standardized equipment was adopted, namely, a Stephan high-speed mixer and a Brabender ball homogenizer.

With these and other modifications to an intervention test, the following procedure was instituted:

Wheat was cleaned and conditioned to 15.5% moisture level, then milled on a Bühler laboratory mill type MLU-202 to give 72% extraction flour with an ash content of 0.5 to 0.6% on a dry basis.

Water absorption was found using the farinograph and the falling number of the flour adjusted to 200 to 250 sec with addition of malt flour. After 1 min mixing, the dough in the Stephan mixer is judged to be nonsticky and machinable as follows: 1) The dough forms a coherent mass that does not adhere to the spindle and sides of the bowl of the mixer. It should be possible to remove it from the mixer in one motion without loss. 2) Sticky and nonmachinable—dough sticks to the bowl and wraps around the spindle. Much remains if an attempt is made to remove it.

If the dough is machinable, it is then divided, rounded, proved, and remolded and its stickiness is again judged. As there has been no further agreement or

TABLE VI
Coefficients of Equivalence (CoE, in units of account per 1,000 kg) for Wheat Imported by the EC for 1977-1978 (EC Regulations 158/67 and 1636/71[43])

Origin	Quality	CoE To Be Deducted
	Soft Wheat	
U.S.A.	Soft Red Winter Garlicky II+III	2.50
	Soft Red Winter I+II	3.75
	Western White II	3.75
	Soft White II	3.75
	Hard Winter/Dark Hard Winter I+II (up to 12.4% protein content guaranteed or without guaranteed protein content)	9.00
	Hard Winter/Dark Hard Winter I+II (guaranteed protein content 12.5 to 12.9%)	9.75
	Hard Winter/Dark Hard Winter I+II (guaranteed protein content 13 to 13.4%)	10.50
	Hard Winter/Dark Hard Winter I+II (guaranteed protein content 13.5 to 13.9%)	11.25
	Hard Winter/Dark Hard Winter I+II (guaranteed protein content 14% and over)	12.00
	Red Spring/Northern Spring/Dark Northern Spring I+II (up to 12.4% protein content guaranteed or without guaranteed protein content)	9.50
	Red Spring/Northern Spring/Dark Northern Spring I+II (guaranteed protein content 12.5 to 12.9%)	10.25
	Red Spring/Northern Spring/Dark Northern Spring I+II (guaranteed protein content 13 to 13.4%)	11.00
	Red Spring/Northern Spring/Dark Northern Spring I+II (guaranteed protein content 13.5 to 13.9%)	11.75
	Red Spring/Northern Spring/Dark Northern Spring I+II (guaranteed protein content 14 to 14.4%)	12.50
	Red Spring/Northern Spring/Dark Northern Spring I+II (guaranteed protein content 14.5% and over)	13.25

improved process, this standard test still holds. Lack of funds and standard equipment have halted further progress towards full development.

ELECTROPHORESIS

The second possibility for testing for bread- and feed-quality wheats is the application of starch gel electrophoresis to wheat gliadin proteins. This test, developed by CSIRO in Australia and Autran and Bourdet at INRA, Paris, enables the user to identify the varieties of grain in a sample within two days.

Basically, the test involves extracting gliadin proteins from grain with alcohol. The extracted protein carried on a slip of absorbent paper is placed in a slab of starch gel. A charge of 400 volts is then applied across the gel and the charged protein units move under the potential gradient. Within four or five hours, the fastest moving protein units will have moved about 15 cm (14). The protein units are stained to make them visible and to give protein patterns as seen in Figure 2.

The Autran and Bourdet procedure has been extensively tested in the United

TABLE VI–*continued*

Origin	Quality	CoE To Be Deducted
Canada	No. 1 Canada Western Red Spring (up to 12.4% guaranteed protein content or without protein guarantee)	10.00
	No. 1 Canada Western Red Spring (guaranteed protein content 12.5 to 12.9%)	10.75
	No. 1 Canada Western Red Spring (guaranteed protein content 13 to 13.4%)	11.50
	No. 1 Canada Western Red Spring (guaranteed protein content 13.5 to 13.9%)	12.25
	No. 1 Canada Western Red Spring (guaranteed protein content 14 to 14.4%)	13.00
	No. 1 Canada Western Red Spring (guaranteed protein content 14.5% and over)	13.75
	No. 1 Manitoba Northern	12.50
	No. 2 Manitoba Northern	12.00
	No. 3 Manitoba Northern	10.50
	No. 4 Manitoba Northern	9.00
	No. 5 Canada	6.00
Argentina	Southern Wheat (Bahia Blanca, Necochea)	9.00
	Up River (Rosa Fee)	9.00
	Down River (Buenos Aires)	9.00
Australia	Faq	6.75
	Hard	9.00
	Prime Hard (guaranteed protein content 14% and over)	12.00
Sweden		0
Bulgaria		2.25
Romania		3.75
Soviet Union	Type 441	9.00
	Type 431	10.50
	Type 121 (SKS 14) (guaranteed protein content 14% and over)	12.50

Kingdom, and the results on 29 cultivars have been reported (15). Most cultivars can be identified with this procedure and variation in site, protein level, and crop year do not influence the identification. The implications for milling and agriculture are considerable, apart from its use as a potential intervention test. It can be used to check that wheat shipments do not contain excluded cultivars. Seed suppliers can use it as a purity check and plant breeders as a test for distinctiveness of breeding lines (15).

V. EC MILLING

The development of the milling industry in Europe has followed a fairly uniform pattern with slight country-to-country variations since the introduction of roller milling. The differences have been mainly in response to the demands of the bakery industry for the type of bread that is produced. For instance, West German millers include some rye in their grists, but the milling processes are basically the same.

A. Milling Principles

INTAKE AND CLEANING

Wheat arrives at the mill by truck or barge and, before intake, usually

TABLE VII
EC Intervention Prices for 1977-1978 (EC Regulation 1152/77 [43])

	1977-1978 (per tonne)		1976-1977 (per tonne)	
	u.a.	DM[a]	u.a.	DM
Baking quality wheat	135.59[b]	462.71	131.00	455.99
Feed wheat	120.06	409.71	116.00	403.78
Durum	203.01	692.79	202.00	703.13
Barley	120.06	409.71	116.00	403.78
Rye	128.96	440.09	124.00	431.62
Maize	118.03	402.79	112.20	390.55

[a] Conversion rate, 1977-1978: 1 unit of account = DM 3.41258.
1976-1977: 1 unit of account = DM 3.46084.

[b] Reference price.

TABLE VIII
Minimum Quality Standards for EC Intervention Wheats

	Bread Wheat	Feed Wheat
Sound basic grain[a]	min. 90%	min. 88%
Moisture	max. 14 to 16%	max. 14 to 16%
Natural weight	min. 72 to 75 kg/hl	min. 68 to 72 kg/hl
Broken grains	max. 5%	max. 5%
Foreign grain admixture	max. 5%	max. 12%
Includes shrivelled grains	...	max. 12%[b]
Foreign admixture	max. 3%	max. 3%
Sprouted grains	max. 6%	max. 8%

[a] Samples of bread-making wheat offered for intervention in the U.K. must have a Hagberg Falling Number of not less than 160 sec.

[b] Fully included in the calculations of grain other than sound basic grain.

undergoes some form of quality testing. The tests are Hagberg for α-amylase, sedimentation for protein quality, moisture test, and protein test with the FMBRA Rapid meter in the United Kingdom. This testing also determines binning of the wheat for gristing.

To meet flour quality requirements, wheat must have all impurities removed before milling.

Impurities can be classified into three groups: 1) poisonous—ergot, smut; 2) flour quality—dust, dirt, mud balls; and 3) mechanical damage—stones, metal.

The characteristics of impurities differ from those of wheat and make their separation possible. These characteristics are size, shape, specific gravity, air resistance, and magnetic properties.

Size. Separation by size is done on sieves of robust construction, usually steel, perforated metal, or brass. The machines are usually referred to as milling separators and remove peas, straw, sand, etc. Such machines are the Simon ripple sifter and Bühler Granostar separator. An aspiration stage is also included.

Shape. This factor is useful to remove from wheat such seeds as oats and barley, which cannot be separated by their size alone. Separation is achieved by using indented discs or cylinders or a combination as in the Simon Disc Cylinder

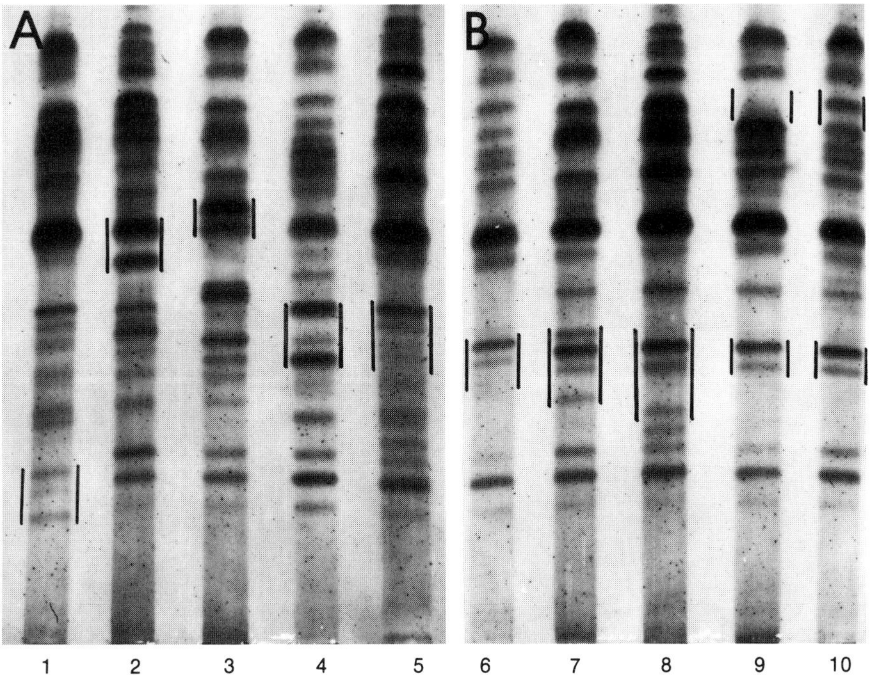

Figure 2 A and B. Electrophoretic patterns of solubilized gliadins from 10 major groups of wheat cultivars. A, Cultivars of: 1 = Maris Dove, 2 = Highbury, 3 = Bouquet, 4 = Clement, and 5 = Maris Ranger. B, Cultivars of: 6 = Maris Nimrod, 7 = Cappelle Desprez, 8 = Maris Freeman, 9 = Atou, and 10 = Maris Huntsman.

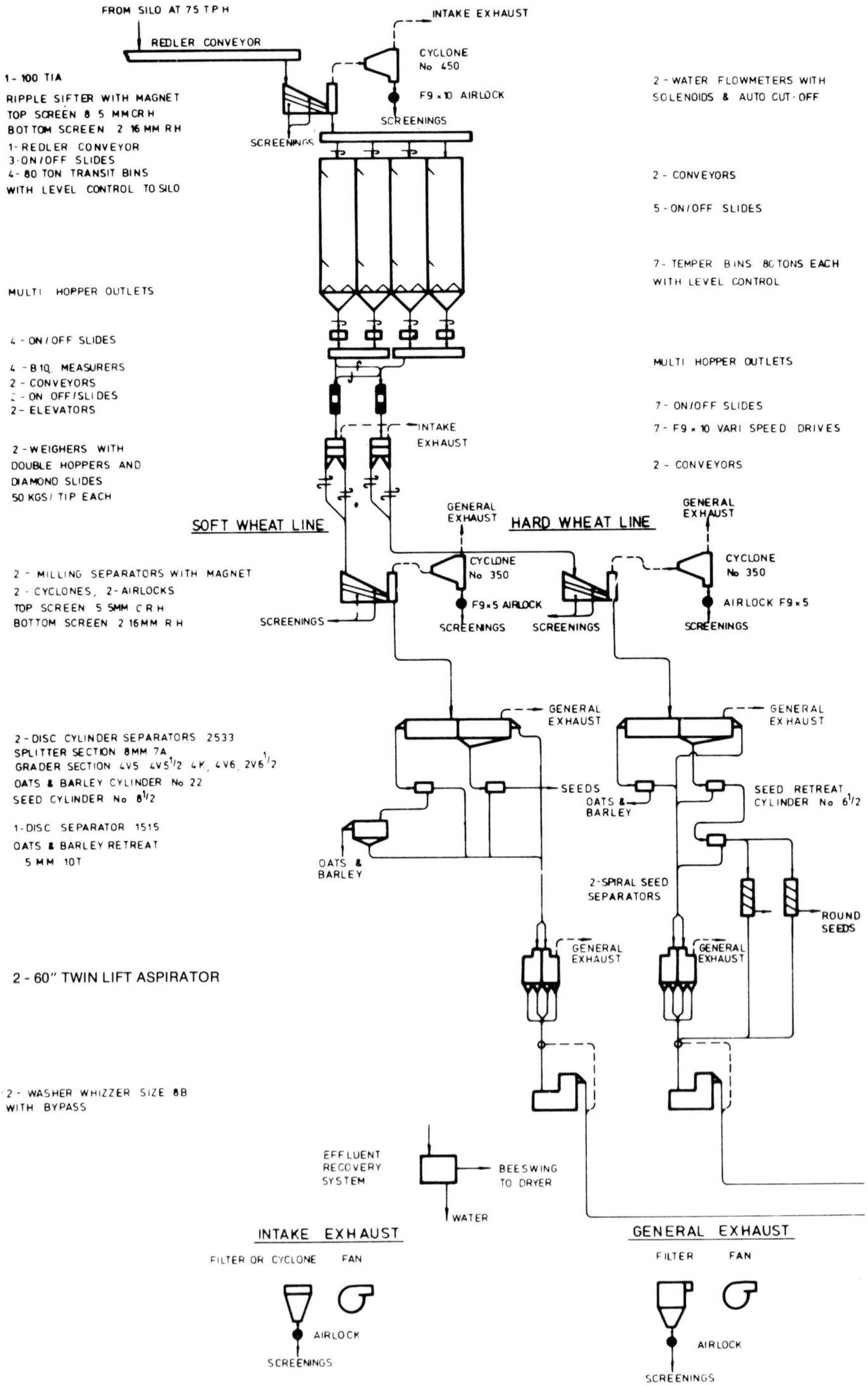

Figure 3. Wheat flow diagram for a typical screenroom. (Courtesy Henry Simon, Ltd.)

TABLE IX

Comparative Terms for British, French, Spanish, German, and American Mill Flow Diagrams [26]

English	French	Spanish	German	American
I Bk.	B1	I Rotura	I Schrot	I Bk.
II Bk. Coarse	B2 Gros	II Rotura gruesa	II Schrot grob	II Bk. Coarse
II Bk. Fine	B2 Fin	II Rotura fina	II Schrot fein	II Bk. Fine
III Bk. Coarse	B3 Gros	III Rotura gruesa	III Schrot grob	III Bk. Coarse
III Bk. Fine	B3 Fin	III Rotura fina	III Schrot fein	III Bk. Fine
IV Bk. Coarse	B4 Gros	IV Rotura gruesa	IV Schrot grob	IV Bk. Coarse
IV Bk. Fine	B4 Fin	IV Rotura fina	IV Schrot fein	IV Bk. Fine
V Bk. Coarse	B5 Gros	V Rotura gruesa	V Schrot grob	V Bk. Coarse
V Bk. Fine	B5 Fin	V Rotura fina	V Schrot fein	V Bk. Fine
X^1	D_s1	X^1	1 Koppen	Chunks 1
X^2	D_s2	X^2	2 Koppen	Chunks 2
Y	D_s3	Y	3 Koppen	Chunks 3
Z	D_s4	Z	4 Koppen	Chunks 4
A	Cl 1	A	1 Griess	1 Sizings
B	Cl 2	B	2 Griess	1st Middlings
C	C 1	C	1 Mahlung	2nd Middlings
B^2	Cl 3	B^2	2 Griess fein	2nd Quality
D	C 2	D	2 Mahlung	3rd Middlings
E	C 3	E	3 Mahlung	4th Middlings
F	Cl 4	F	1 Übergang	1st Tailings
G	C 4	G	4 Mahlung	5th Middlings
H	C 5	H	5 Mahlung	6th Middlings
J	Cl 5	J	3 Griess	2nd Tailings
K	C 6	K	6 Mahlung	7th Middlings
L	C 7	L	7 Mahlung	8th Middlings
M	Cl 6	M	2 Übergang	3rd Tailings
N	C 8	N	8 Mahlung	1st Low grade
O	C 9	O	9 Mahlung	2nd Low grade
P	C 10	P	10 Mahlung	3rd Low grade
Germ I	Germe I	Germen I	1 Keim	Germ I
Germ II	Germe II	Germen II	2 Keim	Germ II
Pollard I	Recoupes	Afrechillo grueso	Kleienmehl grob	Shorts 1
Pollard II	Recoupettes	Afrechillo fino	Kleienmehl fein	Shorts 2
Bran flaking	Aplatissage des sons	Aplastamiento de afrecho	Kleiepresse	Bran flaking
Coarse semolina CS	Grosses semoules G.S.	Semolina gruesa	Grober Griess	1st & 2nd Bk. Coarse middlings
Medium semolina MS	Moyennes semoules M.S.	Semolina mediana	Mittelgriess	1st & 2nd Bk. Medium middlings
Fine semolina FS	Fines semoules F.S.	Semolina fina	Feingriess	1st & 2nd Bk. Fine middlings
Coarse middlings CM	Gros finots G.F.	Semita gruesa	Grober Dunst	}
Medium middlings MM	Moyens finots M.F.	Semita mediana	Mittlerer Dunst	}
Fine middlings FM	Fins finots F.F.	Semita fino	Feiner Dunst	} Midds Re
Coarse middlings duster CMD	Sécheur à gros finots sch I	Clasificador de semita gruesa	Grober Dunst Sichter	}
Medium middlings duster MMD	Sécheur à moyens finots sch II	Clasificador de semita mediana	Mittlerer Dunst Sichter	}
Fine middlings duster FMD	Sécheur à fins finots sch III	Clasificador de semita fina	Feiner Dunst Sichter	}
Semolina grader SG	Diviseur à semoules div I	Clasificador de semolina	Griess-sortierer	Coarse middlings grader
Dunst	Finots	Dunst	Dünste	Dunst
Screenings	Dechets de nettoyage	Residuos	Reinigungs Abfälle	Screenings
Bran	Son	Afrecho	Kleie	Bran
Broad bran	Sons aplatis	Afrecho aplastado	Kleieflocken	—
Coarse weatings	Rebulets	Semitin	Bollmehl	Coarse middlings
Fine weatings	Remoulages	Rebacillo	Futtermehl	Fine middlings

continued

TABLE IX–*continued*

English	French	Spanish	German	American
Bakers' grade flour	Farine de force boulangerie	Harina 000	Vollmehl	Bakers' grade flour
Patent flour	Farine extra 0000	Harina 0000	Auszugsmehl	Patent flour
Straight run flour	Farine boulangerie 00 ou 000	Harina de 00 a 000	Durchgemahlenes Mehl	Straight run flour
Low grade flour	Farine basse 0	Harina 0	Nachmehl	Clears
Breaks	Broyage	Roturas	Schrotung	Breaks
Scalping	Division	Separación	Vorsichtung	Scalping
Grading	Classement	Clasificación	Sortierung	Grading
Dusting	Séchage	Cernir	Trockensichtung	Dusting
Purification	Sassage	Purificación	Griessputzerei	Purification
Dressing	Blutage	Cernido	Sichtung	Bolting
Reductions	Convertissage	Pasajes lisos	Ausmahlung	Reductions
Offal grading	Classement des issues	Clasificación de subproductos	Abfallsortierung	Feed grading
Redressing	Sûreté	Repasador	Nachsichtung	Rebolting

Separator.

Specific Gravity. The main application of this principle is in the separation of stones from wheat. Formerly, it was applied in water immersion equipment such as the Mersey Vat Stoner, which separated stones and sand, or in the form of a stoning tank adjacent to a whizzer. This type of separation is being phased out owing to effluent and hygiene problems. Dry stoners such as the Bühler Vibrostoner and the Simon and Kipp Kelly machines have proved as effective.

Air Resistance. Air currents are applied throughout the screening process to separate impurities lighter than wheat. Air currents initially remove chaff, straw, light grains, and dust and later remove beeswing and dust generated by moving the wheat through the screenroom.

Once the grain has been freed of impurities, it is cleaned to remove mold, crease dirt, and also beeswing by means of brushes, scourers, and aspiration.

A typical screenroom layout is shown in Figure 3. The advantage of this type of system lies in its adaptability and flexibility to meet the cleaning requirements of different types of wheat with widely differing impurities. Where the mill grist is mainly homegrown wheat, there is scope for some simplification.

CONDITIONING

During the conditioning process, wheat is brought to the right physical state for milling. The main objectives are to toughen the bran to reduce its friability so that it can be sieved out of the process easily and to minimize the amount of fine bran particles finding their way into the flour. Conditioning mainly involves addition of water to wheat with or without the application of heat.

Different types of wheat require addition of different quantities of water. Hard North American wheat is generally milled at a higher moisture level and allowed to rest longer than is soft milling homegrown wheat; in the latter category, however, the higher protein wheat needs to lie for a longer time. The lying time is necessary for moisture to penetrate the wheat to toughen the bran and to mellow or increase the friability of endosperm to ensure lower milling power requirements. The addition of water is a compromise, as increasing moisture level reduces the ease of separation of bran and endosperm, and sifting.

To speed the process, warm or hot conditioning, depending on temperature, can be used to increase the rate of water penetration. Because this is expensive, it is not used frequently.

Water is usually added by means of metering into a damping worm (auger) before entry into the conditioning bins. Intensive dampeners such as the Technovator and Bühler Type MOZJ (48) have recently been introduced, which are said to allow more uniform and greater dampening of grain.

MILLING PROCESS

The milling process consists basically of two stages: 1) separation of the bran from the endosperm and 2) grinding of the endosperm into fine particles. However, owing to the structure of the grain and the need to maximize the production of flour free from bran particles, the process is very complex in practice. Before describing the milling process, it is worth mentioning that milling terminology varies widely throughout the world. Table IX, comparative milling terms (26), is included for reference.

First, milling exploits the difference in toughness between bran and endosperm to remove the bran. Second, the cohesiveness of the endosperm allows its production in fairly large particles that are free-flowing and capable of being aspirated to remove fine bran particles. Third, the friability of endosperm allows it to be ground to fine particles, whereas any accompanying bran is not ground and can be separated. The germ must also be removed because of its injurious effect on baking quality.

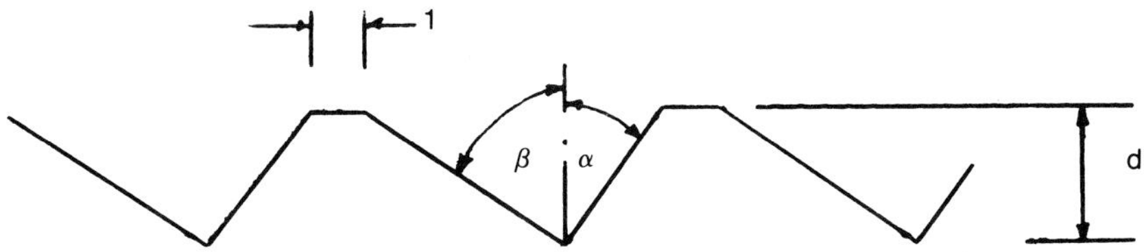

Figure 4. Flute profile. l = land, d = depth, α, β = flute angles.

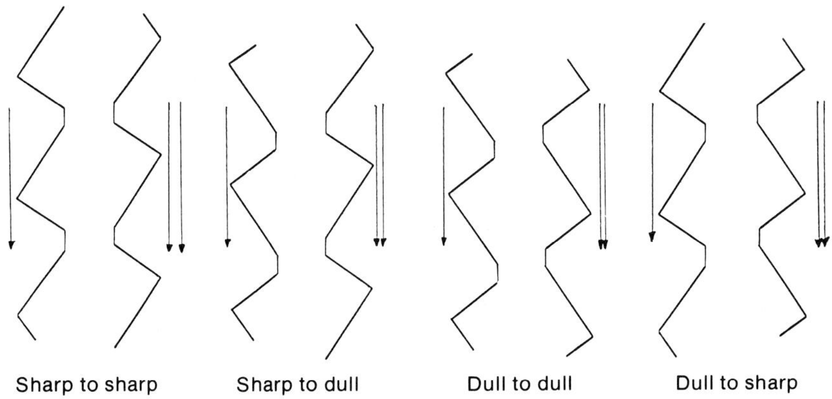

Figure 5. Roll dispositions in break system (double arrow indicates fast speed roll).

The first of these processes takes place on the break system.

The Break System. This was developed to open up the grain from the crease and to remove, by scraping, the endosperm adhering to the bran. The usual number of break stages is four, which normally suffice to clean the bran, but in West Germany, for instance, up to six may be used (33). Each break passage may be divided to mill fine and coarse stock.

The roller mills used on the break passages are fitted with fluted rolls that run at differential speed, usually 2.5:1. This differential is necessary so that one roll acts as a holding roll while the other shears open the grain, or removes endosperm.

Break roll performance depends on the type and number of flutes. A typical flute profile is shown in Figure 4. The section is referred to as a saw tooth shape and is fairly standard.

Scott reported that for U.K. practice (36) $l = 0.010$ to 0.035 in. for scratch, and IBk, $d = 0.02$ in. for scratch, and IBk.

Stanger stated that an optimum land exists in terms of break flour production and ash content (39) and that the depth of fluting is important. It must be deep enough to accommodate released semolina, as used in Simon profile 4 fluting ($\alpha =$ 21 to 25°, $\beta = 60$ to 67°) where from IBk to IIIBkF, excluding IIIBkC, the fluting is designed to produce semolina, whereas IIIBk and IVBks assist the scratch system in removing endosperm from bran (3). For bran cleanup the profile 3 fluting is preferred ($\alpha = 45°$, $\beta = 60°$). Fluting theory and practice have been described for West German milling (33).

The condition of flutes is important. They wear and sometimes are worn out within three months (39), especially on high-capacity mills. If flutes are worn, higher pressures are needed, leading to evaporative losses. Different views may be taken on the desirability of sharp or blunted flutes, depending on whether the bran needs to be cut or unrolled. This is influenced by wheat type (36).

There are four possible ways of placing break rolls together in relation to the

TABLE X
Break System Parameters for U.K. Practice for Roll Speeds Greater Than 500 rpm[a]

Passage	Flutes per: in.	Flutes per: cm	Disposition	Spiral	Profile	Differential
IBk	10.5	4.1	D/D	1:50	4	2.38
IIBk	14	5.5	D/D	1:50	4	2.38
IIIBkC	16	6.3	S/S	1:24	4	2.38
IIIBkF	18	7.1	S/S	1:24	4	2.38
IVBkC	24	9.5	S/S	1:24	3	2.38
IVBkF	26	10.2	S/S	1:24	3	2.38
(VBkC)	(28)	11.0	(S/S)	(1:24)	(3)	(2.38)
VBkF	30	11.8	S/S	1:24	3	2.38
XBkF	30	11.8	D/D	1:24	3	1.8
SIZ 1	24	9.5	D/D	1:24	3	1.8
SIZ 2	26	10.2	D/D	1:24	3	1.8
Bran cutting	18	7.1	S/S	1:8	Collins	2.38
Reductions	32 to 40	12.6 to 15.8	...	1:24	TCC	1.8

[a]Data from Tech. Newsletter (3).

flute angles and fast roll speed, as shown in Figure 5.

U.K. mills use a disposition of either dull to dull or sharp to sharp, depending on the break passage. Other EC millers prefer more variation in cutting angles and dispositions. For equal release settings, D/D produces less bran contamination and S/S produces more bran cutting, although this is preferable for softer U.K. wheats to improve bran cleanup (36).

Scott gives the following flute numbers for a four-break system (36):

Break stage	Flutes (no./in.)
I	12
II	14
III	18
IV	26/30 or 28
Scratch	36

Flute depth is set by the flute angles, if too shallow, excessive crushing results, and if too deep, stock is held. Roll flute spirals can vary from 1 in 40 to 1 in 7 and roll diameters are generally 10 in. Gap setting is a compromise but depends on wheat size and break release.

As every mill has its own requirements, differences in flute design are invariably produced. Table X gives the fluting for a typical U.K. mill. Break release is the stock that does not pass on to the next break passage. The way the break releases are set has important implications on extraction rate and the balance of the mill. Factors that affect the way the release is set include the length

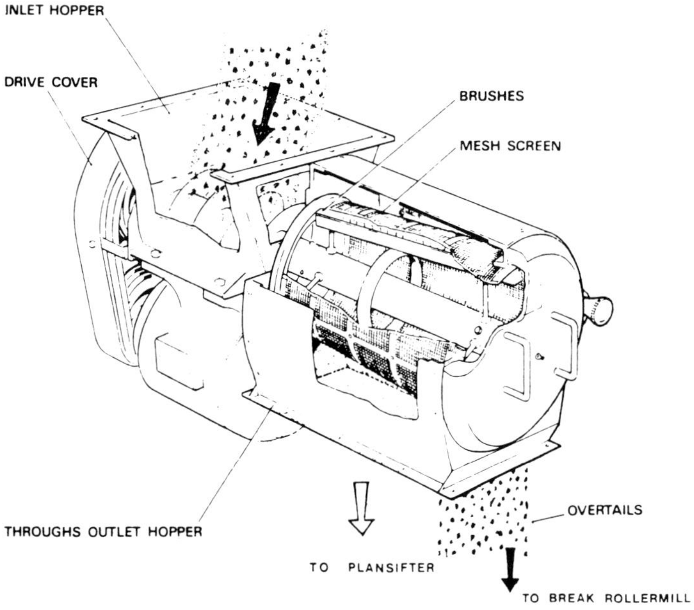

Figure 6. Le Coq prebreak scalper. (Courtesy Henry Simon, Ltd.)

of the milling system, allocation of roll surface, the type of wheat being milled, and the type of flour being produced.

Scalping and grading include the sieving process after the break rolls, scalping being the separation of coarsest stock that passes to the next break. This is usually done on a plan-sifter, although in some French and Belgian mills, a

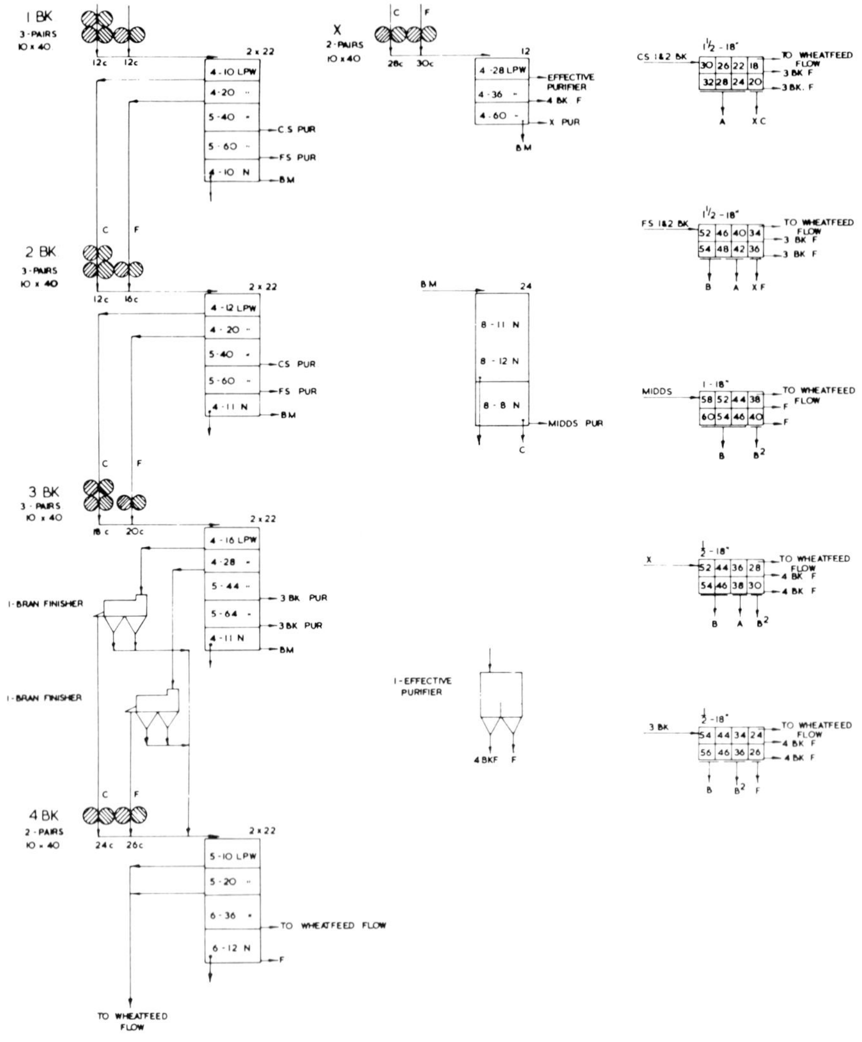

Figure 7. Mill flow pattern for a 40-sack short-surface mill showing stream classification and purification (28).

prebreak scalper unit of the Le Coq type shown in Figure 6 is often used. The advantage of this unit is that it removes bulky stock through more thorough sifting on the plan-sifter by allowing reallocation of sieves.

Subsequent sieves are used for grading stock into material of increasing fineness such as coarse, medium, fine semolinas, coarse and fine middlings, and

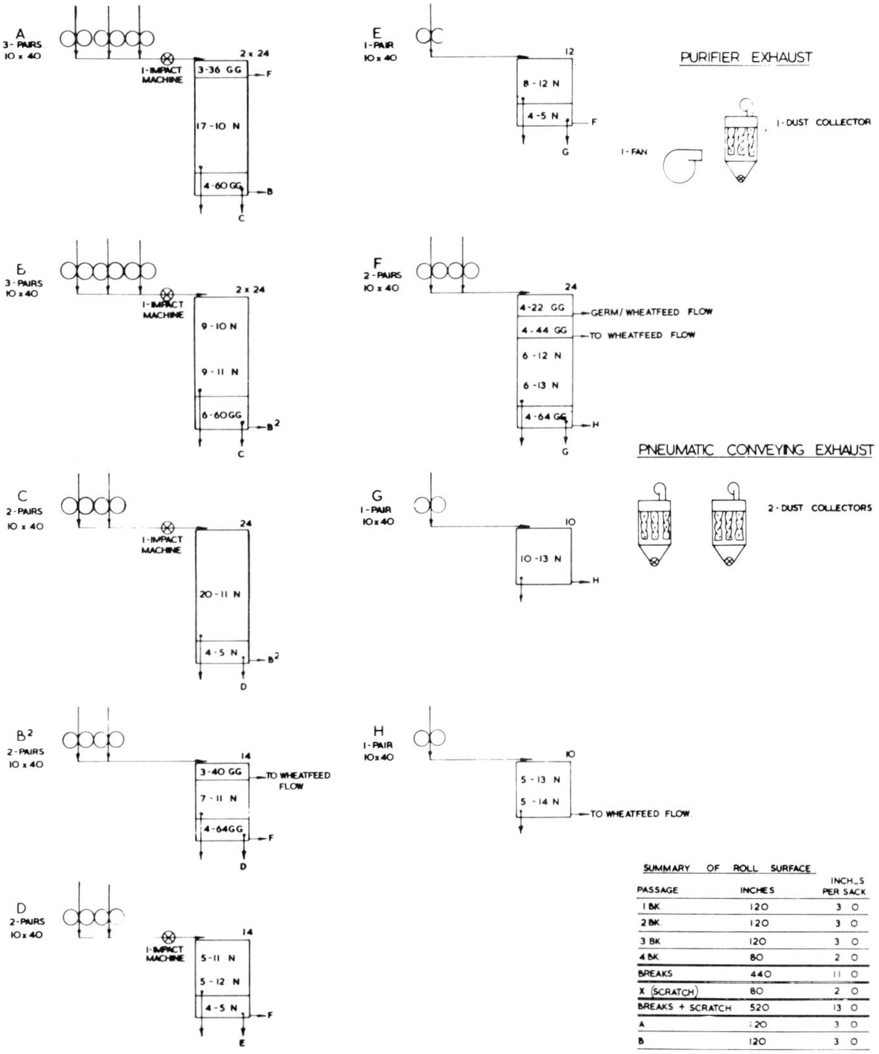

SUMMARY OF ROLL SURFACE

PASSAGE	INCHES	INCHES PER SACK
1 BK	120	3 0
2 BK	120	3 0
3 BK	120	3 0
4 BK	80	2 0
BREAKS	440	11 0
X (SCRATCH)	80	2 0
BREAKS + SCRATCH	520	13 0
A	120	3 0
B	120	3 0
C	80	2 0
B^2	80	2 0
D	80	2 0
E	40	1 0
F	80	2 0
G	40	1 0
H	40	1 0
REDUCTIONS	680	17 0
BREAKS + SCRATCH + REDUCTIONS	1200	30 0

dunst. The latter usually goes straight to reduction rolls, whereas all other stocks go to the purification stages.

In some French mills a sizing roll is included to reduce stock size for purification or reduction grinding.

The semolina and middlings stocks need to be cleaned of their bran before reduction grinding, to reduce the production of unavoidable bran powder that takes place in the reduction system. The purification stage also grades the stock according to particle size for feeding to the various reduction rolls, as shown in Figure 7. Early models of purifiers relied on air currents alone to lift off the bran but were inefficient because particles of bran still attached to endosperm were not separated. Grading was introduced, therefore, so that such particles could be treated further on a scratch system. Grading also ensures that a narrow particle size range is fed to certain reduction rolls, giving more efficient grinding.

Early purifiers were wood but now are gradually being replaced by metal. They operate as a pair, driven off a common drive so the same unit can treat two separate flows. The design feature is a long oscillatory sieve that slopes slightly from head to tail and is divided into four sections with the sieve covers becoming progressively coarser, moving from head to tail.

The sieves are aspirated and each of the four compartments is capable of individual adjustment of airflow. The sieves are supported by hangers, and its characteristic motion ensures that the particles are shaken, undergoing a different motion from that on a plan-sifter. The motion leads to stratification, smaller particles settling below the larger, and the airflow lifts off bran. The following separations are therefore made (26):

Stock	Destination
Small pure endosperm	Coarse reduction rolls
Large pure endosperm	Coarse reduction rolls
Small composite particles	Scratch or break rolls
Large composite particles	Scratch or break rolls
Heavy bran	Wheat feed
Light bran	Wheat feed

The scratch system uses finely fluted rolls (26 to 30/in.) to separate endosperm from particles of bran/endosperm and a sifter for grading. Scratched stock then passes to purifiers and redressers.

The presence of germ oil in bread flour affects its keeping quality and baking performance. The germ has a high commercial value and efforts are made to recover the germ from the process. Most of the germ finds its way to the coarse and fine semolina purifiers and hence to coarse reduction and scratch rolls. Because of its high fat content, germ is flattened easily on the reduction rolls and can be sieved out on the coarse reduction roll sifters. Germ material from the scratch system is recovered from X sifter tails by using aspiration and a germ roll for flattening the germ for separation by sifting.

The Reduction Systems. The purposes of the reduction system are threefold: 1) to reduce the purified stocks to flour particle size, generally less than 150 μm, 2) to remove bran without creating bran powder, and 3) to create starch damage for baking purposes.

The system comprises smooth rolls and sifters. The rolls grind down endosperm particles while the tougher bran and germ are flattened, making separation by sifters easier. As stock cannot be ground completely on one set of rolls, the sifters grade stock for further rolling. In this way, stock works its way through the reduction system, producing flour, stock, and bran overtails. The quality of flour varies at each reduction stage and the better quality flour in terms of least bran content is produced by the head reduction rolls A, B, and C. No attempt is made to grind the stock in one operation, as excessive flow of stock through the rolls causes a cushioning effect; besides, extreme roll pressures would be required, causing excessive bran breakup, heat, and power consumption. The quality of flour produced by various rolls has changed with modifications of the process. The earlier milling systems used a gradual reduction process and ground very lightly on a set of rolls from A to P. This system used A and B rolls as sizing rolls that produced stock for grinding to flour on roll C. In this way, no excessive heat was produced; starch damage was ensured by the hard grists then in use, which included North American wheats; the life of equipment was much longer, too. Owing to wartime destruction in Europe, after 1945 the cost of installing the same equipment was too great and contraction of the system became inevitable. The main problem with high-capacity, short-surface mills was a tendency to produce flakes on the head reduction rolls A, B, C, and D. The centrifugal sieving machines in operation at the time broke down flakes to a certain extent but they did not have the capacity for efficient sifting provided by the development of the plan-sifter. The introduction of the entoleter, or flake disrupter, after the head reduction rolls, improved the performance of flour mills greatly. In addition, pneumatic conveying of milling stocks removed another restriction on high-capacity milling by cooling stock, thereby improving sieving efficiency. High-capacity milling has meant the introduction of high roll pressures and roll speeds up to about 600 rpm; therefore, stronger roller mills had to be developed.

Flour Production. A commercial flour mill produces many flour streams of different quality in terms of bran and protein content. The miller's skill is knowing which streams to combine to produce the most economic final flour. Often a mill may produce two types of flour from the same grist, one a very white patent flour, the other of lower quality but acceptable for baking. If all the streams are combined, a straight-run flour is produced. Nevertheless, quality considerations mean that even in the latter case, some flours may be sent to the wheatfeed stream to maintain final flour quality.

Flour quality differences are caused by the individual streams originating in different parts of the grain. As an example, first-break flour is mainly material from the center of the grain, and the reduction flours contain increasing quantities of bran and subaleurone material, leading to an increase in flour color, ash, and protein content. Table XI shows ash content of various flour fractions from a long-surface mill.

Flour Treatment. Treatment to improve flour quality by addition of chemical substances in the form of powder or gas is not a recent practice and may be done for reasons other than making the flour more acceptable; for instance, it may be legally required.

Individual EC countries have their own regulations governing the additives used in flour for domestic use. To improve trade between countries, the EC

Commission is trying to harmonize the legislation covering flour additives. Table XII shows the allowable additives in the EC countries.

B. Milling Developments

SHORT-SURFACE MILLING

As has been noted, reduction in the number of roller mills and other equipment has led to development of short-surface milling systems (Figure 7). In earlier mills the surface was about 100 in./sk, producing flour at 72% extraction (J. Savory, personal communication). In the United Kingdom during World War II and until 1953, the flour extraction rate was controlled by the government and at one point reached 90%. Some of the lessons learned during this time allowed the introduction of medium systems with reductions going to J and K, eventually short systems with reductions going to G, and in some cases only three breaks. Table XIII shows data on mills in the EC.

Some reduction in surface was achieved by various systems, such as the Bühler system in West Germany (24), which uses roller mills, plan-sifters, bran dusters, and detachers on a fine reduction and dispenses with purifiers. Milling without purifiers as used on soft wheat, has some advantages, but it should be used only in single-grist mills and not for bread flour production. Its use requires (38): 1) good quality wheat to IBk at 16 to 17% moisture, 2) minimal exhaust drying, 3) critical roll fluting, 4) break releases as low as possible, and 5) a good scratch system.

TABLE XI
Color Grade and Ash and Protein Contents of Individual Flour Fractions from a Long-Surface Mill (28)

Flour	% of Feed to First Break	Color Grade	Ash Content (%)	Cumulative Color Grade	Protein Content (%)
A	12.2	−0.1	0.36	−0.1	10.0
B	14.0	−0.1	0.38	−0.1	10.5
C	9.4	0.3	0.38	0.0	10.7
D	7.4	0.4	0.39	0.05	11.2
B2	2.5	0.5	0.40	0.10	11.3
E	2.1	1.8	0.45	0.15	11.1
F	1.2	3.7	0.58	0.25	11.2
H	1.2	4.0	0.60	0.35	11.6
II Break	1.9	4.6	0.69	0.50	12.8
G	3.1	5.0	0.75	0.75	11.3
BMR2	1.6	5.0	0.72	0.90	12.3
K	1.2	5.2	0.77	1.00	11.9
I Break	1.6	5.4	0.79	1.10	12.0
X1	0.5	5.4	0.80	1.10	
BMR1	5.8	5.5	0.82	1.50	12.5
J	0.7	5.9	0.88	1.55	11.5
X2	0.2	6.0	0.90	1.55	
L	1.2	6.1	0.92	1.65	12.8
III Break	0.8	6.5	1.00	1.70	13.3
IV Break	2.4	10.6	1.02	2.00	13.5
IV and Y Red	0.3	13.0	1.20	2.10	13.9
Coarse Offal			5.6		12.4
Fine Offal			4.4		14.8

TABLE XII
Summary of EC Flour Additives Regulations (Number Values in ppm)[a]

	L-Ascorbic	Azodi-carbonamide	Baking Powder	Potassium Bromate	Calcium Carbonate	Chlorine	Chlorine Dioxide	Citric Acid	L-Cysteine Hydro-chloride	Iron	Lactic Acid	Lecithin	Benzoyl Peroxide	SO_2	Tartaric Acid
Belgium	50	P	GMP	P	GMP	P	P	GMP	50	GMP	GMP	2,000	P	P	GMP
Denmark	200	NR	NR	NR	NR	NR	NR	(1)	NR	NR	(1)	NR	NR	NR	NR
France	5	P	GMP	P	GMP	P	P	3,500	P	P	3,000	P	P	P	GMP
West Germany	AWL	P	AWL	P	AWL	P	P	AWL	P	P	AWL	AWL	P	10	AWL
Eire	NR	NR	NR	NR	NR	NR	NR	NR	NR	NR	NR	NR	NR	200	NR
Italy	200	P	P	P	P	P	P	P	P	P	2,000	P	P	P	P
Luxembourg	50	P	P	P	P	P	P	P	P	P	P	P	P	P	P
Netherlands	50	P	GMP	50	GMP	P	P	GMP	75	GMP	GMP	2,000	35	P	GMP
U.K.	GMP	45	NR	GMP	2,350–3,900	GMP	GMP	NR	75	NR	NR	NR	50	200	NR

[a] AWL = allowed without limit.
GMP = good manufacturing practice.
NR = not regulated.
(1) = rye flour only.
P = prohibited.
Source: Food Additives Tables, Food Law Research Centre, University of Brussels, Elsevier Publishing Co.

With this system the roll surface required is 30 to 40 in./sk/hr and that of the sifter, 44 to 60 ft^2/sk/hr; it is similar to the Bellera or Molinostar system. Another system introduced in France was Pratique's system using the "Compacto-Depositif" device, a curved plate that fed stock directly into the nip and allowed a reduction in roll surface down to 20 in./sk/hr (32). Knauff devised an even shorter system for medium and small mills, using high break differentials and special flutings (35). Roncaglia, Italy, introduced a special system for small mills, which dispenses with plan-sifters and purifiers and uses an inbuilt sifter in the exhaust. Most milling is still done on traditional systems, however, and recent developments have suggested the reintroduction of a fifth break roll to improve extraction. Alternatively, bran finishers (or similar vibrosifters) are being used on IIIBk and IVBk overtails to produce extra flour.

Impact mills have been introduced for use on first and second break middlings redressed stock to improve release. Impact milling with a Simpactor, for instance, has been attempted on mill stocks such as C in order to replace roller mills with their higher power requirement.

TABLE XIII
Change in Roll and Sifter Characteristics in EC Mills, 1900-1963

	Roll			Sifter	
Country	**Speed (Bks)**	**Reductions**	**Length (in./sk/hr)**	**Surface (ft^2/sk/hr)**	**Reference**
U.K. prewar	350	210	110-120	135	**(31,49)**
U.K. immediate postwar	350	210	80-85	...	**(31,49)**
U.K. 1963	450	450	60-65	54	**(31,49)**
Germany 1900-1950	300	230	85-120	160-270	**(20,35)**
Western Europe	...	...	35	54-135	**(20)**

Figure 8. View of a mill of Grands Moulins de Pantin, Paris, France. (Courtesy Henry Simon, Ltd.)

AIR-CLASSIFICATION

Air-classification as a protein shifting technique is attractive to European millers. First, low-protein bread wheats can be upgraded by reallocating the fine high-protein fraction that is less than 10 μm in particle size. Second, for certain applications, such as brewing, a low-protein flour is required. Flow characteristics of household flours improve by having fine fractions removed. In the United Kingdom, a pin mill of the Alpine or KEK type and a particle size classifier are part of the system (13). In France, a larger horizontal turbine machine called the ARIA Spinflow classifier has been introduced; it operates with a capacity of more than 3 te per hour.

C. Milling in France

In 1977, French millers milled nearly 6 million te of soft wheat, of which 99% was homegrown and 1% imported. One fourth of the flour produced was exported. Figure 8 gives a view of a mill of Grands Moulins de Pantin, with its capacity of 1,200 te of wheat per 24 hours.

Table XIV shows statistics for the French milling industry.

As in other EC countries, consumption of bread and flour is falling in France. The reduced number of operating mills is a reflection of this, although the

TABLE XIV
French Milling Statistics, 1970-1977

	End 1970[a]	End 1972-1973	1976-1977
Wheat usage (te)	5,317,593	5,673,279	5,908,073
Homegrown (%)	95.6	98.5	98.9
Imports (%)	4.4	1.5	1.1
Flour production (te)	3,537,997	4,278,665	4,402,433
Imported flour (te)	1,472	6,987	2,458
Total, flour (te)	3,539,469	4,285,652	4,404,891
Flour exports (te)	520,526	926,149	1,018,987
Flour usage			
Bread (%)	86	82	79
Domestic (%) }	14	5	21
Pastry (%) }		13	
Rye usage (te)	26,470	26,120	33,778

[a] Data for 1970, 1972-1973, and 1976-1977 from references 1, 2, and 22, respectively.

TABLE XV
Number of Working Mills in France, 1970-1977, Classified by Capacity

Capacity (sk/hr)	1970	1972-1973	1973-1974	1974-1975	End 1977
0.5	2,138	1,629	481	449	372
0.5–2 }			887	814	704
2–4 }			337	314	310
4–10 }			229	217	205
10–20 }	511	481	44	48	55
20–50 }			15	15	19
50+ }			7	8	7
	2,649	2,110	2,000	1,865	1,672

conversion to larger, more economical milling units is also a factor. In 1977, 30% of flour was produced by seven mills. The numbers of working mills are shown in Table XV.

The mills in the 4 to 10 sk/hr and 50+ brackets produced 45% of all flour. The very small mills produce for their local market and the proportion of export flour increases with mill size, so that in 1974 one half of the output of the seven largest mills was destined for export (5).

EXPORT FLOUR

Recently, the amount of French flour exported has risen sharply and now approximates 1 million te per year. The main customers are in the Mediterranean area, i.e., Egypt (398,000 te in 1975), Libya (81,000 te), and Algeria (44,000 te), but a substantial trade exists with Sri Lanka (172,000 te) (29). EC regulations regulate the ash content of export flour. The export contract usually limits moisture content from 13.5 to 14%.

FRENCH FLOUR

French laws on flour are restricted to the purity of flour in terms of bran and germ content. Regulations in force since 1963 define six types with corresponding ash contents expressed in percent on a dry basis: Type 45, 0.50; 55, 0.50 to 0.60; 65, 0.62 to 0.75; 80, 0.75 to 0.90; 120, 1.00 to 1.20; 150, above 1.40.

More than 90% of flour produced is Type 55, used for bread production. Type 45 is used for cake and Type 65 for biscuits. Types 80, 120, and 150 are wheat meals.

The standard for the finest wheat flour in the 45 and 55 groups is governed by alveograph and protein level, the minimum figures being as follows: Type 45—220 W, 11.5% protein d.b; Type 55—220 W, 12.0% protein, d.b. These types are used for rolls and long bread.

FRENCH QUALITY TESTING

Flour strength for baking is usually expressed by the W figure of the Chopin Alveograph plus the swelling figure G. The mills use this test to grade wheats and as a check on grists and flour quality.

Basically the test involves inflating a dough piece and recording the response on a chart. From the curve produced, W is obtained from the area under the curve. Further information is supplied by the ratio between pressure and swelling (P/L). It is assumed that pressure is proportional to the toughness and firmness of gluten and by implication indicates the water-absorbing capacity of the flour. The swelling and length of curve are proportional to gluten extensibility. Other tests used to supplement these quality data are Hagberg Falling Number, protein content, and sedimentation value.

BREAD FLOUR

French bakers use mostly Type 55 for bread production. Bread is produced from a simple dough of flour, water, salt, and yeast. Bread making in France is still a craft industry with about 40,000 bakeries, ranging in flour usage from 60 to 700 te per year. Despite the deep craft traditions, French bread-making techniques have been largely replaced by the so-called intensive mixing method, which still uses bulk fermentation (12) but in addition employs improvers such as ascorbic acid, emulsifiers, malt, and bean flour. The characteristic French bread, such as *baguette* or *Pain Parisien,* uses a highly hydrated dough that is baked unsupported on trays. The resultant loaf has a thick crust and a coarse irregular texture. Good quality gluten is essential, with the correct balance of extensibility, elasticity, and firmness. The introduction of the intensive mixing technique has required flours of increased strength. On average, a W specification of 150, and protein range of 10 to 11% (d.b.) is suitable, but there are slight regional variations.

Other types of flour have the following specifications:

Type	W	Protein (%, d.b.)
Rusks, batch bread	160–180	11–12
Confectionery, biscuits	200–300	13–14
Biscuits	60–120	6–9
Household	140–180	10–12

D. Milling in West Germany

In 1973, of the 7.1 million te of wheat produced, 4.1 million te was milled by flour millers and also 1.1 million te of rye. The total usage of 5.2 million te shows a drop of 1 million te from 1959 owing to the decrease of bread consumption (7). Since 1959, 20% of the mills have closed, assisted by governmental subsidies reducing flour production by 12%. Between 1963 and 1973, the use of North American wheats was reduced from 25 to 13%. In 1970, there were 5,155 mills, of which 6.0% produced more than 2,000 te per year. This number had decreased to 3,861 in 1972-1973 and to 2,572 in 1976-1977. Large mills, as defined here, accounted for 6.9% of the total in the former period and 10.6% in the latter. About 80% of wheat is milled in a few large plants.

Table XVI shows that the greatest number of mills, large and small, are located in Bavaria.

Wheat usage and flour production for West Germany are tabulated in Table XVII.

TABLE XVI
Structure of the Milling Industry in West Germany (1975)[a]

Region	Less than 500 te	500–3,000 te	3,000–10,000 te	10,000+ te	Total
Schleswig-Holstein	60	7	5	9	81
Hamburg	5	1	···	···	7
Lower Saxony	201	44	15	13	273
Bremen	2	1	1	1	5
North Rhine	48	10	4	8	70
Westphalia	153	37	12	9	211
Hesse	322	31	6	3	362
Rhineland-Pfalz	136	27	15	4	182
Saarland	23	7	4	3	37
Baden-Württemberg	491	86	29	10	616
Bavaria	952	131	30	19	1,132
Berlin	···	3	···	4	7

[a] Data from APRIA (5).

EXPORT FLOUR

Like France, West Germany exports a substantial amount of flour. Table XVIII shows the quantitites exported between 1970 and 1975.

WEST GERMAN FLOUR

Flour types in West Germany are classified on the basis of ash content as shown in Table XIX (37).

E. Milling in the United Kingdom

Historically, U.K. millers imported wheats from all over the world in order to produce flour with the characteristics required for bread for the U.K. market. Table XX shows the import figures for 1935-1936, representing about 80% of the U.K. wheat consumption. This situation lasted until the early 1970s when market

TABLE XVII
Wheat Usage and Flour Production in West Germany

	1970[a]	1972-1973	1976-1977
Wheat use (te)	3,799,237	3,810,862	4,009,766
Homegrown (%)	60	68	75
Import (%)	18	32 (Import and Third country combined)	20
Third country (%)	22		5
Flour production (te)	2,915,387	2,930,000	3,103,559
Imported	37,341	35,081	43,633
Total	2,952,728	2,965,081	3,147,192
Exports	419,854	385,615	445,506
Flour use (%)			
Bread	75	75	72
Biscuit	15	15	18
Domestic	10	10	10
Rye use (te)	1,163,935	1,099,671	1,020,476
Flour production	886,984	890,000	869,446
Imports	2,465	1,912	2,023
Exports	80	213	433

[a]Data for 1970, 1972-1973, and 1976-1977 from references 1, 2, and 22, respectively.

TABLE XVIII
West German Wheat Flour Exports and Destination, 1970-1975[a]

	By Destination (1,000 te)		
	1970-1971	1971-1972	1974-1975
Egypt	14	20	73
Saudi Arabia	16	21	49
Yemen	34	61	60
Libya	71	51	45
EC	33	16	...
Other	38	54	49
Total	206	223	276

[a]Data from APRIA (5).

forces were such as to halve the quantity of wheat from third countries, making the U.K. miller rely more heavily on the EC and, more importantly, on homegrown wheat.

The adoption of the Chorleywood bread process has permitted U.K. millers to use more homegrown EC wheats (Table XXI).

The number of mills in the U.K. has declined from about 2,500 before 1939 to 106 in 1975-1976. Closures were necessary owing to overproduction and were assisted by the formation of a voluntary body, The Millers Mutual Association (MMA), which collected a levy, based on flour deliveries from its members' firms. The levy provided money with which to buy mills due for closure. In addition, the large millers, Ranks and Spillers, acquired mills and also bakeries in order to protect their flour markets. In 1961, Associated British Foods (ABF), in order to safeguard its flour supplies for its own bakeries, began to acquire flour mills. RHM, Spillers, and ABF now supply more than 80% of the bakers' flour market.

FLOUR PRODUCTION

More than half of the flour produced in the United Kingdom is made into bread by the Chorleywood bread process. Very briefly, this technique requires a certain quantity of work input during a dough mixing time of less than 5 min, oxidizing agents, fat, and extra yeast and water.

Before this process, U.K. bread was produced from a flour with 12% protein content, necessitating imports of high-protein wheat. With the Chorleywood bread process, the baker can produce bread of similar quality using a flour of from 1 to 1.5% lower protein content.

TABLE XIX
Principal Types of Flour Produced in West Germany[a]

Type	Ash (%) Min.	Ash (%) Max.	Production (%)
Rye			
610	0.58	0.65	1
815	0.79	0.87	9
997	0.95	1.07	30
1150	1.10	1.25	33
1370	1.30	1.45	6
1740	1.64	1.84	1
1800	1.65	2.00	20
Wheat			
405	0.38	0.44	18
530	0.49	0.58	54
630	0.60	0.70	3
812	0.75	0.87	10
1050	1.00	1.15	13
1200	1.16	1.35	...
1600	1.55	1.75	1
2000	1.85	2.20	...
Wheat meal			
1700	1.60	1.90	1

[a] Data from Siebel et at (37).

A typical flour analysis would be: moisture, 14 to 14.5%; protein, 11 to 11.5%; starch damage, 25 to 30 Farrand Units; α-amylase, 8 to 10 Farrand Units; flour color grade, 2.5 to 3.0 units; water absorption, 17½ to 18 gallons per sack; and oxidant level, 10 to 15 ppm potassium bromate.

Grist composition sets the protein level, and moisture levels are determined by conditioning practice. Starch damage and therefore, water absorption, can be controlled by mill adjustment. α-Amylase is adjusted by addition of malt flour if too low, and high-alpha wheats are screened out by the Hagberg test at the quality control stage. Flour color grade is controlled during milling.

U.K. millers also produce flours for traditional bulk fermentation and other baking processes. These flours can vary widely from high-protein, patent types to low-protein, 100% English wheat flours.

There is also a substantial market for cake and biscuit flours that are made from 100% homegrown wheat. In 1975-1976 the types of flour produced were 62% white, 3% brown, and 1% whole-meal bread flours; 11% biscuit, 2% cake, 4% household, 5% self-raising, and 12% other flours.

F. Milling in Other EC Countries

Italy is the largest miller of soft wheat in the EC. Currently, approximately 1,200 mills process more than 6 million te of wheat per year. Seventy

TABLE XX
U.K. Wheat Imports, 1935-1936

Country	Tons
Canada	2,368,000
Australia	1,023,000
Argentina	592,800
France	310,200
U.S.S.R.	176,800
U.S.A.	159,000
Romania	156,200

Source: Lockwood, J. 1945. Flour Milling, 1st ed. Northern Publishing Co., Ltd.

TABLE XXI
U.K. Wheat Usage and Flour Production, 1970-1977[a]

	1970	1972-1973	1976-1977
Wheat use (te)	5,182,110	4,880,000	5,275,000
Homegrown (%)	30	40	45
Imported (%)	70	60	55
Flour production (te)	3,856,100	3,632,000	3,956,000
Imports	79,256	64,412	6,212
Total	3,935,356	3,696,412	3,962,212
Exports	11,661	5,329	30,805
Flour use (%)			
Bread	66	63	66
Pastry	20	21	13
Domestic	14	17	10

[a] Data for 1970, 1972-1973, and 1976-1977 from references 1, 2, and 22, respectively.

percent of the grain is homegrown, and 30% comes from France, Argentina, North America, and West Germany. Of the 4.5 million te of flour produced yearly, 0.25 million te is exported. Of the remainder, 85% is used in bread production.

In Belgium 96 mills currently produce 0.75 million te of flour from about 85% European wheat. More than 20% is exported.

During 1977, Dutch millers ground 1.3 million te of wheat; 75% was EC or homegrown and 25% came from third countries. The Dutch prefer bread similar to the U.K. type and, owing to the high proportion of feed wheat grown in the Netherlands, very little homegrown wheat is used and a large proportion comes from France.

Danish millers produce about 0.25 million te of flour yearly from nearly 100% homegrown wheat. Seven companies own the 14 mills in operation, and 35% of the flour produced is sold for domestic uses.

At present, 17 mills operate in the Irish Republic. In 1900, 217 were in operation (4), and by 1927, the number had fallen to 32. The use of flour has declined gradually so that excess milling capacity is still a problem.

Table XXII presents milling and flour data for the other EC countries for 1972-1973.

VI. OUTLOOK

The continued existence of the Common Agricultural Policy in its present form will force millers to use more EC wheat for bread-making purposes. Encouragement to grow quality wheats for bread making is very necessary if the EC is going to be self-supporting in its milling wheat requirements. Whether this encouragement takes the form of increased premiums paid by millers or lower

TABLE XXII
Milling Statistics for Other EC Countries, 1972-1973[a]

Number of Mills	**Netherlands**	**Belgium**	**Denmark**	**Eire**	**Italy**	**Luxembourg**
>2,000 te/a	**14** }		**16**	**17**	**1,298**	**6**
<2,000 te/a	**60** }	**101**	**3**	**0**	**143**	**3**
Wheat milled (te)	1,300,000	977,138	270,000	327,888	6,600,000	1,300,000
Rye milled (te)	65,000	7,782	110,000	...	...	495
Wheat (%)						
Homegrown	55	38	100	67	89	59
Imported	45	62	...	33	11	41
Flour production (te)						
Wheat	952,000	719,873	200,000	237,272	5,150,000	952,000
Rye	35,000	5,945	100,000	...	...	222
Wheat flour (te)						
Imports	8,000	12,966	...	...	...	25
Exports	151,000	91,846	...	...	539,000	1,221
Flour use (%)						
Bread	80	97	65	63	82	95
Pastry	19	3	9	12	18	5
Domestic	1	...	26	25	...	...

[a] Data from reference 2.

support price arrangements for feed-quality wheats remains an open question.

As far as milling processes are concerned, there will continue to be a demand for standard white flour and the relatively small demand for high-fiber bread will not make serious inroads into current milling operations. In the pursuit of efficiency, millers will seek to reduce the number of machines in operation and hence shorter-surface milling and impact milling will be increasingly examined.

More computer control of the milling process is likely to follow the current rapid development of microprocessors, although complete process control by computer for most milling applications is still far in the future.

With the static consumption of bread and excess agricultural output, other uses for wheat are going to be investigated, and its potential as an industrial raw material will come under closer scrutiny.

ACKNOWLEDGMENTS

We would like to thank the following people for their part in the preparation of this chapter: Dr. J. R. S. Ellis, RHM Research Ltd.; J. E. Savory, P. A. Metaxa, T. Barlow, and W. Powry, RHM Flour Mills Ltd.; G. Campbell, Ranks (Ireland) Ltd,; C. Hawcroft and J. P. Dresti, J. Soufflet Ltd.; E. Amme, ABK-Hamburg; Dr. K. Hertzberg, LUFA, Kiel; S. Cauthery and D. Forder, Henry Simon Ltd., Stockport; J. Gasser, Bühler-Miag (England) Ltd.; Dr. N. L. Kent, FMBRA, Chorleywood. We also thank Mrs. S. C. Edwards and Mrs. A. Head for their work in typing this manuscript.

LITERATURE CITED

1. Anonymous. 1972. Données statistiques sur l'industrie meunière Européenne. Bull. Ancien Elèves de l'Ecole de Meunière, No. 250, July-Aug. pp. 202-206.
2. Anonymous. 1974. Situaçao da moagem em 1973 nos países da Europa occidental, Boletim da FNIM, 17(62):21-28.
3. Anonymous. 1977. Simon fluting profiles. Tech. Newsletter 72-2. Henry Simon, Ltd.
4. Anonymous. 1978. Are the millers on the bakers' backs? Bakery World, Feb. 9, 11, 13.
5. APRIA. Blé et pain dans la CEE (other information not available.)
6. Barlow, K. K., Buttrose, M. S., Simmonds, D. H., and Vesk, M. 1973. The nature of the starch-protein interface in wheat endosperm. Cereal Chem. 50:443-454.
7. Becher, M. 1974. West Germany's flour milling and animal feed industries. J. Flour Anim. Feed Milling 156(6):28.
8. Bingham, J. 1978. Breeding wheat for quality improvement. FMBRA Bull. 4, Aug.
9. Bölling, H. 1975. On the problem of wheats "unsuitable for breadmaking"—A rapid method for recognising them. Mühle u. Mischfuttertechnik 112(15):256-257.
10. Bölling, H., and Zwingelberg, H. 1974. Der französische Weizenseine Verarbeitungseigenschaften in der deutschen Müllerei. Mühle u. Mischfuttertechnik 111:33.
11. Brocklehurst, P. A., and Evers, A. D. 1977. The size distribution of starch granules in endosperm of different sized kernels of the wheat cultivar Maris Huntsman. J. Sci. Food Agric. 28:1084-1089.
12. Collins, T. H. 1978. The French baguette and pain Parisien, FMBRA Bull. 3, June. pp. 107-116.
13. Coppock, J. B. M. 1962. Advances in milling techniques. Biscuit Maker & Plant Baker 13(3):210-220, 255.
14. Ellis, J. R. S. 1976. Variety identification by 'protein prints.' Arable Farming, Aug. pp. 32-33.
15. Ellis, J. R. S., and Beminster, C. H. 1977. The identification of U.K. wheat varieties by starch gel electrophoresis of gliadin proteins. J. Nat. Inst. Agric. Bot. 14:221-231.
16. Evers, A. D. 1978. Type of starch damage. FMBRA Bull. 1. pp. 16-23.
17. Evers, A. D., and Lindley, J. 1977. The particle-size distribution in wheat endosperm starch. J. Sci. Food Agric. 28:98.
18. Farrand, E. A. 1972. Potential milling and baking value of home grown wheat. J.

Nat. Inst. Agric. Bot. 12:464-470.

19. Fisher, N., and Jones, C. R. 1932. The wheats of commerce, Part I, Pamphlet 9, Tech. Ed. Series, NJIC. p. 37.
20. Flechsig, J. 1963. Moderne Mehlverfahren. Mühle 100(35):113-116 of inset supplement.
21. Greer, E. N., and Hinton, J. J. C. 1950. The two types of wheat endosperm. Nature 165:746-747.
22. International Milling Association. 1977. (Other information not available.)
23. Jones, C. R. 1940. The production of mechanically damaged starch in milling as a governing factor in the diastatic activity of flour. Cereal Chem. 17:133-169.
24. Jones, C. R. 1964. Recent European developments in flour milling technology. Milling 142(13):324-327.
25. Jones, C. R., Greer, E. N., Thomlinson, J., and Baker, G. J. 1961. Technology of the production of increased starch damage in flour milling. Part I. Factors in individual reductions. Milling 137: July 21. pp. 58-62.
26. Lockwood, J. 1960. Flour Milling, 4th ed. Northern Publ. Co., Ltd., Liverpool.
27. Marston, H. 1937. The development of the modern British roller mill. NJIC Tech. Ed. Series, No. 7.
28. National Joint Industrial Council. 1966. The Practice of Flour Milling. (Two vols.) Northern Publ. Co., Ltd., Liverpool.
29. ONIC. 1977. Activités de la meunière, 1975. Techniques des industries céréalières, No. 158. Dec. 76-Jan. 77. pp. 15-17.
30. Parish, J. A., and Halse, N. J. 1968. Effects of light, temperature and rate of dessication on translucency in wheat grain. Aust. J. Agric. Res. 19:365-372.
31. Povey, D. W. 1959. Latest machinery designs. Their influence on flour mill capacities and production costs. Milling 133(16):332-333,336,343.
32. Pratique, J. 1960. Technischer Fortschritt und Rentabilität in der Mühlerei. Mühle 97(44):583-598.
33. Rhuss, A. M. 1977. Riffeldaten—Theorie—Praxis—Probleme. Mühle u. Mischfuttertechnik, 7 July.
34. Russell-Eggitt, P. W., and Hartley, A. W. 1975. Extraction rates—Milling process implications. In: Bread. A. Spicer; ed. Appl. Sci. Publishers, Ltd., London.
35. Schneeweiss, R. 1963. Rationalisierungsmöglichkeiten in der Mühlenindustrie. Mühle 100(45):492-495.
36. Scott, J. H. 1951. Flour Milling Processes, 2nd ed. Chapman & Hall.
37. Siebel, W., Bruemmer, J. M., and Stephan, H. 1978. West German bread. In: Advances in Cereal Science and Technology, Vol. II Y. Pomeranz, ed. Am. Assoc. Cereal Chem. St. Paul, MN.
38. Stamford, C. 1969. Unusual approaches to milling in Europe. Milling 151(10):30-34.
39. Stanger, E. A. 1970. Process and mechanical development for the modern mill. Milling 152(7):30-37.
40. Stenvert, N. L. 1974. Grinding resistance. A simple measure of wheat hardness. J. Flour Anim. Feed Milling, July. pp. 24-27.
41. Stenvert, N. L., and Kingswood, K. W. 1977. The influence of the physical structure of the protein matrix on wheat hardness. J. Sci. Food Agric. 28:11-19.
42. Symes, K. J. 1969. Influence of a gene causing hardness on the milling and baking qualities of wheats. Aust. J. Agric. Res. 20:197.
43. Toepfer, A. 1977. The EEC grain market regulations 1977/78. Gebr. Hoesch, Hamburg.
44. Vince, J. 1978. Mills and millwrighting. Shire Publications, Ltd.
45. Wanzenreid, H. 1970. Grinding with smooth rolls. Milling 152(11):18,20-22.
46. Williams, P. C. 1969. Nature of mechanically damaged starch, its production in flour. Part I. Northwestern Miller 276:8-12.
47. Willm, C. 1977. Contribution a l'étude de l'endommagement de l'amidon en mouture de blé tendre. Bull. Ancien. Elèves de l'Ecole Francaise de Meunière 277:13-25.
48. Wyss, E. 1975. Bühler Diagram No. 59, (K.P. Wild, ed.), Bühler Bros., Ltd., Uzwil, Switzerland.
49. Youdale, L. M. 1963. Effects of developments in milling technology on mill designs and layout. III. Modern mill machinery. Milling 140(14):348-349.
50. Ziegler, E. 1940. The effect on diastatic activity of surface, pressure, differential, and temperature of the reduction rolls in milling. Cereal Chem. 17:668-679.

CHAPTER 5

SOFT WHEAT USES IN THE UNITED STATES

H. J. LOVING, *Retired*
Mennel Milling Co.
Fostoria, Ohio, U.S.A.

L. J. BRENNEIS, *Deceased*

I. INTRODUCTION AND HISTORY

In early U.S. history locally grown wheat was customarily taken to a nearby mill and exchanged for locally milled flour that was used for all household needs—bread, cakes, cookies, pancakes, gravy, and all other home-prepared foods containing flour. Because climate has considerable influence on the type, strength, and variety of wheat grown, the people of some geographic regions used a flour type poorly suited to some of the products made, for example, cakes from Kansas and North Dakota wheats and bread from the softer wheats grown east of the Mississippi River. Gradually the development of the commercial baking industry, modern transportation, improved wheat varieties, better technologies, and the many influences leading to specialization brought about the characterization and use of the various wheat types according to the best applications for each. So today the baker of bread and other yeast-leavened baked foods draws upon the higher protein, stronger wheats whatever his location, and the cake, cracker, and cookie baker purchases flour milled from the lower protein, soft wheats suitable for the intended use. To this day, however, remnants of the earlier custom endure. Household or family flour tends to be similar to that from wheat grown nearby or in the natural supply area serving a given locale. Thus, the Plains states and the north-central and northeastern states use hard wheat family flour, whereas the southeast prefers flour milled from soft or a blend of soft and low protein hard wheats for its traditional hot biscuits. In spite of this classification of flour types for specific uses, there is some interchange of hard and soft wheat flours in commercial and industrial use. This, however, is likely to occur only where the flour or wheat type used is not critical and under stress brought about by shortages or price.

The emerging complexity of our diet from one based on bread to one in which increasing quantities of flour are used in pies, cakes, cookies, crackers, pretzels, donuts, cones, snack foods, and the like has spurred the development of flours for specific purposes. Rising incomes and an improved standard of living have been paralleled by development of a baking industry capable of accommodating the larger purses and new appetites. Significantly, the direction of change has been toward greater consumption of foods based on soft wheat. The single largest use of flour is still for baking bread, but soft wheat usage makes up for sheer tonnage in its broad diversity of applications. Soft wheat technology is much more varied than that for hard wheat (57); most early studies focused on hard wheat applications.

U.S. Department of Agriculture economists (49) have indicated flour consumption trends as shown in Tables I and II. The distinct decline in all-purpose or family flour is being partially made up by gains of other types, particularly soft wheat flour products, and in variety breads. However, the gains are not sufficient to halt a steady downward trend in overall per capita flour consumption. Translating this into total consumption, thus adjusting for population growth, produces the data in Table II.

By the mid-1970s, U.S. annual wheat production had surpassed 2 billion bushels. Less than one fourth of our wheat is soft red winter and soft white; these varieties provide our soft wheat requirements. There are two major soft wheat-growing areas; the largest is centered around the east-central states of Ohio, Indiana, Michigan, Illinois, and Missouri, with less production in adjacent states, east to the Atlantic Seaboard, and south to the Gulf. The other is the Pacific Northwest, comprising Washington, Oregon, and Idaho, with some production in Arizona, California, and Utah. Most soft wheat mills are either within these producing states or along the path from the wheat growing areas to

TABLE I
Per Capita Flour Consumption (pounds, by types)

	1959-1960	1965-1966	1973	Trend
White pan bread (hard)	50.69	44.37	44.91	Down, to steady
All-purpose (mostly family flour)	38.75	29.57	21.50	Significantly down
Soft	20.37	25.85	28.70	Significantly up
Whole wheat and variety breads (hard)	0.48	4.90	6.30	Significantly up
Semolina	7.32	6.53	6.32	Steady, to slightly down
All flour	117.61	11.24	107.74	Down

TABLE II
Total Flour Consumption (1,000 cwt, by types)

	1959-1960	1965-1966	1973	Trend
White pan bread (hard)	88,860	84,569	92,919	Steady, to up
All-purpose (mostly family flour)	67,929	56,360	44,484	Significantly down
Soft	35,709	49,270	59,380	Significantly up
Whole wheat and variety breads (hard)	841	9,339	13,076	Significantly up
Semolina	12,832	12,446	13,076	Steady
All flour	206,170	212,023	222,914	Up

major population centers or seaports. Of all the western soft white wheat produced, 85 to 90% is exported, chiefly to the Far East, diminishing the importance of this wheat for domestic use.

II. BAKED FOODS FROM SOFT WHEAT FLOURS

A. Cakes

Probably the most characteristic and unique product from soft wheats is cake, traditionally the centerpiece of festive and joyous celebrations—birthdays, weddings, holidays, and homecomings. In the early days the cake served at the dinner table was made from scratch from flour, sugar, shortening, eggs, milk, leavening, and flavorings by a laborious and by no means foolproof procedure including tiptoeing past the stove while the cake was in the oven. But in the past few decades the bake-from-scratch tradition has been largely superseded by two major innovations. First was the trend toward buying finished, iced, and sometimes decorated cakes from the retail bakery or from the retailer of wholesale baked cakes. Second was the perfecting of boxed cake mixes to the point that excellent freshly baked cakes can be served by the proud housewife and even children, with the likelihood of failure almost totally banished (15). This revolution started in the 1930s, before which no more than one fifth of the cakes and pastries was commercially produced.

In the 1920s and 1930s, a few milling companies developed specially milled cake flours based on grinding selected plump, soft red wheat or red and white blends on a finely clothed mill, rejecting all but the best patent streams, grinding and sifting to a very fine particle size, then treating with chlorine for cake baking performance and benzoyl peroxide for optimal whiteness (1,20,21,29,31, 34,41,59). The milling of short patent cake flours opened the way for combining the unused streams into selected strong or weak fractions to be merchandised for other specific commercial applications, such as cracker sponge, donut, and cookie production (35). Early promotions were directed to housewives and then to commercial cake bakers, as more mills offered specialized cake flour. Then the shortening companies developed new processing techniques and the monoglyceride and diglyceride emulsifiers, followed by other types (8,16,19,38). This made possible richer, more moist, longer keeping cakes than ever before. Whereas sugar to flour ratios had been 1:1 or less, it now became possible to increase the sugar to 1.4 parts to 1 part flour, or even more. Also increased were shortening, eggs, and especially liquid, to greatly improve eating quality and lengthen shelf life. All this came at no loss (indeed gains in some cases) in volume, symmetry, crumb structure, and freedom from fallen cakes, the certain sign of a too-rich recipe. Fluid-emulsified shortening also contributed in modernizing formulations.

With the improved ingredients (41,43,47) also came simplicity and time saving in mixing. The older method was to cream sugar, other dry ingredients, and fat until light. This was followed by stepwise addition of eggs and other liquid,with continued mixing. This was followed by addition of flour, with minimum mixing at this stage to avoid gluten development, it was said. Ingredients that were too cold or too rapid additions of liquids could produce curdled batters and questionable results in the oven. Now a simpler two-stage method could be used; all the solids plus part of the liquids were mixed to form a slurry; mixing was continued with addition of the remainder of the liquids, then final mixing until smooth. Even simpler was the single-stage method, where all ingredients were dumped into a bowl and mixed until smooth. There was no need for concern over gluten development in this high-sugar, high-liquid environment (43,47).

In the 1950s further flour technology advances were made through commercial development of subsieve sizing by air-classification (29,56). These studies made clear that "the finer the flour the better" was not necessarily the cardinal rule for cake baking; indeed, a selected intermediate, rather uniform particle size gave best layer cake performance, after rejection of the very fine (5 to 10 μ, 15 to 20% protein) and the chunky or coarse fractions. Studies at about this time also revealed the relationships between particle size, starch damage, and cake-baking quality (31). Reduction in particle size of soft wheat flour by pin milling (accompanied by increased starch damage) improved cake quality to a point. Excessive starch damage (over about 5% as determined by an enzymatic procedure) brought about a decline in cake-baking quality.

Improvements in flour and shortening were paralleled by improvements in the processing of eggs (27,40,41) and the selection of leavening acids (3,17,28,41), all of which helped in perfecting modern formulations and mixing methods.

The shortening-based cakes constitute the most popular type of formulation, with structure depending mostly on fat-liquid emulsification (8) and the leavening action of baking powder. The first rule of formula balance for a high-ratio cake (43) is that sugar exceeds flour weight, usually by a ratio of 1.2 to 1, or more. Eggs exceed shortening, the latter being in the range of 40% of flour weight, depending on the richness desired. Total liquid (formula water plus the water in liquid milk and eggs, if used) equals or exceeds the weight of sugar, so the latter may go into solution. Milk solids can vary within fairly generous limits, their proportion usually being about the same as that of egg solids. Baking powder is commonly 5 to 6% of flour weight.

Sugar, shortening, egg yolks, water, and baking powder are customarily classified as tenderizers, as opposed to the tougheners or structure builders, comprising flour, egg whites, and milk solids (41,43). Formula balance avoids dominance of the tenderizers over the tougheners and vice versa.

This type of cake includes white, yellow, spice, chocolate, and devil's food formulas, which may be baked in layer, loaf, sheet, and cup cakes (43,47,54,55) (Fig. 1). The pound cake is an important variant in certain regions, particularly in the east; raisins and other fruits may be included in the batter.

Foam cakes make up the second major type of cake and depend primarily on the whipping capacity of eggs and the air incorporated therein for structure and volume (5,60). Angel food cake quality is sensitive to flour type even though this ingredient constitutes a relatively small percentage of the total batter. This type of cake is formulated from equal parts of sugar and egg whites, each being about 2.75 times flour weight, plus salt to taste, flavoring, and about 0.4% (flour basis) cream of tartar; the latter assists foam volume and stability, adjusts batter pH, and gives bright color and fine, uniform grain and texture to the crumb. Monocalcium phosphate monohydrate is sometimes used in place of cream of tartar. The liquid egg white with salt, acid leavening agent, and sometimes part of the sugar is whipped to a soft peak. The rest of the sugar and the flour are added; mixing after the flour is incorporated must be restricted to the minimum to avoid gluten development and loss of whipped volume. Critical points to watch are

Figure 1. Wholesale loaf and iced layer cake.

Figure 2. Slurry mixer, holding tank, and continuous cake mixer. (Courtesy of Oakes Machine Corp.)

strict freedom from fat in ingredients or on utensils, correct "peak," and specific gravity of batter (5). Angel food cakes are usually baked in deep, tube pans, so that baking can proceed both from the outside and the center.

Sponge cakes also belong to the foam structure type, the whipping starting with whole eggs, sometimes fortified with egg yolks. The formula consists of

Figure 3. Depositing cake batter into baking pans. (Courtesy of *Bakery Production and Marketing*)

approximately equal parts of flour, sugar, and eggs. The eggs are usually warmed to enhance whipping. The leavening supplied by the whipped eggs may be supplemented with baking powder.

Automation and development of bakery equipment have contributed in no small way to the progress of the baking industry (16,43,47). Continuous mixers are being used in the commercial production of both shortening and foam cakes (16) (Fig. 2). Rotor speed and air pressure controls of the continuous mixing operation provide a convenient means for maintaining optimal specific gravities of batter at the depositor (Fig. 3).

B. Pies

Pie dough can be made from almost any flour, but shortening can be used most efficiently with a low-protein soft white or soft red wheat flour; tenderness is thus achieved at lower shortening levels, a factor important to commercial pie bakers for whom shortening is the most expensive ingredient of the crust. On the other

Figure 4. Battery of mixers used in mixing pie doughs. (Courtesy of Excelsior Industrial Corp.)

hand, if a rich, fatty crust is the objective and cost is unimportant, stronger, higher protein flours will "carry" more fat into the product, but this situation is seldom encountered (32,44).

An important objective in mixing pie doughs is to avoid gluten development. Commercial pie dough is usually mixed in slow-speed mixers with jointed arms that gently lift the dough and set it down without stretching or kneading (Fig. 4). Another objective is to use as little water as possible, which also harmonizes with the avoidance of gluten development, which causes toughness.

Pie dough has a simple formula, comprising flour, water, shortening, and salt; sometimes a small amount of dry milk and/or sugar are used for flavor and browning of the crust. Baking powder and soda are optional ingredients that can give some expansion, structure, and tenderness and help reduce shrinkage. The critical factors contributing to pie crust quality are the quantity of shortening and the mixing procedure. Pie doughs are mixed cool and may be stored for a time before rolling out to improve working properties, probably by relaxing the dough and overcoming any gluten development that may have occurred. The amount of plastic shortening can vary widely depending on desired richness, tenderness, structure (mealy or flaky), and the economics of the situation. Pie doughs can be made with as little as 35% as much fat as flour, but the proportion may run twice as high. With a low fat percentage, mixing before adding liquid is continued for a longer time, until the fat is finely divided; the resultant mealy crust takes maximum advantage of the low fat percentage. However, dry mixing must not be continued until the fat and flour form a creamed mass. As the fat level is increased, the mixing time is shortened to the point that some raw flour is present and chunks of fat remain marble size or even larger. After the liquid is added, mixing proceeds only as long as necessary to make a cohesive dough. In the rolling or sheeting process, the fat chunks are flattened out to form layers, and the result is a flaky crust (Fig. 5).

Pie dough formulation is not critically dependent on flour type or ash level, but a coarser, rather than fine, granulation is desired. A flour with high diastatic activity is considered undesirable in the crust of frozen, fruit-filled pies. A typical commercial pie dough flour is a soft white or soft red of 0.40 to 0.45% ash, 8.0 to 9.5% protein, unbleached. Soft clears up to 0.65% ash are sometimes used. Lard is the traditional fat for pie doughs, since it has good plastic properties and high shortening value and contributes to flavor in a way esteemed by many. The hydrogenated vegetable fats are also suitable for pie dough production and are likely to be first choice for pie dough mixes, pie doughs for frozen pie manufacture, and for some ethnic foods.

Of the two types of pies generally made, the fruit-filled pies usually have both bottom and top crusts. The other type has a cream, custard, or pudding filling (Fig. 6), does not have a top crust, and may be topped with an eggwhite meringue. Typical of the latter are chocolate, lemon, custard, coconut cream, and pumpkin pies (44,50).

A variation of the usual full-size baked pie, 20 to 25 cm in diameter, is the small individual pie or tart, which lends itself to mass production and is sold in supermarkets, service stations, and corner groceries and through vending machines.

Another variation is the fried pie, which is deep fat fried (10,44). The pie dough is sheeted and formed around a precooked, starch-thickened fruit filling. After cooking, a sugar glaze usually is added for both appearance and taste. This product is attractively packaged and may be eaten as a snack or dessert while being held in the hand. Fried pies are popular in some regions, particularly the southeast.

C. Cookies and Crackers

One of the largest outlets for soft red and soft white wheat flours is the wholesale cookie and cracker industry. Usage runs about 30 million cwt annually, equivalent to approximately 70 million bushels of wheat. The soft wheat flours are superior for these products because the lower protein levels provide tenderness at a lower level of the expensive enriching ingredients; processing and machining qualities are excellent; high cookie spread is available if desired; and eating qualities are attractive to consumers (51,53). Large wholesalers utilize low product moisture levels and protective packaging to preserve shelf life for a relatively long time. They enhance the demand through strong advertising and promotional activities.

Soda crackers are leavened with yeast and are thus an exception to the rule that soft wheat flours are used for chemically leavened and nonleavened

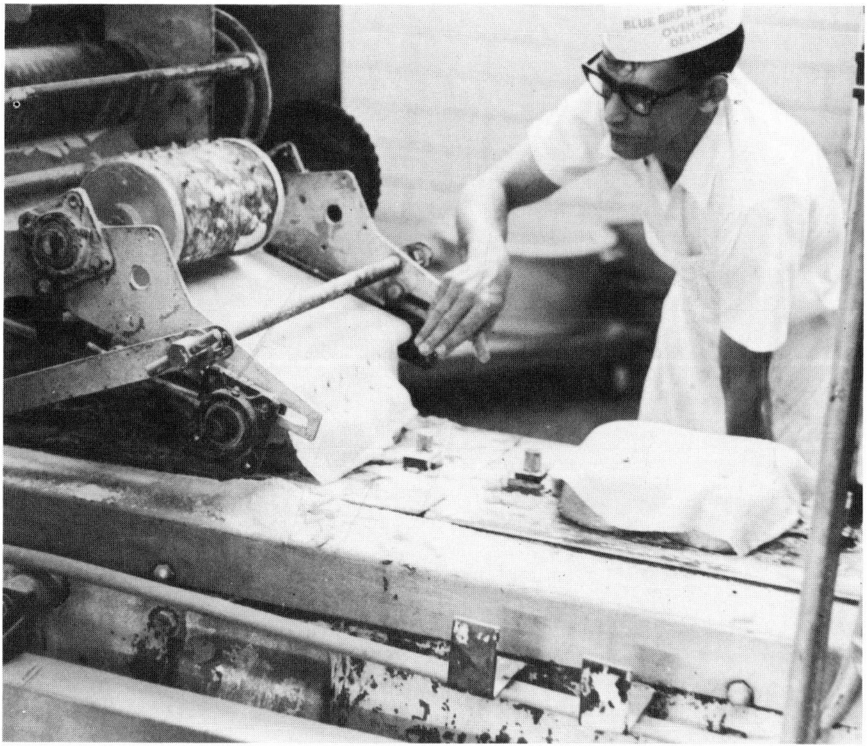

Figure 5. Sheeting pie doughs. (Courtesy of Blue Bird Baking Company)

applications. A low level of yeast (about 0.2%, basis flour weight) is added to the sponge phase of cracker production, which commonly extends 18 to 20 hours before doughup and further processing. Actually, the sponge fermentation of

Figure 6. Pie filling operation. (Courtesy of Blue Bird Baking Company)

crackers is a combination of yeast and bacterial fermentation (30) and is also influenced by the enzyme provided by the flour and other ingredients.

Cracker sponge flour is one that needs to be strong (usually soft red, unbleached) with high protein (9 to 10%), a high no-time viscosity[1] (65 to 85° MacMichael), and low cookie spread (7.2 to 8.0 width to thickness [W/T] ratio). This flour type withstands the long sponge fermentation but retains enough strength to provide the oven spring and open structure of the saltine (25,53). Where the available soft wheat is not strong enough, a percentage of hard wheat may be introduced into the blend. Extensigraph and recording dough mixer curves are sometimes used to measure the strength and uniformity of cracker sponge flours.

After the lengthy fermentation, at about 27° C and 80% relative humidity, the sponge is returned to the spindle mixer and a weaker flour is used for doughup (8.0 to 8.8% protein, 40 to 55° MacMichael no-time viscosity, and 8.0 to 8.8 cookie W/T ratio), plus shortening, salt, and sufficient sodium bicarbonate to overcome the acidity generated during sponge fermentation—hence the name "soda

[1] MacMichael no-time viscosity here and throughout this chapter indicates that the 1-hour digestion time is omitted.

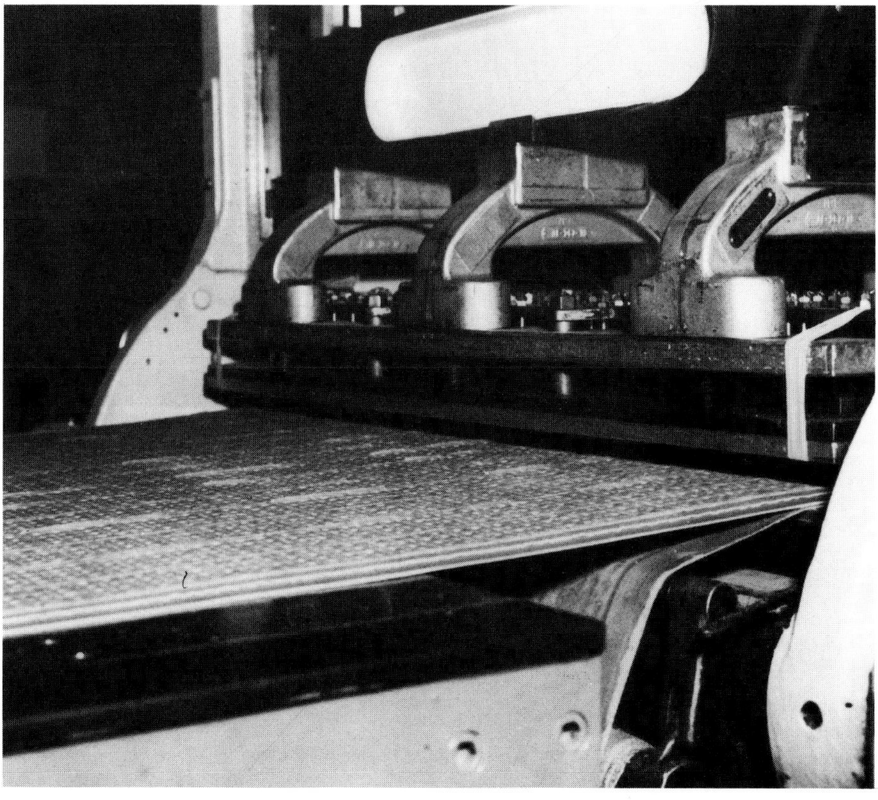

Figure 7. Cutting and stamping soda cracker dough. (Courtesy of Nabisco, Inc.)

cracker." After a resting period, the dough is fed to equipment that rolls, folds, and rerolls it to build up many layers and achieves gradual reduction to a very thin dough sheet (about 1.6 mm). This ribbon of dough is stamped or cut and docked into a square or oblong shape (Fig. 7), salted, then conveyed on endless belts through a hot oven where it is baked rapidly (2 to 4 minutes at 230 to 260° C) to the delightful color, flavor, and texture we favor. The final moisture of the crackers is below 3%, and they must be protected from moisture pickup so that they retain their appealing crispness and flavor. This is achieved with a double protective wrap, the inner wrap holding 1/8, 1/4, or 1/2 of the usual pound package (454 g); these smaller quantities are then enclosed in an outer package.

An unfermented, sweeter dough is similarly processed and baked into chemically leavened "snack" crackers that are usually cut into round or oval shapes; they may be sprayed with fat for greater tenderness and richness. A soft wheat flour of intermediate strength is used, similar to the doughing flour in the production of soda crackers. Soft wheat flour and whole wheat flour are used in the production of graham crackers, which are also chemically leavened.

Size, shape, color, and flavor of the cookies produced by wholesalers are legion (24,36). Wholesalers may seek high spread in cookies that are extruded or wire-cut (Fig. 8) or may mix a heavy, relatively stiff and dry dough that is

Figure 8. Wire-cut cookie depositor. (Courtesy of Nabisco, Inc., Pilot Plant, Fair Lawn Research Center)

pressed into the dies of cylinders that release the formed pieces onto a canvas apron for transport to the oven (Fig. 9). The stiff doughs in this case assist in mold release. Through flour choice and formulation, the manufacturer of rotary cookies arrives at an end-product in which spread is minimal, the baked piece emerging from the oven with an intricate design embossed upon it by the dies of the rolls. This is a large-volume item and is typified by many varieties of sandwich cookies. High-speed sandwiching machines apply the fillings and assemble the cookies, which are then packaged and distributed. Cream fillings must be smooth, tasty, and virtually without moisture, which would soak into the cookies and shorten shelf life (26). Formulas for rotary doughs are characterized by low water and sugar levels but higher fat percentages than those of spread-type cookies. The flour selected for this kind of cookie should be stronger than for high-spread types but need not be as strong as flour for cracker sponges. However, formulation can be adjusted to harmonize with differing flour strengths.

Ingredient and processing variations are more readily sensed in wire-cut and deposited cookies (24,36). Typical of this type are the thin, flat or slightly rounded vanilla wafers and chocolate chip cookies. Highest spread and tenderness with crispness are achieved by starting with soft white or soft red wheat flour with low protein (7.0 to 8.5%), low no-time viscosity (30 to 45° MacMichael), and high cookie W/T (8.8 to 9.5) (Fig. 10).

Figure 9. Forming rotary-molded cookies. (Courtesy of Nabisco, Inc.)

Ice-cream sandwiches are popular, particularly along the East Coast. They are often dispensed in vending machines. A special product of the cookie-cracker baker is the ice-cream wafer used for these sandwiches. They are made from an unfermented dough, rolled and sheeted like a cracker dough, fed to reciprocating cutters, and stamped into a docked oval or rounded oblong shape for baking.

There are two varieties: dark or chocolate, and light or vanilla. Besides a satisfactory appearance, texture, and eating quality, the ice-cream wafer is characterized by close dimensional tolerances—length, width, and thickness—and by hardness or resistance to breaking. Both of these important characteristics are related to successful passage through the high-speed machines that assemble ice cream and wafers into sandwiches. The fingers of the machine accept only dimensionally correct wafers, with small margin for variation. A wafer that is too tender also contributes to excessive breakage at the machine.

A flour for making ice-cream wafers may be anything from a high-ash soft red wheat straight grade of 0.44 to 0.48% ash, to a stuffed straight or clear of 0.50 to 0.70% ash. Close control of cookie spread by the miller is thought to be the best contributor to dimensional uniformity. Mills have at times used chlorine bleach to limit cookie spread. It is not well understood which factors help provide high breaking strength for ice-cream wafers. It has been theorized that higher ash flours may be best because of slightly higher levels of damaged starch, fiber, and pentosans. Furthermore, the higher pH in unbleached, high-ash flours may cause less initial reaction with the soda in the dough, resulting in fewer bubbles, better machinability, and a stronger structure in the baked wafers.

A single integrated cookie-cracker production line can make several tons of product per hour, and many plants have five or more complete production units side by side that often work both day and night. The consumption of flour, the highest tonnage ingredient, can reach many truckloads or carloads per day during heavy production seasons. To assure an uninterrupted supply of flour, the buyer, in close coordination with the sales and production departments, must

Figure 10. Sugar cookies from soft wheat flour.

purchase well in advance of need, and schedule flour to be shipped in orderly sequence, so that it can be ground and delivered to the carrier on a carefully structured schedule. He must avoid either running short or overfilling supply bins, and the flour must be protected at every step from deterioration owing to exposure to moisture or to animal or insect contamination. The materials handling revolution of the 1960s that led to conversion from sack to bulk shipment and storage has aided in this protection, although many sanitation precautions still must be taken.

The very fact of high-speed production demands uniform supplies of all ingredients to avoid making significant amounts of off-quality merchandise. This has led to development of flour specifications that are usually expressed in terms of commonly performed analyses such as moisture, protein, ash, MacMichael viscosity, and in cookie spread. Another factor that may be included is amylolytic enzyme, i.e., amylograph, gassing power, Hagberg falling number, or maltose value. Limits in fragment counts also may be made a part of flour specifications.

Unfortunately, an analysis does not guarantee product line performance; performance testing in the laboratory has not been developed to the point that it is always meaningfully applied to commercial practice. Hence we use the practices we know today to help provide uniformity and seek better methods of measurement and control.

Recently, alongside the large wholesaler of crackers and cookies, smaller regional wholesalers have established themselves. They bake a type of cookie that is more moist, thicker, and more chewy and that often contains fruit, oatmeal, or nuts. These may be called "home-made" type cookies with a shorter shelf life owing to higher moisture and usually less sophisticated packaging. This lends itself to limited operations where shelf life can be observed and stale products removed by the driver-salesman of the smaller (often franchise) operator. The flour used is of two types—a high- and a low-spread flour, at least one of these being chlorine-bleached, unlike the large wholesaler's flour, which is unbleached. The bleached flour yields a thicker, smaller-diameter cookie, with the moist, chewy consistency and peaked profile of the home-style cookie. The pace of production in the franchise bakery is slower and the equipment less automated than at large wholesalers. Marketing efforts focus in a different direction and over smaller territory. These operators do not make crackers.

The cookie-cracker industry is a large user of flour. Operating on small mark-ups on large tonnage, it must procure ingredients at the lowest possible price, but the usually cheaper, high-ash by-product is not practical for two major reasons. First, large-quantity requirements mean that the plant may at times need more of this type flour than is available and therefore may have to substitute other available types. This switching of type, grade, or class can interfere with smooth, trouble-free operation. Second, by-product flours do not run as uniform as patents and straight grades, which are usually available in unlimited quantity, subject only to mill running time and available supplies of grain. Specified flour ash levels mostly run in the 0.40 to 0.45% range, where uniformity can be best achieved by the miller. In some cases a blend of straight grade and by-product may be used, with ash levels up to 0.50% or even higher, where a limited quantity of flour is needed and some variability in quality can be tolerated. Patent grades of less than 0.40% ash are infrequently specified, because of cost and because there is no compensating gain in processing uniformity or product quality.

D. Sweet Goods

The term "sweet goods" is applied to a class of baked foods that includes cinnamon rolls, coffee cake, Danish pastry, and various fruit-filled and nut-topped breakfast rolls that may be baked in individual servings or in full pan-size units to be cut to serving size. These products may be made with baking powder but are usually yeast-leavened. They differ from bread doughs by their higher levels of sugar and fat, plus eggs. Their makeup often requires considerable handwork to achieve attractive finished shapes, appearance, and varied flavors. Wholesale bakers have introduced equipment innovations to bring a degree of automation not practical in the retail bakeshop (44,47) (Fig. 11). Retailers usually feature selections of attractive individual sweet rolls demonstrating skill and imagination.

Fillings and toppings used in sweet goods production add variety in eating quality, appearance, and flavor to the same basic formula and make-up.

Danish pastry is a special variation, involving the introduction of part of the fat by the roll-in method. A low-fat dough is lightly mixed and allowed to rest.

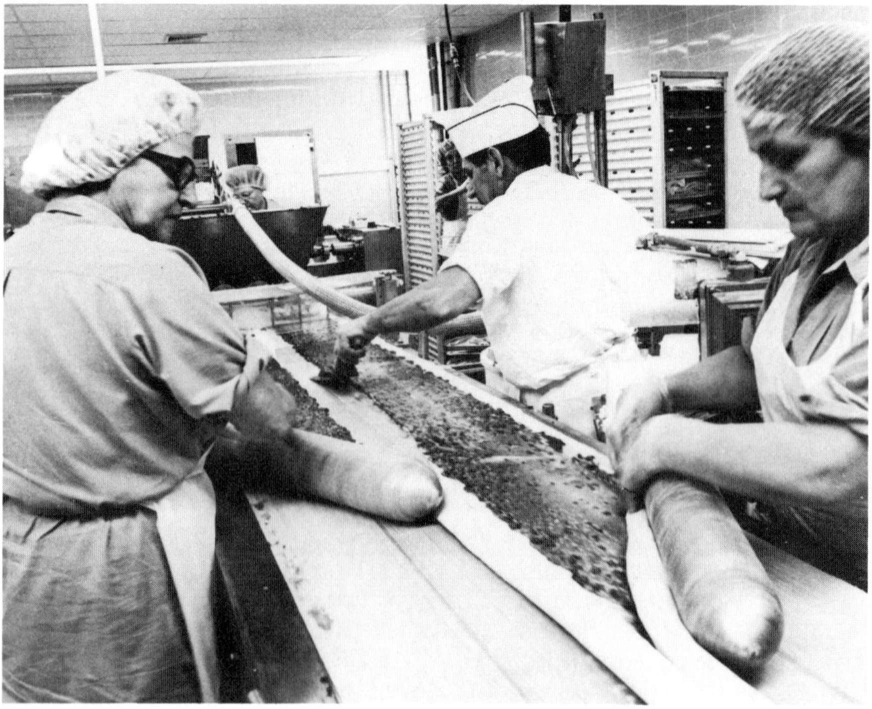

Figure 11. Continuous production of fruit-filled coffee cake. (Courtesy of Food Service Group, Durkee Foods, Division of SCM Corp.)

After the dough piece is rolled into an oblong shape, tempered shortening (up to 25% or more of flour weight) is spread over two thirds of the dough; it is then folded over to form three layers of dough and two layers of fat, rolled out thin, again folded into three layers, and rolled out; this is repeated to give a multilayered structure (up to 108 layers of dough, 72 layers of fat). After again relaxing, the layered dough is ready for make-up, proofing, and baking. The alternate layers of fat and dough make a rich, short, tender pastry with distinctive and highly esteemed eating quality. Because the layers of dough separate and open up during baking, Danish pastry is said to be fat-leavened, and it is important to select a fat suitable for that purpose. The consistencies of the dough and the roll-in shortening must be nearly the same. If the shortening is too firm, it will break through the dough layers; if too soft, it will run out between the dough layers or "bleed," and tend to be absorbed by the dough. The desired plasticity must be maintained throughout the roll and fold operation (44).

Bakery mixes for making sweet goods are available in which all the dry ingredients are premixed, thus simplifying the inventory and scaling steps for the baker. Rich or lean formulas can be selected according to market conditions. A blend of soft (pastry) and hard (bread) wheat flours is commonly used in sweet goods production.

E. Ice Cream Cones and Sugar Wafers

Ice cream cones come in two different kinds (Fig. 12). Best known and most widely distributed is the cake or molded cone—tender, crisp, symmetrical, and usually golden yellow. Another closely related item is the molded cup, which is

Figure 12. Molded or cake ice cream cone (left) and sugar cone (right).

widely available and differs only in shape. The traditional flour for this product came for many years from Pacific Northwest white wheat. As freight rates increased and shipboard flour traffic through the Panama Canal ceased, Eastern and Midwestern molded-cone producers reluctantly turned from their favorite source and converted to eastern soft wheat flours.

A thin fluid batter is mixed to desired consistency, as measured by a viscosimeter. Batters are kept cool during mixing and before use, to give stability and discourage growth of microorganisms. The batter is next pumped to a machine resembling a ferris wheel. Around the circumference of the wheel is a series of two-piece molds with cores, which are injected with a charge of the batter, then transported through the gas-heated oven to expand the batter, then bake and dehydrate the finished cone. At the completion of the circular ride (about 2 minutes), the mold separates into its component parts, dropping the cone into a trimmer that removes excess scrap, then onto a conveyor where it is cooled, and taken to the packaging operation. Cones must be very dry in order to provide the crisp eating quality responsible for their appeal; however, they must not be so tender that they collapse when handled or when the ice cream is deposited in them. On extended storage, cones sometimes shatter even in the absence of any notable physical force. The cause of this phenomenon is not well understood. Flour for molded cups and cones may be milled from either soft red or soft white wheat, patent or straight grade, unbleached. It may be dried to a predetermined moisture level, such as 10 to 11%, for better mixing and performance characteristics.

The second type is the sugar cone, quite different in appearance as well as in its manufacture. The batter is poured onto baking irons that produce a flat, waffle-patterned sheet when a second iron is pressed down onto the batter and the irons are passed through the oven. While still hot and flexible, this sheet of baked material is rolled into a cone shape, then cooled, and conveyed to the packaging line. This cone is harder, more brittle and chewy, and is usually dark in color. Its appearance is that of a flat sheet rolled into a cone shape, with the outer layer wrapped, sometimes loosely, around the inner folds of the cone. Flour for sugar cones is often of a higher ash (by-product) type than that of molded cones (up to 0.55% ash) and has a moderate chlorine bleach treatment (5.1 to 5.4 pH) to prevent sticking during baking and to enhance cone strength.

Sugar wafer batters contain little or no sugar and are baked between heated metal plates similar to sugar cones (Fig. 13). A sweet filling is spread thinly over several baked slabs, which are sandwiched together and cut to size with high-speed saws. Sugar wafers are commonly served with a dish of ice cream or other dessert.

F. Pretzels

A most interesting, historic, and unique product of soft wheat flours is made by the pretzel industry—interesting and historic because of the stories of the origin of pretzels in the monasteries of the Middle Ages and pretzel making in America, mostly in Pennsylvania in the early days. There are still many of the original pretzel factories in that state, although pretzels are now more widely produced and consumed. Insofar as modern technology is concerned, the pretzel industry was slow to emerge, many of the early plants being family owned and operated and seldom having a laboratory. Little has been published on flours for pretzels or on process control, although some steps are being taken to study the methods of manufacture. Automation (45) is replacing the traditional "hand-made" concept, typified by the long line of human "pretzel benders" stretched out along a conveyor belt.

Hard pretzels are unique in that after the doughs are formed, they are first cooked in a lye bath, coarse salt is applied, and then they are baked. This unusual treatment, plus drying to a low level, gives a color, sheen, flavor, and texture that is greatly appreciated as a snack and party food widely served in America.

Because the industry developed in Pennsylvania, the hard pretzel manufacturers learned to use locally ground soft red winter wheat flour; this is still being used. As there seems to be no advantage to using short patent grades, the trend has been to procure higher-ash, soft red winter flour types, from 0.40 to 0.45% ash straight grades to 0.60% ash stuffed straights or soft clears, untreated. Because the final color of pretzels obscures the original color of flour used, bleach is inappropriate.

A stiff dough is mixed (about 40% water, flour weight basis), usually in a horizontal mixer, comprising flour, water, shortening, nondiastatic malt or other sweeteners, yeast, and salt. Small amounts of chemical leavening may be included. Bread flour is sometimes blended with the soft wheat flour to achieve certain objectives in the finished product.

The dough next goes to the forming equipment, which may be of many kinds. The old hand-twisted hard pretzel is now made by machine (Fig. 14). Shapes may also be varied: sticks, rods, rings, nuggets, or other familiar styles. After a proofing period to gain some volume and form a crust (Fig. 15), the pieces are carried by a mesh belt into and through the lye cooking solution, usually about

Figure 13. Sugar wafer oven. (Courtesy of Weidenmiller Company, Inc.)

1% sodium hydroxide at about 90 to 95° C for up to half a minute. While pretzels are still moist and sticky from the cooking bath, either evaporated or coarse rock salt is applied before baking. Temperatures up to 260° C are applied in the fore part of a tunnel oven, to fix high volume and rich color and develop inner structure, with lower temperatures in the rest of the oven to complete baking and dehydrate to about 15% moisture. From the oven, the pieces pass through a drying tunnel to bring about the crisp and brittle texture characteristic of the finished product (less than 3% moisture). Packaging must hold moisture at this low level; well-protected pretzels retain their flavor and crispness for a year or more.

G. Prepared Mixes

A large use of soft wheat flours, and one closely related to family flour use, is in boxed cake mixes (15). The manufacturer provides a wide range of formulas and flavors of cakes, representing an intermediate stage between the ultimate convenience of buying the finished baked product and the older tradition of

Figure 14. Pretzel-twisting machine. (Courtesy of American Machine & Foundry Co. and Nabisco, Inc.)

starting from scratch with all individual ingredients. Premixed products assure almost foolproof results, thus permitting housewives to serve oven-fresh, home-baked, and iced cakes. The formulators of prepared cake mixes have taken

Figure 15. Conveyor carrying pretzels from forming machines. (Courtesy of American Machine & Foundry Co., and Nabisco, Inc.)

advantage of modern developments in flour, fats, emulsifiers, stabilizers, leavenings, and packaging to bring top performance to their products. Forceful merchandising has also helped to achieve widespread acceptance.

Cake flours for mixes are commonly dried, usually to about 10% moisture. This helps to stabilize and protect the mix from premature leavening action and consequent loss of baking performance. Drying may be targeted at lower moisture levels for longer shelf life or high-stress climatic conditions. Packaging to bar entry of moisture is likewise important.

The earliest example of prepared mixes, dating back to the 1850s, was the manufacture of self-rising flour, in which the proper amounts of salt and leavening (sodium bicarbonate and acid reactant) were premixed with the flour; although not a complete mix, this did form the foundation for making the hot biscuits of the south (Fig. 16), for home-made cakes, pancakes, and dumplings. This mix even today comprises much of the "family flour" of the southeast; there the nonleavened or "plain flour" contains about 0.5% of monocalcium phosphate, which keeps the hotbread slightly acid, contributes to the much treasured pure white crumb color, and also adds needed calcium to the diet. The biscuit mix, complete with shortening and dried milk solids, reflects further development of the self-rising flour idea. This can form the basis for making hot biscuits, shortcake, dumplings, pancakes, waffles, and coffee cake.

Self-rising flour is so well recognized that it appears in the Food and Drug Administration's Definitions and Standards of Identity for Cereal Flours (9). From this can be derived the formula shown in Table III.

Preference for a bright white crumb dictates use of soft wheat short patent flour, usually 0.33 to 0.36% ash, bleached with chlorine to 5.2 to 5.4 pH for baking performance, and with benzoyl peroxide for optimal color removal.

Figure 16. Baking powder biscuits from soft wheat family flour or self-rising flour.

TABLE III
Formula for Self-Rising Flour

Ingredient	%
Flour	95.00
Anhydrous monocalcium phosphate	1.64
Sodium bicarbonate	1.32
Salt	2.04

Monocalcium phosphate is usually added for vigorous and sustained leavening action. Sodium acid pyrophosphate and sodium aluminum phosphate are also acid-reacting substances permitted in self-rising flours. Manufacture may be by either batch or continuous system. Blends of soft and low-protein hard wheat flour may be used in this application.

The more complete mixes require more complex manufacturing equipment and procedures, especially in the methods for incorporating fat; it may be added as a dry shortening (a spray-dried mixture of fat and dry milk), by spraying in liquid fat, or by cutting in plastic fat.

Cake and other prepared mixes are made for and sold to housewives, institutional users (hotels, restaurants, schools, factory in-plant cafeterias, prisons), and bakers, the chief differences being in the package sizes and marketing efforts. Mixes may be complete, requiring only addition of water, or may require addition of water and eggs or eggs and milk.

Cake flour for mixes is generally specified in the 0.33 to 0.36% ash range, with flour pH adjusted for best performance, usually 4.5 to 4.8. Considerable latitude is possible in the choice of protein level, influenced by natural year-to-year crop variations. Soft red winter or a blend of soft red and soft white is the choice for cake flours used in the eastern United States. The cake flours in mixes manufactured in the western United States are milled from the western soft white wheats.

Granulated sugar (sucrose) is the largest ingredient of cake mixes in percentage by weight, with flour next. An emulsified shortening follows in proportion. The formulator must provide baking performance and also must design a product that will give long shelf life, which must be demonstrated in undiminished performance, freedom from rancidity or other breakdown, and resistance to discoloration over time and under unfavorable environmental conditions. The shortening manufacturer has helped meet these needs by careful selection of raw materials, deodorizing, hydrogenation, controlled crystalline structure, and the use of additives—emulsifiers for performance and antioxidants for keeping quality.

Cake mix formulas contain dried eggs and nonfat milk solids in roughly similar proportions, their selection again being guided by performance and flavor criteria. In the interest of these same objectives, the choice of leavening acid is critical; much progress has been made to steadily improve performance and shelf life. Chemical composition, particle size, and the use of coatings have all contributed. Soda of suitable granulation is added to properly balance the acid and to govern final pH. Salt and flavorings complete the usual ingredient list for a typical layer cake formula.

H. Toaster Pastries

A recent development has been that of toaster pastries, a filled dough containing fruit or chocolate, sugar, and spice, sized and shaped to fit into the household toaster; it is toasted to serve at breakfast or as snacks or dessert. A soft wheat flour dough forms the envelope for the filling; it is baked enough to set the dough, sterilize, and reduce moisture for shelf life and stability but short of full bakeout. It is often iced to enhance appearance and taste.

The toaster heating browns the exterior, completes the baking, and refreshens the interior, giving the product a fresh-from-the-oven flavor. Toaster pastries did not go through the long period of hand fabrication or small-shop production of many baked foods; they virtually sprang into existence as a highly automated, large-scale process during the 1960s and 1970s. Some of the leading breakfast cereal manufacturers also make and distribute this product.

Ingredients include soft wheat flour, sucrose and other sugars, shortening, salt, baking powder, flavorings, colors, vitamins, and other additives, plus the distinctive varietal filling.

A similar product is the frozen waffle. A square of frozen, cooked waffle is dropped into the toaster and comes out piping hot, with its crust crisp and interior steaming, ready for butter and syrup. This product has been on the market for a number of years. Freezer storage and protective packaging preserve moisture and freshness until use.

III. INGREDIENTS USED IN BAKED FOODS

A. Sweeteners

Ingredients other than flour are as diverse as the products baked with soft wheat. These formula components help provide structure, texture, color, and appearance as well as sweetness, aroma, richness, flavor, and nutrition. In quantity, the sweeteners are usually most important in determining the characteristics of baked foods and in some instances exceed the weight of flour in the formula (37). Granulated sugar (sucrose) is the dominant form used. It contributes sweetness, which has universal appeal to the human palate, tenderness, color, and flavor through crust caramelization; its crystal size and shape affect creaming properties and spread in cookies. Dextrose and corn syrups may be substituted in some formulas to reduce cost and modify performance, but they are less sweet than granulated sugars (48). Invert syrup affects crumb softness and moisture retention, contributes to crust color and increases shelf life and sweetness. Malt (syrup or dry) has a low sweetening value but gives a distinct flavor and dark color. Use of a diastatic type also adds to the enzyme systems.

B. Shortening

An entire industry has been built around supplying the baking industry with shortening (2,7,38,41,44,46). Although lard, butter, and other animal fats are still used, the big swing has been to the vegetable oils and hydrogenated shortenings, rising with the phenomenal growth of the soybean economy in the last generation. Each class of baked food has available specially selected and processed fats formulated for its needs. The shortening industry research and service laboratories have developed specific products for every baking purpose and have pioneered and promoted new concepts to help bakers perfect their products to improve consumer acceptance. An outstanding example is the now universal acceptance of the high-ratio layer cake, a development begun in the 1930s. Investigations have covered raw materials selection, deodorization, bleaching, hydrogenation and chemical rearrangement, addition of emulsifiers and antioxidants, and specific crystalline structure. Characteristics provided include creaming, aerating, and emulsifying properties; specific consistency; selected melt point; stability for long shelf life and high smoke point; and deep fat cooking quality. All fats provide richness and high nutritive value. The newest focus is on how to help provide the greatest enchancement of the diet in light of evolving studies of food's effect on health and longevity.

Raw materials range from lard and beef fat to the domestic oils from soybean, cottonseed, corn, safflower, and peanuts and imported coconut and palm kernel oils. Their processing skills enable manufacturers to exercise a degree of interchangeability that allow raw material shifts to offset price rises and shortages.

Shortening's function in baked foods is to tenderize, to supply softness, nutrition and richness, sometimes crispness, and structure and volume where emulsification takes place, and to extend shelf life. The importance of fat's contribution to the diet is attested to by its rationing during World War II.

C. Eggs

Eggs substantially contribute to many baked products and are available in various forms including fresh shell eggs in home baking, refrigerated bulk eggs, frozen whole eggs, egg whites and yolks, and dried egg products that closely match the performance quality of the liquid forms (5,40,41,60). The dried products have advantages offsetting higher original cost including lower shipping costs, simplicity of storage, superior keeping quality, and advantages in handling at scale and mixer. Much progress has been made recently in perfecting dried egg products to high performance standards through studies of drying methods, temperature levels, use of additives (mainly sugars and whipping aids), and pasteurization to ensure a salmonella-free product (27). The baking industry uses more frozen eggs than any other form, but dried products and chilled bulk eggs are gaining in use.

As for the specific contributions to baked foods, color is notable. The golden yellow of pound cake, yellow layers, sponge cakes, and donuts is the traditional sign of richness and wholesomeness. Eggs also help give volume, structure, grain, and texture and improve that nebulous attribute called "eating quality." The lecithin of egg yolk is an emulsifier; hence it aids in providing better structure and stability. In the classification of the toughening or tenderizing effect of ingredients, eggs play a dual role, whites being a toughener and structure builder, yolks a tenderizer through high fat content.

Outstanding examples of function as part of the leavening and structural system are angel food and sponge cakes (5,60), where the egg or egg white is whipped to considerable volume, with other ingredients added to the foam before baking in such a way as to maintain lightness and good cell structure. Not as self-evident to the casual consumer is the superior nutritional contribution of eggs.

D. Milk

Whereas housewives use whole milk in recipes, this is impractical for bakers for several reasons (39). Storage, sanitation, and keeping quality would pose special hazards to use of liquid milk in quantity. Extra refrigeration and many handling precautions would be required to prevent spoilage. Shipping costs of liquid milk, which is seven-eighths water, are high compared with that for other forms. Furthermore, it is more economical to substitute other fats for butterfat. Therefore, the form of choice is nonfat dry milk solids, with use of minor quantities of evaporated milk (from whole or skim milk) and sweetened condensed milk. Dried whole milk is available but not widely used because of cost and the possible development of rancid flavor.

Nonfat dried milk may be spray dried or roller dried, the former being much more popular. Where reconstitution is anticipated, dried milk is usually supplied in instant form.

The heating of milk before drying is adjusted to suit the ultimate use of the product; for example, high-heat processing is necessary to condition dried milk for use in yeast-fermented doughs (58).

The amount of milk in a cake batter is usually based on all the liquid in the formula being in the milk and eggs. When nonfat dry milk and dried eggs are used, cake formulas tend to maintain about these same levels of milk and egg solids, substituting in each case solids and water for liquid milk and eggs. Formula balance is more directly influenced by the percentage of eggs, shortening, sugar, flour, and liquid; hence there is somewhat more tolerance to the amount of milk solids. Under these conditions changes may be made according to cost and availability.

Nonfat milk solids function as a structure builder and a dryer, contribute significantly to nutrition, and help develop crust color, through the presence of lactose, which caramelizes in the oven.

E. Leavening

An outstanding feature of most baked foods is lightness, or a high ratio of volume to weight. Some of this property derives from wheat flour's unique gluten-forming protein, which envelopes the evolved gases and characteristically expands to produce the greatly esteemed light-textured products. Leavening brings about this evolution of gases during batter and dough formation and in the oven. Air whipped or creamed into a batter is a form of leavening; so is water converted to steam in the oven; eggs and shortening that whip or emulsify are vital parts of the leavening system. The natural acidity of some ingredients in the presence of soda releases carbon dioxide in batters and doughs (17,25). Ammonium bicarbonate is a leavening agent sometimes used in cookies for spread control.

The two basic classes of added leavening are yeast and chemical leavening. Products that are baked with soft wheat flour and that utilize yeast fermentation are soda crackers, pretzels, and sweet goods (coffee cake, sweet rolls, etc.). Most soft wheat bakery products are chemically leavened with baking powder (3,15,17,28,41). Baking powder consists of balanced proportions of sodium bicarbonate and an acid leavening agent. To provide the acid leavening, many chemical compounds are used, according to the particular needs of the end-product. Housewives used to provide leavening, particularly in baking powder biscuits, by using sour milk (clabber) and soda. However, depending on the degree of acidity in the sour milk, a different quantity of soda is required for balance, leading to significant variability in baking results, especially in flavor and color of both crust and crumb.

Cream of tartar (potassium hydrogen tartrate), sometimes with tartaric acid, was long the standard acid reactant, but it was a fast reactor and lacked the tolerance of many acid reactants that have now largely superseded it; moreover, it was costly. Monocalcium phosphate is widely used today; it is inexpensive and fast in reaction except when treated to give delayed effects. Dicalcium phosphate dihydrate, sodium aluminum phosphate, and sodium aluminum sulfate are also used in leavening systems for certain desired effects. Because of its slow-reacting property, sodium aluminum sulfate is frequently used in combination with monocalcium phosphate in household baking powders. Besides cream of tartar and tartaric acid, other organic acids more recently developed and employed are fumaric acid and glucono-delta-lactone. The latter has special application in some mixes where a breadlike flavor is sought. The organic compounds suffer a price disadvantage. Sodium acid pyrophosphate is a slow-acting leavening acid, which is much used in layer cakes and in cake donuts. Some applications call for early evolution of gases in the batter phase and further gas production under oven heat. This is achieved by making a "double-acting" baking powder, consisting of the usual sodium bicarbonate plus two or more acid components with different reaction rates.

F. Salt, Flavors, and Colors

Salt (sodium chloride) is incorporated into baked foods primarily for its contribution to flavor, but it has other effects, particularly in fermented products where it slows the rate of gas evolution and strengthens the gluten. These side effects are useful in some special situations but the consequences on taste limit the degree to which salt can be used as a controlling agent; it is, of course, deleted in baked foods for salt-free diets.

Spices and flavoring agents seldom have any function other than to affect aroma, flavor, and taste (42). Label statements must indicate the use of artificial flavors.

The preferred way to provide attractive color to baked goods is with the use of natural ingredients that possess the desired color; e.g., yellow (denoting richness) from egg yolks and red from fruit jams or jellies. The use of artificial colors in baked foods is generally restricted by law (18), but they may be used if three criteria are met. First, artificial color must be an approved additive, signifying that it is harmless to man. Second, it must be labeled properly. Third, it must not be used to cover up inferiority otherwise visible and must not lead the consumer to infer greater richness or value (e.g., use of yellow to suggest the presence of eggs or brown to imply the use of cocoa or chocolate).

IV. NONBAKED FOODS FROM SOFT WHEATS

A. Cake Donuts

The basic flour for making cake donut mixes or batters is a strong soft wheat flour—soft red winter in the east, soft white in the west. The desired strength may be obtained through wheat selection, stream selection (35), or in some instances by blending hard with soft wheat flour to the desired strength and performance level. Cake donut flour for mixes is often dried to aerate and oxidize the flour and to help achieve desired performance and stability. The flour is usually unbleached, although light treatment with a maturing agent, such as chlorine dioxide, has been used by some mills.

Cake donut mixes are composed of 55 to 65% flour, plus sugar, shortening, nonfat milk solids, leavening, egg yolk solids, salt, flavorings, and sometimes other ingredients such as soy flour, corn sugar, starch, gums, coloring, and surface-active agents as emulsifiers. They may be formulated for hand-cut or mechanized production. Desired characteristics are controlled water absorption in the batter for desired yield of donuts; tolerance or the ability to maintain uniform performance from beginning to end of the batch; appropriate leavening action to deliver good volume, desired shape, and smooth skin; tenderness and good eating and keeping qualities; and frying fat absorption sufficient to enhance appearance and eating character but not enough to make the finished donuts greasy, overly rich, or expensive from the standpoint of frying fat loss (2,44,46). This calls for what may be termed "balance" of ingredients, to match the flour and each of the other formula components.

Protein and ash often appear in donut flour specifications, but they are not important performance criteria. The farinograph curve is often used to judge the proper strength and absorption. Performance testing is the ultimate arbiter of flour quality; fortunately, a small production machine can be operated in or in conjunction with the laboratory.

Selection of leavening is important; a delayed-action acid leavening agent is used so that gas evolution is not completed before the second side has expanded and been cooked or fried to its final shape. Sodium acid pyrophosphate with a controlled reaction rate is usually employed, although others such as sodium aluminum phosphate, monocalcium phosphate, dicalcium phosphate, and glucono-delta-lactone are also used in cake donut formulations.

A modern large-scale production unit (Fig. 17) turns out hundreds of dozens of cake donuts per hour. The operator changes temperature, mixing time, water absorption, and extrusion pressure to maintain correct size, shape, crust color, fat absorption, and finished appearance of the product. Temperature and care of the frying fat are also factors important to successful operation (2,38,46).

After cooling, the donuts may be coated with powdered sugar, mixed with starch and fat to help resist soakage and discoloration (Fig. 18). The flavoring of donuts can be quite varied but has traditionally included a spice, such as mace or nutmeg, often combined with a citrus companion such as lemon, and vanilla.

B. Breakfast Cereals

Breakfast cereals are usually classified into those that must be cooked and those that are ready to eat. The soft wheats infrequently appear in the former but are quite important in the latter. There are a few exceptions; for example, a soft wheat cereal for cooking is made as a co-product of the milling of soft red winter wheat. In effect, the base material is farina plus bran and germ; it is roasted until the germ is brown, giving a desirable flavor and extending shelf life. Bran is held at about 12%, flour at 1% or less. Granulation is maintained within fairly close tolerances.

Figure 17. Cake donut cutting and frying operation. (Courtesy of Belshaw Bros., Inc.)

Figure 18. Cake donuts, plain (left) and sugared (right).

TABLE IV
Fiber in Ready-to-Eat Cereals

Product	Crude Fiber (%)
All bran, 100% bran cereals	7.0–8.0
Raisin bran, 40% bran cereals	2.5–3.5
Shredded wheat, wheat flakes	1.0–3.0

In wheat-based, ready-to-eat cereals, the soft wheats play a prominent role, particularly soft white wheat (13,14,23,52). The bran's whiteness provides a bright, attractive color in the finished product and may have flavor advantages, the reasons for its choice over red wheat bran as raw material. Varying amounts of bran and fiber are found in prepared cereals; they doubtless are the most readily available means of restoring dietary fiber lost from the American diet in the past 100 years. On a crude fiber basis, the wheat-based, ready-to-eat cereals rank as shown in Table IV.

The flaked cereals start out as discrete particles that retain their individuality throughout processing, each day granule becoming a finished flake (23,52). For example, each wheat flake in the package started out as a clean, plump kernel of soft white wheat. The charge of ingredients going to the cooker comprises the wheat, sugar and/or malt, salt, water, and coloring material. After cooking to about 50% moisture the mass is discharged, cooled, and partially dried in preparation for flaking (Fig. 19). Intermediate steps are taken to separate any granules that have adhered to each other. Mineral and vitamin additives are sprayed onto the product, and heavy flaking rolls compress each particle into a thin, flat plate that is then baked in ovens to toast and dehydrate each flake to the proper color and texture. After oven treatment and cooling, the product is ready to package and ship. Each kernel of wheat has become a thin, crinkled, slightly blistered, irregularly shaped, and tasty flake up to 15 mm or more in diameter.

The all-bran and 40% bran products are similarly processed. Raisins are added after flaking to make raisin bran. Shredded wheat consists of whole wheat only, not augmented with sugars, malt, salt, or other flavoring materials. The wheat is cooked as in the preparation for wheat flakes. Then the dough is formed or spun into strands that are wound by machines into the familiar biscuit shape to be further cooked and dried, with the final moisture being near the normal equilibrium point for wheat and flour. This calls for a less sophisticated package than is necessary to maintain condition in flakes, with their much lower moisture (2.5 to 3.0%) (6).

The flaked cereals have become an important vehicle for introducing some of the micronutrients to the American diet (13). Promotions are primarily aimed at the children's market, through sponsorship of Saturday morning television cartoons, in-the-box and send-away-for premiums, and other activities designed to attract children. We think of the cereals as the logical place to find the "B" group of water-soluble vitamins, but with progress and competition in the ready-to-eat cereal field, as many as 10 vitamins and five minerals are being added—not only the "B" group of vitamins, but also C, and the fat-soluble group, vitamins A, D, and E, plus iron, calcium, phosphorus, copper, and zinc. More than 90% of the ready-to-eat breakfast cereal brands on the market in 1976, excluding granola types, were fortified with added nutrients (13). Shelf space for these products in the supermarkets has multiplied many times since the days when cornflakes were the big item. The convenience and high nutritive levels in

Figure 19. Flaking operation producing bran flakes. (Courtesy of Kellogg Company)

the ready-to-eat cereals, combined with forceful marketing programs, have carved a notable place for them in the American diet. The high sugar levels in some of these products have been criticized.

C. Soup Thickeners

Soft wheat flour contributes body and consistency to canned processed soups (Fig. 20). Hard wheat flours can also be used, but soft wheat flours are more readily available in the areas where most of the canned soup is produced. Soft wheat flours are usually lower priced than hard wheat flours, and the thickening value of soft wheat flours is most often greater than that of hard wheat flours, owing in part to the higher starch content of soft flours but also to the inherent characteristics of soft wheat starch. There are exceptions to these generalizations, specifically when rains at harvest in soft wheat-producing areas induce high and undesirable enzyme levels through sprout damage. The thickening value of a flour for soup ingredient use is measured by the amylograph, which is very sensitive to the presence and effect of alpha-amylase. Generally, flour for soup thickening is best milled from soft red winter wheat, selected for high test weight and soundness. It must be free of sprout damage and should be milled to achieve a good separation of endosperm from by- product, especially those fractions having high enzyme activity (e.g., germ). Amylograph value can be elevated by the application of strong chlorine bleach even to the point of overcoming modest levels of alpha-amylase. Longer extraction flours exhibit lower amylograph peaks and less thickening value owing to the inclusion of more enzyme-active fractions. Because of the greater susceptibility to sprout damage and the consequent greater likelihood of high levels of α-amylase, the eastern white wheats are not usually milled into soup flours (11). There have been exceptions to this rule.

Ground cracker meal is another way in which soft wheat flour is used as a soup ingredient. A simple dough of flour, water, and salt is mixed and run through the

Figure 20. Soda crackers from soft wheat flour; soup thickened with soft wheat flour.

rolls and cutters of a soda cracker line, baked, then ground, and sifted to a uniform meal. This is included in the soup formula to give a different kind of body and appearance.

D. Batters and Breaders

To enhance their appearance and palatability, cooked fish, shrimp, scallops, poultry, some meats and meat patties, and onion rings are often first coated with a batter or breader consisting of cereal-based preparations, usually starting with soft wheat flours, which may be combined with spices, leavening, stabilizers, colors, and flavorings. The coating may comprise as much as 30% of the finished product.

The widely differing nature of these food coatings is demonstrated in the following descriptions:

CRACKER MEAL BREADING

A soft wheat flour, straight to stuffed straight to clear, usually bleached with benzoyl peroxide, sometimes with chlorine as well, is mixed into a cracker-type dough, sheeted, cut, baked to a creamy white color, then ground and sized. The formula is simple, consisting of flour, water, salt, and perhaps reducing sugars for fast browning. The flour most often used is milled from soft red winter wheat, 0.45 to 0.55% ash. Use of this breader contributes a distinctive granular texture and appearance.

RAW FLOUR BLENDS

These are simply soft or hard wheat flour, mixed with spices and seasonings and used to coat poultry pieces before deep fat frying, sometimes under pressure.

BREAD CRUMBS

Bread ground into suitable particle sizes is used to coat meat or fish before cooking. Stale bread returns are the source for some of this breader, but the diminishing supply of returned bread has brought about the practice of baking a standardized bread specifically for grinding and stuffing purposes.

ORIENTAL CRUMBS

A proprietary baking process is used to process wheat flours into elongated, white, fragile crumbs resembling grains of rice. This product is enjoying growing popularity in the preparation of seafoods.

BATTER COATINGS

A dry mix is made of flour, leavening, flavors, and colorings. This is made with water into a batter that is applied to meat or seafood in preparation for cooking or precooking, resulting in a crisp, golden brown, appetizing finish. With the proper level of leavening, the familiar *tempura* effect is achieved (swelled, blistered). A complex and specialized system is required to convey the product into the batter, then out, and through the cooking fat.

The rapid growth in production of batters and breaders is related to the growing acceptance of frozen foods and the expanding numbers of fast-food outlets. The finished-product weight of items on which batters and breaders are used increased from 504,561,000 lb in 1963 to 1,187,471,000 lb in 1969 and 1,535,112,000 lb in 1975 (22).

Small packages of breader mix are available on the supermarket shelf for use in the home.

E. Fried Noodles

The fried noodles sprinkled over chow mein or other oriental dishes are processed from a dough usually made with soft red winter clear flour. Besides the flour, ingredients are shortening, salt, water, and sometimes yeast. The dough is lean and very stiff and may be mixed in a continuous mixer like that used in making macaroni and spaghetti. After a relaxation period to give workability, the dough is sheeted to a thin film, cut into long fine strands by a rolling circular cutter, then cut crosswise to short lengths. The dough filaments then go through a wet steam-heated tunnel for conditioning before cooking in hot fat; the individual pieces here achieve greatly increased volume and become covered with blisters, contributing to the appealing crunchy texture that complements the other elements of an oriental dish. The product can also be eaten alone as a snack food.

F. Bulgur

Most Americans are unaware of the processing of wheat into bulgur, a product that is more than 4,000 years old. The Bible records the serving of this dietary staple in the Middle East. U.S. exports of bulgur go to many points in southeast and southwest Asia and South America, under government-sponsored programs. Domestic consumption is minor, chiefly in Armenian restaurants (e.g., *Pilaf*).

Bulgur may be made from soft or hard, red or white wheat. An estimated 3 to 4 million bushels of western soft white wheat was used in making this product in 1975 (12).

Bulgur was traditionally made in the homes in the Middle East by soaking wheat in water, parboiling, and drying in the sun. After rubbing to remove some of the bran, it was crushed into a meal. The coarse cereal was cooked and consumed much as rice is.

In the 1950s and 1960s, great strides were made in America in modernizing and mechanizing the technology of bulgur manufacture, culminating in 1970 in the U.S. export of about 6 million cwt. The half dozen or so producers of bulgur are concentrated along the Pacific Coast and in the hard red winter wheat growing areas. Commercial development came through the activity of flour milling organizations and in the facilities of the Western Regional Research Laboratory at Albany, California.

Hot water is added to cleaned wheat to about 40% moisture, then heated and held until penetration and starch geletinization are complete (at about 88° C). Advantage is taken of the time-temperature relationship to speed the process but not so rigorously as to rupture the bran and produce handling problems. Steeping under pressure would further shorten processing time but is not commercially practiced. After hydration and gelatinization are complete, the kernels are dried to about 10% moisture, near the original state of the wheat kernel before processing. The next step is pearling to remove some of the bran, then grinding and sifting to produce a coarse and a fine product. The keeping quality of bulgur is considered good because of its low moisture level and hard texture.

The modern manufacture of bulgur was developed to help implement a government food export program. Domestic consumption shows no tendency to rise to significant totals, although some promotion has been undertaken.

V. NONFOOD USES OF SOFT WHEAT FLOURS

A. Glue Extenders

A considerable tonnage of low-grade soft wheat flours is shipped to the furniture and plywood industries, where it becomes an ingredient in the urea-formaldehyde-resin glues used in assembling furniture and bonding the laminates of plywood (33). Flour specification is not critical, with the lowest grades (clears) being satisfactory. Use of wheat flour in this application permits cost reduction without reducing bonding strength below a specified level. Other advantages stem from the favorable effects of gluten on glue viscosity and the contribution of color to a desired light or white "glue line." Flour for this use is untreated.

B. Feed Ingredients

When by-product flours are abundant and prices low, a significant quantity may displace some of the traditional feed grain or industrial by-product in pet foods and in animal and poultry formula feeds, particularly when the feed grain-based ingredients are scarce and dear. This usage comes and goes in response to prices; in times of normal supply and price relationships, this is a distinctly minor outlet for by-product flour.

C. Miscellaneous

A small tonnage of soft wheat and flour is used in the metal casting industry, as textile and paper sizing material, in making adhesives and wallpaper paste, in book binding, and as one of the ingredients for brewing beer. During World War II, when alcohol was in short supply for explosives and synthetic rubber and there were severe dislocations in some traditional basic raw materials, wheat was diverted to the fermentation industry.

VI. COMPARISON OF SOFT RED AND SOFT WHITE, EASTERN AND WESTERN SOFT WHITE WHEATS

Many flour specifications in the eastern United States call for a given soft wheat flour to be milled strictly from "soft red winter" or "soft white winter" wheat, based on impressions that there are more or less profound differences between the two classes. There are indeed modest differences, but they tend to be exaggerated. In pie dough production and for high-spread cookies particularly, many think white wheat flour is superior. In fact, it is difficult to distinguish between red and white for these purposes and a high degree of interchangeability is possible. Soft red wheat flour is thought to be superior for cracker sponge and cake donut production, and there is undoubtedly a better excuse for this impression. For soup thickeners, soft red wheat flour is most often preferred, because of one of the true inherent differences between red and white wheats; this is the relative susceptibility to sprout damage, where there is a higher resistance, or postharvest dormancy, accompanying the red pigmentation in the bran of the soft red wheats (11). Even here, under favorable harvest conditions, white wheat produces perfectly satisfactory soup thickener flour.

Close laboratory examination will divulge these characteristic differences between soft red and soft white wheat flour in most cases: protein, MacMichael viscosity, farinograph absorption and curve strength, and hot paste viscosity (amylograph) tend to be greater in red than in white wheat flour. However, specific instances can be cited where exactly the reverse is true, which only emphasizes the substantial interchangeability. Soft white second clears tend to be lower in protein than the red counterpart, when wheat proteins are the same. Unbleached white wheat flour is usually creamier in color than similar grades milled from soft red wheat.

Red and white functional performances in the baking industry differ little. Laboratory performance tests of cookie spread and pie dough baking show virtually no characteristic differences. Both red and white wheat flours are accepted for home baking of baking powder biscuits. The greatest difference occurs only when a wet harvest season favors sprout damage; then white wheat tends to suffer more; but even then the flour user will be conscious of the problem only if the application is sensitive to low hot-paste viscosity. Some products such as cakes and cookies will be little affected. The greatest problem with flour milled from sprout-damaged wheat is in the making of soup thickeners. In 1968 and 1969 a problem arose with importers of western white wheats in Japan, the flour being used in making noodles (4) and sponge cakes. The highly unusual incidence of amylolytic activity caused "droopy" noodles and fallen cakes. The low moisture of western-grown wheats usually guards against sprout damage.

Besides the geographical gap between the soft white winter wheats of the east (Michigan and New York) and the western whites (Washington, Oregon, and Idaho), there are other significant differences. First, the western white wheats may be winter or spring, hard or soft, with soft winters predominating. A variety classified as soft sometimes develops vitreous tendencies under unfavorable weather during ripening. Second, the western white wheats fulfill for the west the functions of both reds and whites in the east. Third, up to 90% of western white wheat goes into export, chiefly to the Orient. Further, western white comes to market at distinctly lower moisture levels than those of the eastern soft wheats. Some of the western wheat is grown under irrigation, as in the Snake River Valley of Idaho and in Washington.

About 70 to 75% of the western soft white wheats are of the short-straw Gaines type varieties; about 20% is white club, a distinctive subclass noted for its soft texture and good quality for cakes and pastries. Club wheats are grown in Washington and Oregon.

As for functional qualities, the western whites produce excellent cake and pie dough flours, as well as molded-cone flours of superior quality. Western white wheats characteristically yield flours somewhat higher in ash than the corresponding grades of eastern soft wheats; they also react less in pH increment drop with a given application of chlorine gas in cake flour milling.

ACKNOWLEDGMENTS

We gratefully acknowledge the comments and technical assistance of Charles Ardinger, Golden Dipt Company; Richard Baum, Western Wheat Associates, U.S.A., Inc.; Samuel Berlin, Maryland Baking Company; David Downs (retired), Krispy Kreme Donut Corp.; Ralph Fishkin, Modern Maid Food Products, Inc.; Edward Glinski, Standard Milling Company; George Minor, Peavey Company; Chester Ofelt (deceased, formerly retired), Sterwin Chemicals, Inc.; Wilbur Schaefer, Northern Regional Research Laboratory, USDA; Donald Sundberg, Fisher Mills, Inc., and Leland Thomson (retired), Keebler Company.

We are also grateful to W. T. Yamazaki for arranging for the reproduction of some of the photographs at the Ohio Agricultural Research and Development Center and to The Mennel Milling Company for resources and use of facilities.

LITERATURE CITED

1. Ash, D. J., and Colmey, J. C. 1973. The role of pH in cake baking. Bakers Dig. 47(1):36-42, 68.
2. Baeuerlen, R., Brody, H., and Erickson, D. 1968. Frying fats and their uses. Bakers Dig. 42(6):51-55, 62-63.
3. Barackman, R. A. 1954. Chemical leavening agents. Trans. Am. Assoc. Cereal Chem. 12:43-45.
4. Bean, M. M., and Fullington, J. G. 1970. Suppression of excess alpha-amylase in sprout-damaged wheats. In: Report of Sixth Nat. Conf. Wheat Utilization Res., Nov. 5-7, 1969, (ARS 74-54), USDA, ARS, Berkeley, CA. pp. 49-61.
5. Borders, J. H. 1968. A look at foam cakes. Bakers Dig. 42(4):53-55, 66.
6. Bridges, C. H. 1976. Protective packaging for cereals. Cereal Foods World 21: 542-545.
7. Carden, J. B., and Meisner, D. F. 1970. Margarine for baking. Bakers Dig. 44(2):58-60.
8. Carlin, G. T. 1944. A microscopic study of the behavior of fats in cake batters. Cereal Chem. 21:189-199.
9. Code of Federal Regulations. April 1, 1976. Title 21, Parts 10-199. U.S. Government Printing Office, Washington, DC. p. 24.
10. Downs, D. E. 1971. Basic aspects of fried pie production. Bakers Dig. 45(3):62-65, 68.
11. Everson, E. H., and Hart, R. B. 1961. Varietal variation for dormancy in mature wheat. Mich. Agric. Exp. Stn. Q. Bull. 43(4):820-829.
12. Fisher, G. W. 1972. Technology of bulgur production. Bull. Assoc. Operative Millers. pp. 3300-3304.
13. Franta, R. 1976. Impact of ready-to-eat cereals on breakfast nutrition. Cereal Foods World 21:550-552.
14. Gravani, R. B. 1976. Breakfast cereals today. Cereal Foods World 21:528-530.
15. Harrel, C. G. 1959. Manufacture of prepared mixes. In: The Chemistry and Technology of Cereals as Food and Feed. S. A. Matz, ed. Avi Publishing Company, Westport, Ct. pp. 321-368.
16. Johnson, N. F. 1973. Continuous cake production using hydrated surfactants. Bakers Dig. 47(4):30-33, 72.
17. Kichline, T. R., and Conn, J. F. 1970. Some fundamental aspects of leavening agents. Bakers Dig. 44(4):36-40.
18. Kinnison, J. W. 1974. Color additives—Current status. Cereal Sci. Today 19:308-318.
19. Knightly, W. H., and Lynch, M. J. 1966.

The role of surfactants in baked foods. Bakers Dig. 40(1):28-31.

20. Kulp, K. 1972. Some effects of chlorine treatment of soft wheat flour. Bakers Dig. 46(3):26-32.
21. Loving, H. J. 1962. Cake flour—Its quality requirements. Bakers Dig. 36(1):64-66, 78.
22. Martin, S. 1976. Quick Frozen Foods. Sam Martin Company, Publisher & Editor, 757 Third Ave., New York.
23. Matz, S. A. 1959. Manufacture of breakfast cereals. In: The Chemistry and Technology of Cereals as Food and Feed. S. A. Matz, ed. Avi Publishing Company, Westport, CT. pp. 547-566.
24. Matz, S. A. 1968. Base cakes and plain cookies. In: Cookie and Cracker Technology. S. A. Matz, ed. Avi Publishing Company, Westport, CT. pp. 119-136.
25. Matz, S. A. 1968. Fermented dough products. In: Cookie and Cracker Technology. S. A. Matz, ed. Avi Publishing Company, Westport, CT. pp. 137-149.
26. Matz, S. A. 1968. Coatings, fillings, cremes, and marshallow. In: Cookie and Cracker Technology. S. A. Matz, ed. Avi Publishing Company, Westport, CT. pp. 163-179.
27. McWhinney, G. H. 1966. Pasteurized eggs. Bakers Dig. 40(1):53-56.
28. Mendelsohn, S. 1939. Baking Powders. Chemical Publishing Company, New York.
29. Mertz, L. C., and Nordstrom, K. L. June 21, 1960. Flour milling process. U.S. Patent 2,941,730.
30. Micka, J. 1955. Bacterial aspects of soda cracker fermentation. Cereal Chem. 21:125-131.
31. Miller, B. S., Trimbo, H. B., and Powell, K. R. 1967. Effects of flour granulation and starch damage on the cake making quality of soft wheat flour. Cereal Sci. Today 12:245-247, 250-252.
32. Miller, B. S., and Trimbo, H. B. 1970. Factors affecting the quality of pie dough and pie crust. Bakers Dig. 44(1):46-55.
33. Miller, D. L. 1974. Industrial uses of wheat and flour. In: Wheat—Production and Utilization. G. E. Inglett, ed. Avi Publishing Company, Westport, CT, pp. 398-411.
34. Myhre, D. V. 1970. The function of carbohydrates in baking. Bakers Dig. 44(3):32-35, 38-39, 60.
35. Nelson, C. A., and Loving, H. J. 1963. Mill-stream analysis—Its importance in milling special flours. Cereal Sci. Today 8:301-304, 326.
36. Pieper, W. E. 1968. The role of ingredients in wire-cut and deposited cookies. In: Proc. 44th Ann. Meeting, Am. Soc. Bakery Eng. pp. 265-268.
37. Pyler, E. J. 1973. Sweetening agents. In: Baking Science and Technology, Vol. I. Siebel Publishing Company, Chicago. pp. 396-434.
38. Pyler, E. J. 1973. Shortening products. In: Baking Science and Technology, Vol. I. Siebel Publishing Company, Chicago. pp. 435-486.
39. Pyler, E. J. 1973. Milk and milk products. In: Baking Science and Technology, Vol. I. Siebel Publishing Company, Chicago. pp. 487-513.
40. Pyler, E. J. 1973. Eggs and egg products. In: Baking Science and Technology, Vol. I. Siebel Publishing Company, Chicago. pp. 514-545.
41. Pyler, E. J. 1973. Cake ingredients. In: Baking Science and Technology, Vol. II. Siebel Publishing Company, Chicago. pp. 898-939.
42. Pyler, E. J. 1973. Flavor products. In: Baking Science and Technology, Vol. II. Siebel Publishing Company, Chicago. pp. 940-965.
43. Pyler, E. J. 1973. Cake baking technology. In: Baking Science and Technology, Vol. II. Siebel Publishing Company, Chicago. pp. 966-1002.
44. Pyler, E. J. 1973. Miscellaneous bakery products. In: Baking Science and Technology, Vol. II. Siebel Publishing Company, Chicago. pp. 1003-1038.
45. Reisman, H. 1970. Modern methods of pretzel production. In: Snack Foods Blue Book. Gary L. Larson, ed. Harcourt Brace Javanovich, New York. pp. 115-117.
46. Roth, H., and Rock, S. P. 1972. The chemistry and technology of frying fat. I. Chemistry. Bakers Dig. 46(4):38-45, 66-67.
47. Rumsey, L. A. 1959. Commercial baking procedures. In: The Chemistry and Technology of Cereals as Food and Feed. S. A. Matz, ed. Avi Publishing Company, Westport, CT. pp. 242-273.
48. Schanefelt, R. V. 1977. Corn sweeteners; products and properties. Cereal Foods World 22:44-47.
49. Sosland, M. I. March 23, 1976. Strong baked foods market analyzed. Milling & Baking News 55(5):7,33-35.
50. Stephens, C. 1972. Automation in modern pie making operations. Bakers Dig. 46(4):56-59.
51. Tanilli, V. H. 1976. Characteristics of

wheat and flour for cookie and cracker production. Cereal Foods World 21:624-628,644.

52. Thompson, J. J., and Raymer, M. M. May 27, 1958. Production of ready-to-eat composite flaked cereal products. U.S. Patent 2,835,495.
53. Thomson, L. S. 1976. Flour needs for the commercial cracker process. Cereal Foods World 21:642-644.
54. Trimbo, H. B., Ma Shao-Mu, and Miller, B. S. 1966. Batter flow and ring formation in cake baking. Bakers Dig. 40(1):40-45.
55. Trimbo, H. B., and Miller, B. S. 1973. The development of tunnels in cakes. Bakers Dig. 47(4):24-26,71.
56. Wichser, F. W. 1958. Air-classified flour fractions. Cereal Sci. Today 3:123-126.
57. Yamazaki, W. T. 1969. Soft wheat flour evaluation. Bakers Dig. 43(1):30-32,63.
58. Yamazaki, W. T., and Lord, D. D. 1971. Soft wheat products. In: Wheat Chemistry and Technology, 2nd ed. Y. Pomeranz, ed. Am. Assoc. Cereal Chem., St. Paul, MN. pp. 743-776.
59. Yamazaki, W. T., and Kissell, L. T. 1976. Soft wheat: Source of pastry flours. Cereal Foods World 21:630-632, 650.
60. Zabik, M. E. 1968. Foam cakes prepared with dried eggs. Bakers Dig. 42(6):32-35.

CHAPTER 6

SOFT WHEAT USES IN EUROPE

C. T. GREENWOOD[1]
University of the South Pacific
Suva, Fiji

R. GUINET
Ecole de Boulangerie et de Pâtisserie de Paris
Paris, France

W. SEIBEL
Institute of Baking Technology
Federal Research Center of Grain and Potato Processing
Detmold, Federal Republic of Germany

PART ONE: SOFT WHEAT USES IN THE UNITED KINGDOM

I. BAKED PRODUCTS IN THE UNITED KINGDOM

In the United Kingdom (U.K.), the grower tries to produce the maximum quantity of wheat without reference to its ultimate use: varieties are judged according to their ability to yield reliably and resist diseases. Legally in the U.K., any one of about 60 wheat cultivars can be grown,[2] although the number is restricted to about 30. Agricultural husbandry in the U.K. entails the wheat crop being grown in about 150,000 fields (each an average of 17 acres), and these "packages" of wheat vary in baking quality from the hard North American type to the very soft type. A voluntary wheat grading system was introduced by the Home Grown Cereals Authority to encourage farmers to produce wheat of the best baking quality (32).

In terms of baked products, the U.K. is a wheat-importing country; it produces

[1]Formerly, Flour Milling and Baking Research Association, Chorleywood, England.

[2]A list issued by the Ministry of Agriculture, Fisheries and Food, in accordance with Seeds Regulations of the Plant Varieties & Seeds Act 1964 and amended by the European Communities Act 1972.

an average 5 million tonnes (te) of wheat, but the flour milling industry uses about 5.1 million te per year. Milling products can be broadly classified into four types, i.e., bread flour, confectionery flour, biscuit flour, and domestic culinary flour. Table I shows the estimated annual demand for these types, and the estimated wheat requirements for homegrown wheat and imported hard and soft wheat (32). Soft wheat is used almost exclusively in the preparation of confectionery and biscuit flours, but increasing amounts are now included in the bread grist.

Baked goods still make an important contribution to nutrition in the U.K. The National Food Survey for 1977 showed that bread, flour, and other cereals provided an average of about 18% of the calories, 19% of the protein, 33% of the carbohydrates, 2% of the fat, 19% of the calcium, 27% of the iron, 30% vitamin B_1, and 17% of the nicotinic acid consumed per person per day.

II. BREAD MAKING

A. Introduction

Before 1900, bread was made by hand by master bakers and sold from shops attached to the bakery. Since then, plant bakeries have increased dramatically which is related to technological advances in production and introduction of mechanized wrapping and slicing of bread, which occurred when the market widened as better transport developed. The new plant bakeries built up their own chains of retail shops, which remained essentially local until World War II. After this, U.K. companies in the bread-making industry amalgamated, and currently two large groups produce about half the bread in the country. A few independent plant bakeries and about 7,000 small bakers make the rest of the country's bread.

The large companies operate highly mechanized bakeries (39), usually situated near centers of population; the other bakeries are distributed throughout the country. The independent plant bakers distribute their products in relatively small areas, and the small bakers usually have only a bakery behind the retail shop. Modern bread-making practice uses carefully controlled raw materials and high-output processing equipment and methods.

In the 1977-1978 wheat crop season, about 2.5 million te of wheat flour was used for bread making in the United Kingdom.

The industry thus consists of two types of bakeries:

TABLE I
Estimated Current Annual Wheat and Flour Requirements of the United Kingdom (in million tonnes)

	Flour Production	Wheat Equivalents	Wheat Sources: Homegrown	Wheat Sources: Imported Soft	Wheat Sources: Imported Hard
Bread	2.48	3.35	1.25	0.90	1.20
Biscuit	0.49	0.66	0.63	0.03	0
Confectionery	0.07	0.10	0.36 (Confectionery and Domestic)	0.14 (Confectionery and Domestic)	0.10 (Confectionery and Domestic)
Domestic	0.37	0.50			
Total	3.41	4.51	2.24	1.07	1.30

• The highly mechanized plant bakery principally produces sliced and wrapped loaves and a wide range of crusty, fancy, and proprietary breads. The bread is distributed and sold over a wide area through the bakeries' shops and retail vans and through wholesale outlets such as supermarkets, grocery, and small general stores.

• The small to medium-sized, family-type bakery primarily produces bread, crusty type bread and rolls, probably by labor-intensive methods, for its own sales outlets.

The degree of mechanization in a bakery varies with the bread output, but in recent years all types of bakeries have tended to use more semiautomatic equipment.

Output of bread bakeries is described in terms of the number of sacks of flour used; for historic reasons, a sack of flour is 127 kg (280 lb). Mixing machines are designed to mix a sack or a portion of a sack of flour into dough. Bread-making plants are classed as having a throughput of a certain number of sacks of flour per hour, each sack producing 220 to 230 loaves of bread. A large plant bakery may have a weekly output in excess of 2,000 sacks; that of the family-type bakery may be only 2 to 100 sacks.

The net weight of bread sold in the United Kingdom is controlled legally by the Weights and Measures Acts, and a whole loaf must weigh less than 300 g, 400 g, or multiples of 400 g. The 800-g loaf has the highest volume of sales.

B. Essentials of Bread Making

Figure 1 shows the order of the various processing operations in bread making.

DOUGH DEVELOPMENT PROCESSES

In the United Kingdom, several dough development processes are in common use, but the two main ones are bulk fermentation and the Chorleywood bread process (CBP). Bulk fermentation was used to produce nearly all U.K. bread until the introduction of the CBP in 1961 (3).[3] Currently, the CBP is the established method in plant bakeries and is used in about 85% of bread production in the country.

In bulk fermentation, the basic bread ingredients—flour, yeast, salt, and water—are mixed in a low-speed mixer fairly gently for 20 to 30 min to produce a smooth dough. The total work expended on the dough by the mixer is about 8 kJ/kg. After this mixing, the dough is resistant to stretching and is relatively inelastic with poor gas retention properties, and so it is set aside for 3 to 4 hr to ferment in bulk.

The physicochemical changes that occur involve the following sequence:

When water is first added to flour to make the dough, the proteins hydrate and swell. Forces holding the proteins coiled begin to weaken, and a series of sulfydryl-disulfide interchange reactions are initiated. When one disulfide group breaks, one half forms a new sulfhydryl group, and the other forms a new disulfide bond to relax the stress.

[3] Developed by the British Baking Industries Research Association, the forerunner to the Flour Milling and Baking Research Association at Chorleywood.

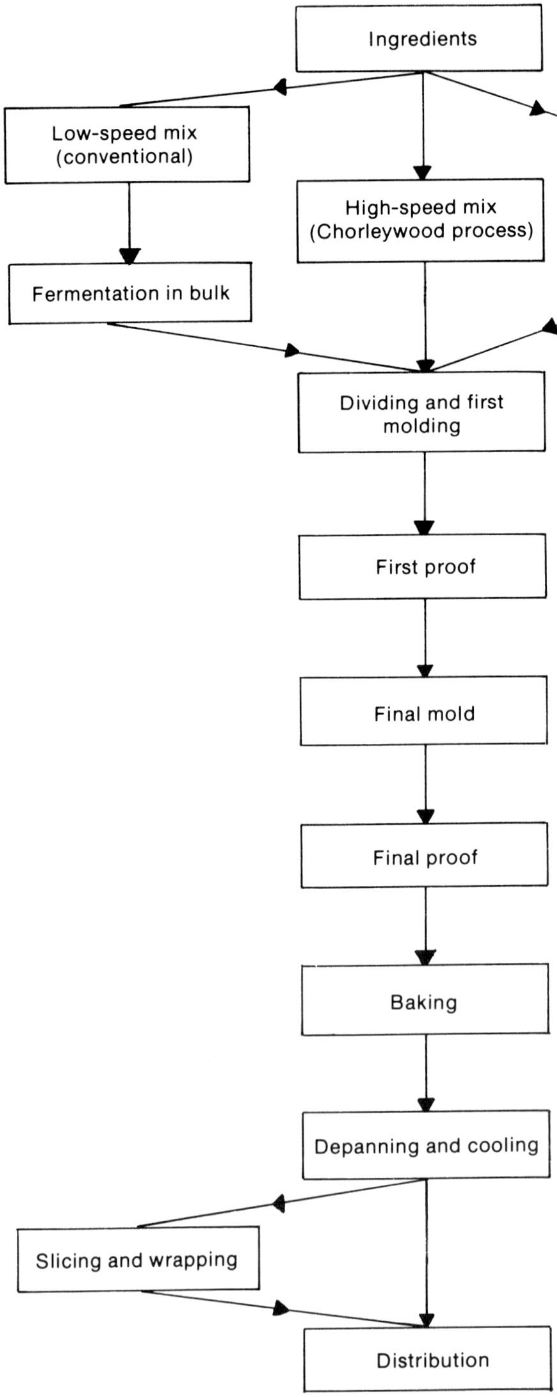

Figure 1. Flow diagram of the order of processing in bread production.

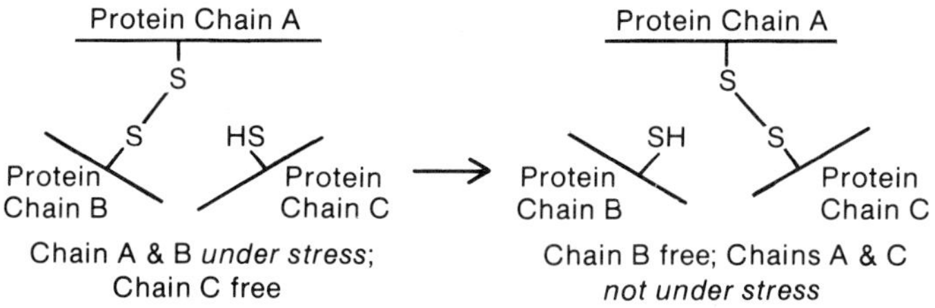

Two adjacent sulfydryl groups may also link oxidatively to form a disulfide bond, i.e.,

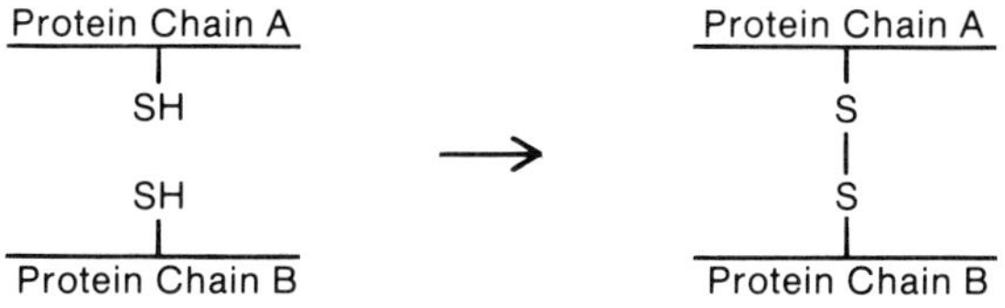

At the same time, water absorbed by damaged starch granules enables these to swell, and they help build a three-dimensional structure, i.e., an essentially protein matrix in which starch granules of varying degrees of hydration are held by secondary forces (30). Again, this structure does not exhibit true elasticity, because many of the protein chains are still capable of sliding past each other when stress is applied, but during the bulk fermentation, gas produced by the yeast stretches the protein matrix and the chance interaction of reactive groups allows the cross-linking reactions to occur more readily, thus enabling a three-dimensional protein matrix to be built. This process is aided by proteinases from the yeast acting on the gluten. Such a protein matrix exhibits true elasticity (31), and this "ripening" of the dough can be achieved with proper control of yeast levels and dough temperature.

These chemical changes in the protein can be aided by the addition of chemical oxidizing and reducing agents. In bulk fermentation, low levels of an oxidant, such as potassium bromate at 10 to 20 ppm, are commonly added.

In contrast, the Chorleywood process enables the dough to be fully developed and ready for processing immediately after mixing. The essential feature of this process is that dough is mixed in a specially designed, high-speed mixer that imparts to the dough a work input of 40 kJ/kg in the presence of an oxidizing agent (e.g., ascorbic acid,[4] 75 ppm) or an ascorbic acid-potassium bromate mixture (75 ppm). Under these conditions, dough development is complete less than 5 min after mixing.

The formation of the three-dimensional elastic gluten matrix in this mechanical dough development occurs because the intense mixing and the chemical oxidant accelerate the essential sulfhydryl-disulfide interchange reactions (31).

The CBP lends itself to mechanized production (Fig. 2). It saves time and

[4]Ascorbic acid acts as an oxidizing agent because it is transformed into dehydro-ascorbic acid by the ascorbic acid oxidase present in the wheat flour. The fate of this additive (61) and potassium bromate (62) has been determined in bread making.

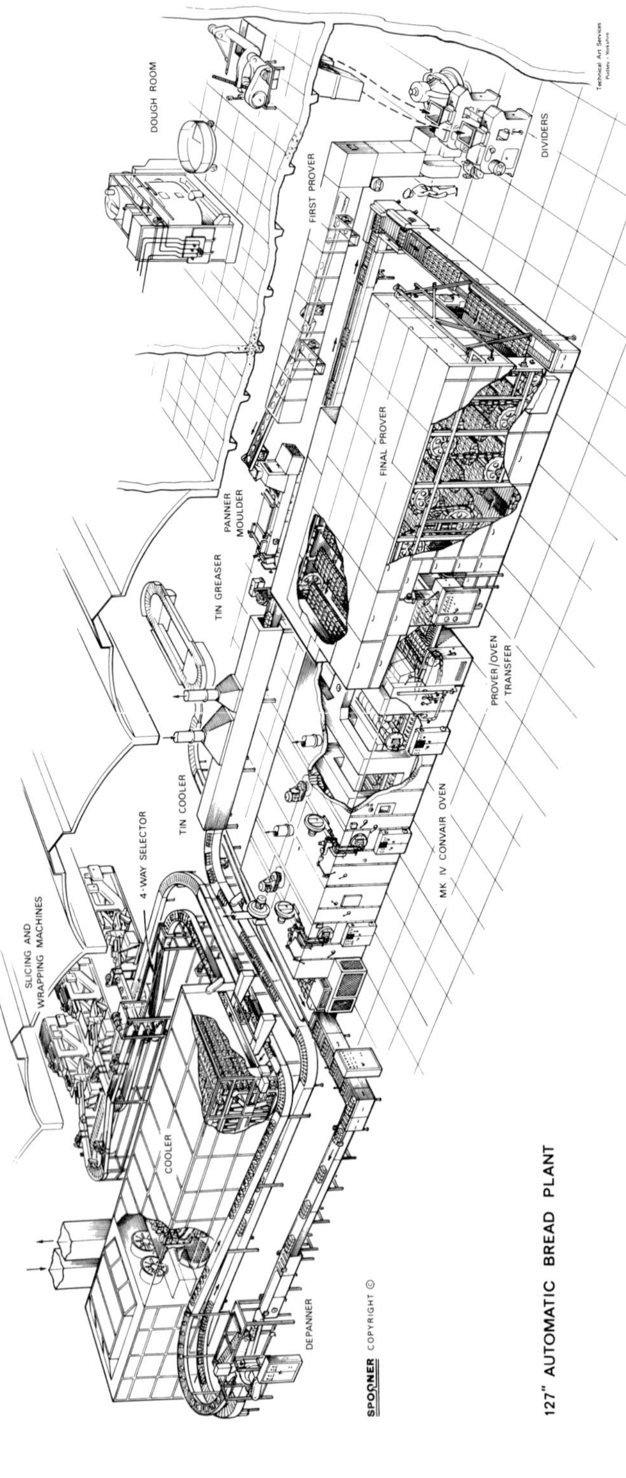

Figure 2. Diagram of an automatic bakery plant. (Courtesy of Spooner Food Machinery Engineering Co. Ltd.)

space; the process is simpler than bulk fermentation; the denser, ungassed dough can be divided more accurately; the yield of bread is higher; and a very high proportion of soft wheat can be incorporated into the grist.

Extra water (3.5% based on flour weight) must be added to the mix to compensate for the negligible fermentation losses. Fat (approximately 0.7%) must also be included in the recipe, and its melting point should be higher than the final dough temperature to ensure that at the end of the final proof period, a proportion of unmelted fat remains in the dough. The function of this fat is not fully understood (5,6,25,27). Extra yeast must also be included to ensure a good gassing rate during final proof, and to give a final proof time of 40 to 50 min.

The high-speed mixer in general use is of the batch type (Fig. 3) where weighing of flour and water into the mixer is automatic, and mixing stops when the required work input has been applied to the dough. Chilled water is generally necessary to maintain the final dough temperature at about 30° C. Because bread of a finer texture is attained when the amount of air taken up by the dough during mixing is reduced, most high-speed mixing machines work the doughs under a pressure of 305 mm of mercury.

Another related and more recently introduced process is that of activated dough development, which allows the dough to be fully developed and ready for further processing about 10 min after completion of mixing in a low-speed mixer. The process depends on ripening the gluten network by first chemically reducing disulfide bonds and then using a similar concentration of oxidizing agent to stabilize the network: L-cystine hydrochloride (35 ppm) is used as the reducing agent and a mixture of potassium bromate and ascorbic acid (70 mmp) is the oxidizing agent. For the same reasons as in the CBP, the recipe must contain a high-melting fat and extra water and yeast.

DIVIDING AND FIRST MOLD

The developed dough is then passed through dough-dividing machines that generally work on a "volume displacement" principle. For a 400-g loaf, about 900 g of dough is used to allow for moisture loss during baking and cooling. After dividing, the dough pieces are formed into a sphere by a first molder.

FIRST OR INTERMEDIATE PROOF

After the first mold, dough pieces are allowed to rest for 6 to 16 min which allows recovery of the elastic tensions built up during dividing and first molding and production of carbon dioxide (24).

In the CBP, further oxidation of the dough protein occurs during this period.

FINAL MOLDING

The dough piece is formed into the required shape, i.e., a cylinder used for all bread baked in rectangular tins, by reducing the sheet from the intermediate proof to a thickness of 6 to 13 mm by passing through rollers. It is then formed into a "Swiss roll" cylinder and passed under a pressure plate to reduce its diameter and length to a required dimension.

SECOND OR FINAL PROOF

This proofing allows the gradual evolution of carbon dioxide from yeast fermentation within the final dough piece structure (24). To achieve a steady fermentation rate, the temperature in the final proofer is controlled (35 to 43 °C), and the humidity is maintained at 70 to 85%. The normal final proofing time for most loaves is 45 to 60 min.

BAKING

As heat is conducted through the dough piece during baking, the fermentation rate initially increases with consequent rapid expansion of loaf volume known as

Figure 3. Typical high-speed batch mixer for the Chorleywood bread process. Dough piece being discharged into bowl. Automatic hopper is above mixer; control panel with work time meter on right. (Courtesy of the Flour Milling and Baking Research Association at Chorleywood)

"oven spring." As the internal dough temperature rises further, however, yeast activity decreases at about 43°C and the yeast is killed at 54°C. As the temperature rises further, starch granules begin to gelatinize at about 66°C, enabling the amylolytic, α- and β-amylases to react to form maltose and dextrins. This enzyme activity continues until temperatures of 74 to 77°C are reached. The gluten network begins to coagulate at 74°C, a transformation that continues until the end of baking when an internal temperature of 93 to 100°C is reached.

Water and alcohol vapors are driven off during baking, and consequently the internal temperature of a loaf only just reaches 100°C. In contrast, the outside of the loaf reaches the full oven temperature, and as a greater amount of moisture is driven off, the protein coagulates to form a crust. The thickness of the crust varies with baking temperature and time. At the elevated temperatures of the crust, complex chemical reactions take place that are responsible for the color and flavor; browning effects are thought to result from caramelization of the sugars present as products of amylolytic enzymes and from the interactions between these sugars and proteins. The ovens used in bread production are principally of the traveling tunnel and setter types.

Traveling ovens consist of a horizontal conveyor moving through an insulated baking chamber in which the dough pieces in tins or on trays are loaded at one end, and the loaves are baked as they travel through to the other end. In the case of crusty bread, the dough pieces are loaded directly onto the traveling belts. These ovens can be gas- or oil-fired, and the method of heat transfer is usually a combination of radiation, forced convection, and conduction (38).

In setter ovens, the bread is baked directly on a fixed base or sole (varying in area from 0.9 to 9 m^2). The oven may have several decks of baking chambers situated above each other; each chamber has a door the full width of the sole, which allows special "setters" containing the fully proofed bread to be pushed into the oven, and then these are retracted to leave the dough pieces on the oven sole. Bread baked in this way is called "oven-bottom bread" and includes crusty types such as Vienna, bloomers, French sticks, and cottage loaves. Bread is often unloaded from these ovens with the aid of a wooden peel.

Steam is normally introduced into the oven during the first few minutes of baking. The effect of steam on the cool dough pieces entering the oven is to cause the starch on the surface of the dough pieces to gelatinize and so produce a pleasing glaze on the finished product. This effect also retards the drying and setting of the crust, which permits the dough piece to expand fully. The steam is then drawn off, and baking is completed in a relatively dry atmosphere.

Most types of bread require baking times between 25 and 45 min at a temperature of 220 to 250°C. In general, crusty bread is baked longer than bread that is to be sliced. An interesting exception is Scotch batch bread, which is produced by placing dough pieces directly side by side on the oven sole or on the traveling conveyor with each piece supporting its neighbor. After baking at 193 to 295°C for 75 to 90 min, the bread is tall, regular, with thick top and bottom crusts and no side crust (Fig. 4).

COOLING, SLICING, AND WRAPPING.

Bread must be cooled until the internal temperature falls to 27°C under controlled conditions of temperature and humidity, to ensure that the crumb structure is stable during slicing (23). Most bread-slicing machinery in the United Kingdom consists of a horizontal conveyor that carries the load into a vertical slicing head set at right angles to the conveyor. The slicing head consists of two frames carrying scalloped-edge blades. The two frames enmesh so that the blade of one frame is adjacent to that of the other, 6 to 9 mm apart, depending on the slice thickness. Slicing occurs by each frame moving rapidly up and down in the opposite direction to the other. Most bread is wrapped in waxed paper, although thermoplastic bags are also popular. Some wraps are heat-sealed, others enclosed by a twist tie. Perforated polypropylene wrapping with high moisture permeability to ensure the crispness of the crust is used for wrapping crusty type breads.

C. Properties of Bread

The criteria of good bread may be taken as: satisfactory external appearance and crust color; adequate volume; good crumb color; softness of crumb; fine, regular crumb-cell structure; and long shelf life.

The constituents responsible for the flavor of bread are not fully understood. Crumb of the ordinary white U.K. loaf is bland, and consumer tasting panels have determined that the flavor of crumb from the center portion of such a loaf is independent of the way in which the loaf was prepared or baked (21). The flavor

Figure 4. Scottish batch bread unloaded from traveling oven. (Courtesy of the Flour Milling and Baking Research Association at Chorleywood)

of white bread arises from flavor constituents in the crust, diffusing into the center of the loaf. For example, any difference in flavor between a crusty white loaf and a sliced wrapped white loaf arises solely from differences in thickness of the crust and the corresponding increased amount of flavor constituents in that crust.

Enough salt is in the loaf to add to the flavor; bread made without salt is regarded as insipid. Fermentation of the dough contributes little to the flavor of the bread, as the fermenting products are volatile and are lost during baking (21). In wheat-meal bread, the bran may influence the flavor, whereas in whole-meal loaves, there is the additional contribution of the wheat germ.

DETERIORATION OF EATING QUALITY

Bread deteriorates in storage by staling and microbiological spoilage. The phenomenon of staling is complex and is not yet understood (22,70,71). Staling can be defined as progressive firming of the crumb as the loaf ages (41), and it occurs even when moisture loss from the loaf is inhibited. Staling is reversible to some extent, for a stale loaf may be freshened up by reheating. It has been suggested that staling is related to recrystallization of the starch material present. Another factor is the specific volume (4); a loaf with a high specific volume stales at a slower rate than one of lower specific volume (the specific volume is related to the protein level of the flour; the lower the protein level, the lower the specific volume). The rate of staling depends on temperature and can be arrested by freezing, i.e., bread can be kept fresh by storage in a deep-freeze at −20 C.

Mold can grow in bread as a result of postbaking contamination. Such mold problems occur mainly in sliced and wrapped bread where the normal mold-free shelf life is four to five days at a temperature of 21° C.

D. Bread Varieties

Bread is produced in great variety in the United Kingdom, and loaves vary in size and shape, specific volume, recipe formulation, and degree of baking, as well as whether they are sold sliced or unsliced, wrapped or unwrapped (Fig. 5). A legal definition of bread varieties is given in the current Government Bread and Flour Regulations.

White bread, the most widely sold variety, is made with flour of 72% extraction rate, supplemented by the legally required nutritive additives. All U.K. flours must contain at least 0.24 mg of vitamin B, 1.6 mg of nicotinic acid, and 1.65 mg of iron per 100 g of flour. Except for whole meal, wheat meal flour, and some self-rising flours, all flour is enriched with calcium carbonate at no less than 235 mg and no more than 390 mg per 100 g of flour.

Soft white bread with a thin crust arising from a light baking, is sold sliced and wrapped in an impermeable material, whereas the more highly baked crusty white breads have a thick, crisp crust and are sold unwrapped or wrapped in a high-moisture-permeable material. Whole-meal bread is made from whole-meal flour and is generally sold as a 400-g loaf; wheat-meal bread is made from 80 to 90% extraction flour; wheat-germ bread is made from wheat meal or white flour with 10 to 25% of wheat germ added (wheat-germ flour and bread are supplied under proprietary names); high-protein bread must contain 22% d.m.w. of protein—obtained by addition of dried gluten—to legally qualify for this

Figure 5. Breads made in the United Kingdom: (Courtesy of Flour Advisory Bureau) **1,** Malt loaf—made with wheat-meal flour. **2,** Split tin—top surface of dough is slashed down the middle before baking. **3,** Cottage loaf—usually handmolded and baked on the oven sole. **4,** Cholla—rich plaited Jewish bread made from white dough enriched with eggs, sugar, and fat; its glazed crust is normally sprinkled with poppy seeds. **5,** Wheat-germ loaf. **6,** Poppy seed plait—made from a plain white dough, decorated with poppy seeds, and normally baked on the oven sole. **7,** Coburg—a crusty white loaf baked on the oven sole or on flat trays; dough's top surface is slashed twice to form a cross before baking. **8,** Fruit and malt—a malt loaf baked in a tin that contains fruit, normally sultans. **9,** Wheat meal tin loaf—may be sold sliced and wrapped. **10,** French sticks. **11,** Danish loaf—crusty white loaf baked on the oven sole; top is slashed lengthwise and dusted with flour before baking. **12,** Standard sliced tin loaf. **13,** Farmhouse loaf—white loaf baked in a tin, slashed lengthwise, and sometimes floured before baking. **14,** Bloomer—dough is slashed in a decorative fashion several times before baking. **15,** Whole-meal cob. **16,** Bap—small white loaf with soft crust and floured top. **17,** Milk loaf—high protein breads are also baked in this distinctive shape. **18,** Rolls—wide variety of types, brown or white, crusty or soft. **19,** Fruit loaf.

definition: gluten breads are similar but must contain at least 16% protein d.m.w.; starch-reduced bread must contain less than 50% of carbohydrate, which is achieved by adding an equal weight of gluten to the flour; milk bread is made with flour containing no less than 6% of whole milk solids; malt bread is a sticky, sweet, dark loaf containing malt and can be made from white, wheat meal, or whole-meal flour and can be plain or fruited; rye bread is made from a mixture of rye and white flour.

III. FLOUR CONFECTIONERY

A. Introduction

In 1977, 70,500 te of confectionery flour was produced in the United Kingdom. The industry is dominated by about six companies, that together account for some 60% of the industry's production. These large units produce a small number of nationally sold lines, using highly mechanized plants with relatively few workers. The remaining flour confectionery production is provided by small, independent companies and the owner-operated bread bakeries, that supply their own shops with a large variety of goods produced by a highly skilled labor force with little mechanization.

The list of U.K. cakes and teabreads is very large; for example, there are Banbury cakes, Chelsea buns, Dundee cake, muffins, and crumpets. Cakes are often distinct to certain regions, with large variations of flour confectionery among them. There is also a regional variation in the consumption: More cakes are eaten in Scotland than in England.

B. Ingredients

The specific characteristics of confectionery products—structure, texture, flavor, and appearance—are influenced by the ingredients in the recipe. Typical confectionery product recipes are shown in Fig. 6.

Strong (hard wheat) flour is used for sweet fermented products (doughnuts, buns, Danish pastry, and puff pastry), where the gluten network holds the carbon dioxide gas produced by fermentation in sweet fermented doughs, and holds the steam formed between the laminations in puff pastry.

Soft flour is used for rich fruit cake, scones, shortcakes, and short sweet pastry for tarts. Chlorinated soft flour is used in sponges, Madeira cake, sultana cake, Swiss rolls, and angel cake, where the recipe includes high levels of liquid and sugar (29).

Many other ingredients are included in flour confectionery recipes. Eggs, for example, contribute to the structure of the product by coagulation of their protein during baking and to the lightness and volume by their ability to absorb air when mechanically beaten; they also provide moistness, flavor, and nutritional value. Separated egg white (albumen) is used for meringues and macaroons. Bakers use deep-frozen (−20° C) eggs, defrosted before use; chilled liquid (0° C) eggs stored in refrigerated containers and used within 48 hr of delivery; or freeze-dried eggs, which have a long storage life and are reconstituted with water, as required. Milk enriches baked products such as scones or custards and can be added as a spray-dried solid to the ingredient mixture or reconstituted before use.

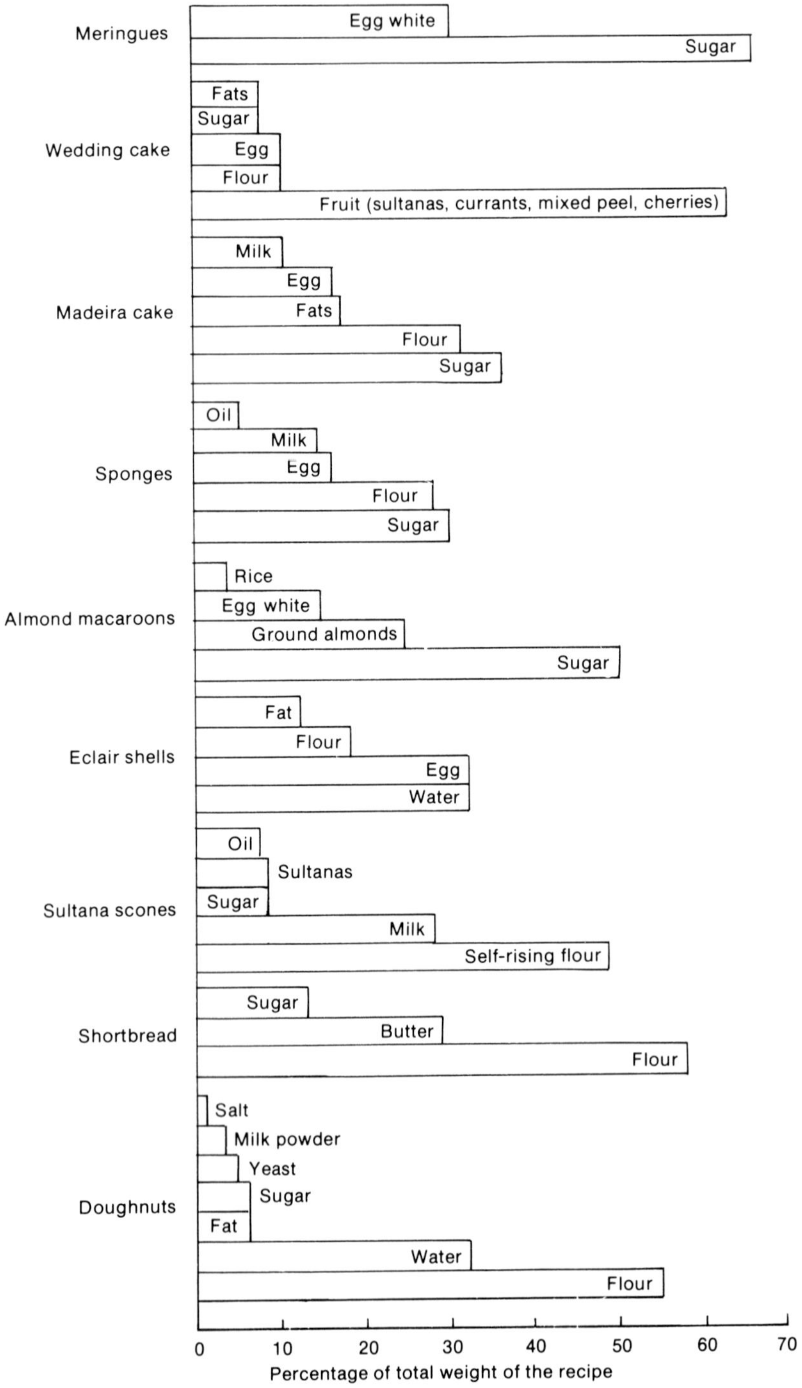

Figure 6. Main ingredients of flour confectionery recipes.

Fats add shortness to cake crumb and pastry and also contribute to the volume of cakes by absorbing air during mechanical beating. Butter, which contributes greatly to the flavor, has only average creaming properties; margarine, which also adds flavor, has good creaming properties; and white compound fat, which is neutral in flavor, has good creaming properties in cake and good shortening properties in pie pastry. Pastry margarine (a high-melting fat, 32° C) is used in the production of puff pastry. Sugar sweetens and adds to the crust color of the baked product. Icing sugar (very finely ground sugar) is used for icings and the top decoration of cakes. Emulsifiers and stabilizing agents (for example, partial polyglycerol and partial glycerol esters) are often used because they greatly increase the speed at which air is incorporated into a sponge or other batter during mixing, and thus they reduce the mixing time.

C. Aeration

The light cellular structure of most confectionery products is achieved with aeration by mechanical, chemical, or fermentation techniques.

In the mechanical method, air is incorporated into the batter by beating or whisking. In a typical machine, the mixing tool rotates around its own axis, and the drive shaft simultaneously carries it in a circle around the bowl to ensure thorough mixing. For sponge production (34–36), aeration can be either by the egg-sponge or all-in method. In the first technique, only the egg and sugar are mixed together until the correct volume is reached, before the other ingredients are slowly mixed in. In the all-in method, all the ingredients are mixed at the same time. For cake mixtures, a sugar/fat, or a flour/fat mixture is often aerated first before egg and other ingredients are mixed in. All-in methods require some alterations in recipe, i.e., baking powder and emulsifying agent. There are also pressure whisk mixers available for sponge batters in which a compressor builds up the air pressure in the enclosed mixing chamber with a rotating whisk in the center; mixing times are reduced and distribution of air cells in the batter is claimed to be more uniform. High-speed mixers similar to those used for the Chorleywood process are also available. In large bakeries, with high outputs of cake and sponges, a continuous cake mixer is likely to be used. This equipment has a holding tank for a premixed slurry of all ingredients; a pump to transfer the slurry to a mixing head; an air supply to the mixing head; and the mixing head, which consists of the rear stator, the rotor, and the front stator. The internal faces of the stator and both faces of the rotor are filled with concentric rows of blades, and the rotor revolves between the two stators. Ingredients are pumped into the center of the rear stator and flow between the blades and this stator and the rotor to the outside of the mixing head. This effective shearing action also evenly distributes air that is being pumped into the mixing head. The rate of airflow is adjusted to give a mixed batter of the required specific volume.

Chemical aeration with baking powder is used in products such as scones, doughnuts, and rock cakes. Yeast fermentation is used for goods such as tea cakes, doughnuts, fruit buns, and hot cross buns. A bulk fermentation process may be used in which—after shaping—the dough pieces are placed in a temperature- and humidity-controlled environment for 40 to 50 min of proofing.

In all flour confectionery products, the evolution of steam during baking is probably one of the most important contributing factors for aeration. But in the production of puff or flaky pastry used for apple and jam turnovers, vol-au-vent, sausage rolls, custard, and vanilla slices, steam is the sole raising agent. The principle is to produce a dough structure with alternate horizontal layers of pastry margarine and dough. During baking, steam is produced within the layers of dough and so pushes them apart, while the fat layer melts and lubricates the layers of dough so that they do not stick together. One technique to produce this lamination involves passing the dough of flour, salt, fat, and water through a roller to reduce the thickness and to form a rectangular shape, two thirds of which is then covered with sheets of pastry margarine of the same thicknes. The uncovered pastry is folded over to the center of the covered pastry, and then folded again to form a lamination of three layers of dough and two of margarine. The dough product is passed through the reducing rollers, folded in three, rested for 20 to 30 min before the whole procedure is repeated another four times before baking.

The lamination method is used in production of Danish pastry, but there are fewer laminations, and the dough also contains yeast.

D. Flour Confectionery Manufacture

The wide variety of confectionery products necessitates a correspondingly large number of processing methods and equipment. A typical flow diagram of the order of processing in large flour confectionery bakery is shown in Fig. 7.

A wide range of depositors is available for placing cake batters and sponges directly into baking containers, and meringues, eclair shells, sponge fingers, or Japanese biscuits in their required shapes directly on the baking trays. In smaller bakeries, these operations are done manually.

Sweet fermented dough can be divided and molded by a variety of equipment. One type divides the dough by volume displacement, and the divided pieces are placed in inverted cups set on a frame. The frame rotates, and the rolling action produces a round dough shape inside the cups. The dough pieces are either placed straight onto baking sheets for round buns or given a rest period before being molded into a special shape. They then pass into the final proofer (40 to 50 min) before baking or, in the case of doughnuts, before frying in a 182° C oil bath. Alternatively, the molded dough pieces may be placed in a dough retarder (refrigerator at 2 to 4° C), to stop fermentation for up to 48 hr. This method has advantages for small bakers, who can prepare goods in the afternoon that will then be immediately ready for proofing and baking the following morning.

Where the sales volume of products is very high, for example, Swiss rolls, angel cake, and fruit pies, semiautomatic production lines have been installed. Figure 8 shows the schematic layout of a mechanized in-line fruit pie plant that produces 16,000 fruit pies per hour. Swiss roll production can also be mechanized by mixing the batter continuously and feeding it to a depository head over a moving steel band (1 m wide). The continuous ribbon of batter that is carried through a tunnel oven (baking time, 4 min at 232 to 260° C). On emerging from the oven, the baked sheet is mechanically removed from the belt, turned over so that the baked side is underneath, and passed through a cooling tunnel. Jam or cream is deposited on the cooled sheet; it is then cut and rolled into individual Swiss rolls that are then wrapped. This plant can be adapted to various products using the basic sponge sheet and applying different finishes.

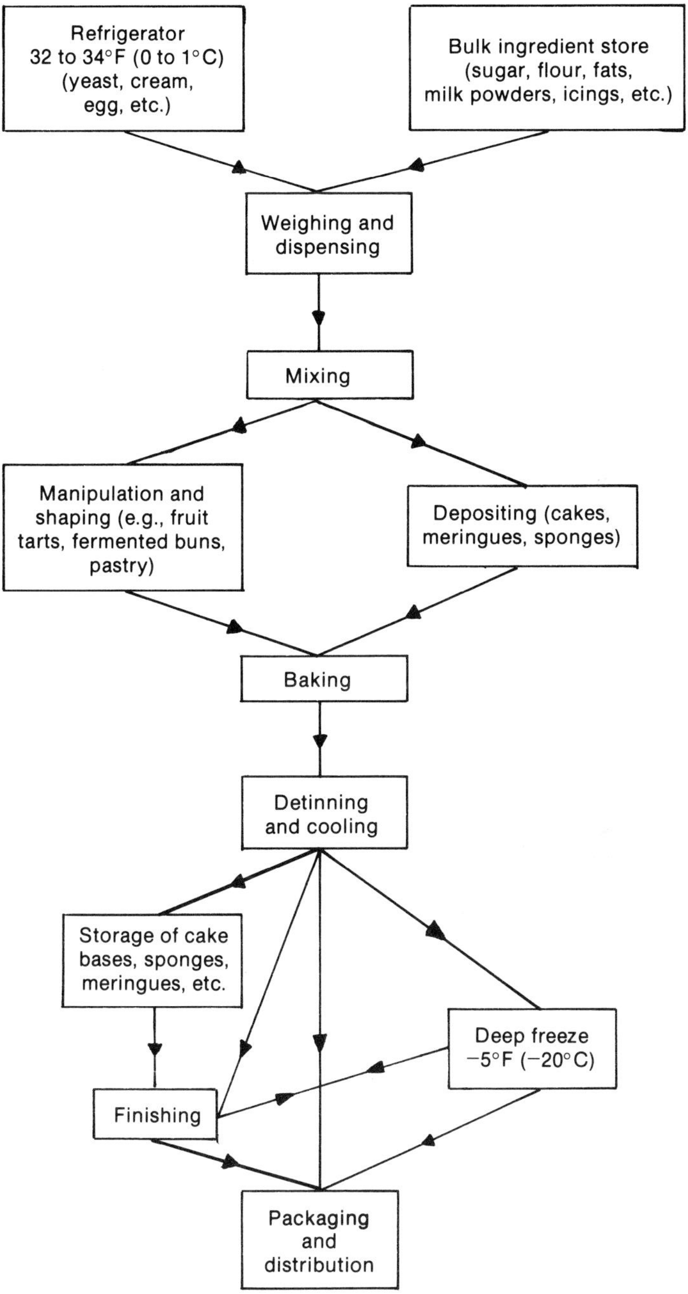

Figure 7. Flow diagram of the order of processing operations in flour confectionery production.

OVENS AND BAKING

A wide variety of ovens is used in flour confectionery. Multideck peel-type ovens are popular with small- to medium-sized bakeries, as are rack ovens, which consist of a large walk-in compartment into which the mobile rack of products to be baked can be pushed; heating is by forced convection, and the rack normally rotates to ensure even heat distribution.

Baking times and temperatures can vary greatly among products depending on their size, the amount of egg and fat, and the sugar content.

FURTHER PROCESSING

Further processing operations in flour confectionery production include the application of creams, icing, marzipan, jellies, fruit fillings, decorations, chocolate, and fondant coatings. Mechanization of these processes is difficult and expensive, although enrobing machines that coat with fondant chocolate are fairly common in large- and medium-sized bakeries. Finishing conveyors are also used in larger establishments in which a continuous flow of a particular product moves along on a conveyor and operatives situated along its length apply different finishes; the product arriving at the end of the conveyor is complete and ready for packaging.

Packaging protects flour confectionery from foreign matter, taints and odors, and mechanical damage during transport, controls moisture loss; and aids sales appeal by design and color. In retail bakers' shops, many products are packed at the point of sale either in paper bags or cardboard cartons. The current trend, however, is toward more prepacking before distribution—essential in self-service stores. One packaging method involves direct wrapping and heat-sealing of the product in cellulose polypropylene or polyethylene. In another technique, the product is placed in a cardboard former with an open top, and the whole is overwrapped with polymer film and heat-sealed. In large bakeries, vacuum-formed transparent polyvinyl containers are also widely used. The medium- to smaller-sized baker often uses a self-sealing polyvinyl chloride film. Machinery is often available for such packing operations.

SHELF LIFE

The shelf life of flour confectionery products is affected by factors such as staling, changes in moisture content, and microbiological spoilage. Staling of cake is usually slow and not a problem, but staling is rapid with fermented buns,

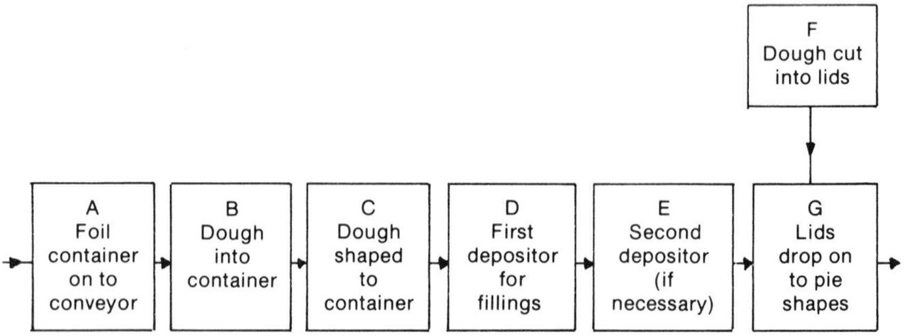

Figure 8. Schematic layout of plant for semiautomatic production of fruit tarts.

scones, and doughnuts so these products are normally baked and sold daily. Factors affecting the rate of staling are still not fully understood (33), but this phenomenon occurs in the absence of moisture loss. Moisture loss from cakes and sponges reduces their eating qualities, and moisture gain in pastry products causes loss of crispness. Such moisture changes can be controlled by the correct selection of wrapping materials. The principal factor limiting the shelf life of most confectionery products is microbiological spoilage. The rate of spoilage largely depends on the equilibrim relative humidity (ERH). Table II shows the ERH of some confectionery products and the possible contaminating microorganisms. Products with dairy cream and high-ERH-imitation cream fillings, custards, and jellies are capable of producing food poisoning and spoilage bacteria and must be made and sold daily. The high fruit level of rich fruit cakes produces a low ERH for these products and consequently a mold-free shelf life of many months. Microbiological contamination may be prevented or reduced: Postbaking contamination is the problem since molds, yeast, and most bacteria are destroyed during baking. To keep contamination to a minimum, good hygiene must be maintained in a bakery (58–60), and many ingredients such as dairy cream and coconut are pasteurized by a mild heat-treatment before delivery. Cakes baked in a can that has the lid hermetically sealed afterwards have an indefinite mold-free shelf life. Deep-freezing (−20° C or below) successfully preserves a wide range of flour confectionery products, and refrigeration of certain perishable products (dairy cream-filled cakes) at 0 to 5° C, is used for short-term storage before distribution; at the point of sale these products are often displayed in refrigerated cabinets (47).

IV. BISCUITS

A. Introduction

In 1977, about 500,000 te of wheat flour was used in the production of biscuits in the United Kingdom. Biscuit manufacture is highly mechanized (45), with a relatively small labor force involved in the actual production, although biscuit quality still depends on the subjective judgment of the operators. Two large

TABLE II
Equilibrium Relative Humidity (ERH) of Flour Confectionery Products and Possible Contaminating Microorganisms[a]

Product	ERH (%)	Microorganisms[b]
Plain cakes	80–90	Normal molds (80)
Fruited cakes	73–83	Xerophilic molds (65)
Christmas puddings	70–77	Xerophilic molds (65)
Jam, marzipan, fondant marshmallow, butter-cream	70–80	Osmophilic yeasts (60) Xerophilic molds (65)
Dairy cream, imitation cream, custard, jellies	97–99	Food poisoning and spoilage bacteria (91)
Fruit fillings	75–95	Normal molds (80) Xerophilic molds (65)

[a] Data from Seiler (58).
[b] Figures in parentheses give the minimal ERH that will support growth.

groups produce 70% of the U.K. biscuit output, most of the remainder being produced by 12 other manufacturers. These concentrated production units supply biscuits to a relatively wide area of the country; some biscuit factories produce 20 or more high-output lines in three daily shifts, and others produce only a few types of product.

Because biscuits have a very long shelf life compared with that of bread, fairly long-term production can be planned to optimize the manufacturing facilities. The biscuit industry has developed essentially independently of the bread and flour confectionery industries in both financial and technical respects.

B. Types of Biscuits

Flour, fat, and sugar are the principal ingredients of biscuits. Their relative proportions determine the handling properties of the dough and the eating quality of the finished biscuits (Fig. 9). Water is an important ingredient at the dough stage, but most of it is removed during baking; the finished biscuit has a low residual moisture content of about 1 to 3%.

Variations in the proportions of fat, sugar, and flour give the following types of dough during manufacture:

- Hard dough, used to make butter puffs and cream crackers, has a well-developed gluten network with marked elastic properties, and the flour used has a relatively high protein content of 10 to 11.5% (68).

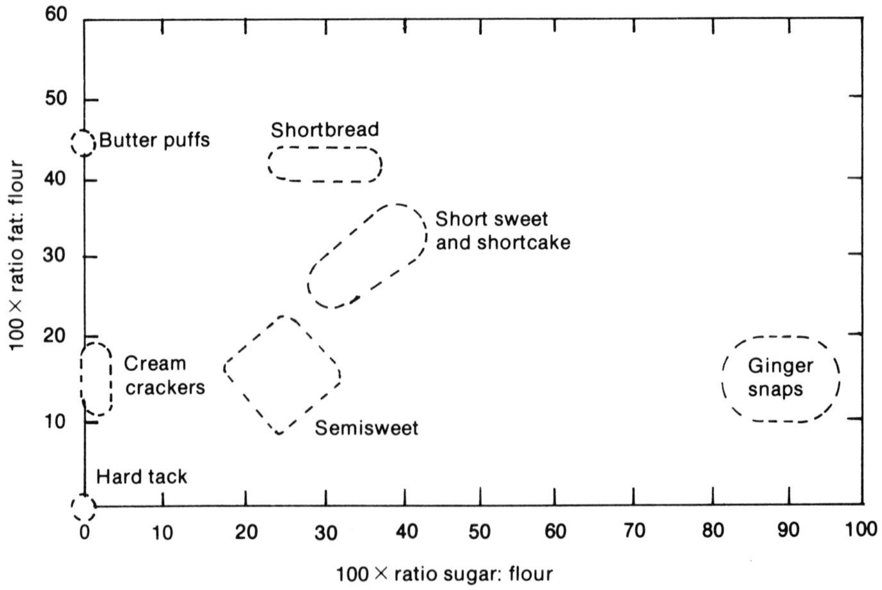

Figure 9. Proportions of fat and sugar associated with 100 parts by weight of flour in various biscuit doughs.

• Hard sweet doughs, used to manufacture semisweet biscuits such as Marie, Rich Tea, Osborne, Petit Beurre, etc., have properties similar to the hard dough, but the flour contains less protein (7.5 to 10 %) and may contain sodium metabisulfite (200 ppm) (63); the recipe also contains sugar (64,65).

• Short doughs contain high fat and sugar levels, the flour used has a low protein content, and only a small quantity of water is necessary to give the required consistency; no gluten network is formed and the dough is inelastic; short doughs produce shortbread, shortcake, and short sweet biscuits.

C. The Biscuit-Making Process

A flow diagram of the processing operations in a biscuit bakery is shown in Fig. 10.

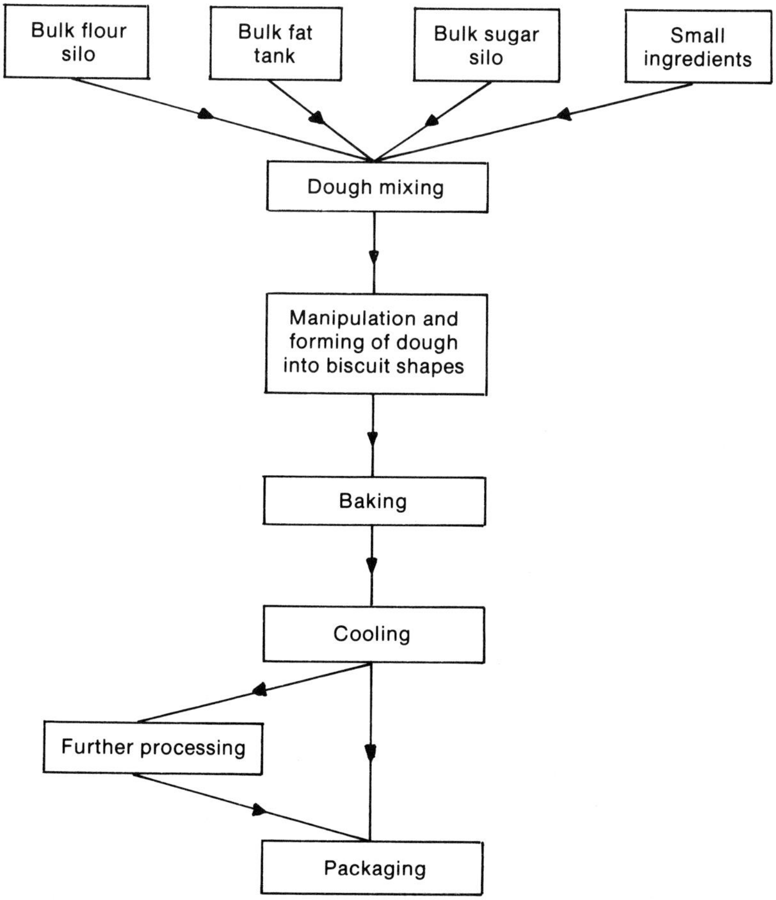

Figure 10. Flow diagram of the order of processing operations in biscuit production.

As outlined, the type of biscuit to be made determines the characteristics of the flour to be used. Butter, margarine, oils, and various hydrogenated fats are all used in biscuit manufacture, the type and amount depending on the recipe. The degree of subdivision (i.e., fine, caster, or granulated) of the sugar appreciably affects dough and biscuit properties. Often sugar in the form of syrups or molasses is added to affect color and flavor. The amount of water added to the dough depends on the water absorption capacity of the flour. Other ingredients that can be added in small amounts include malt flavor and milk powder to contribute to flavor and color; aerating agents such as sodium bicarbonate, sodium acid pyrophosphate, and ammonium bicarbonate, to raise the dough structure during baking.

In most factories, the principal ingredients are stored in bulk-handling systems that feed directly to the mixers. Systems have been developed to control this operation automatically on a batch basis.

MIXING OF THE DOUGH

Mixing conditions differ according to the type of biscuit dough. Mixers in general use are of the batch type, either a twin vertical-spindle mixer or a horizontal-beater, high-speed mixer; the latter mixes dough in a much shorter time. Recently, time control has been introduced to govern the mixing cycle. If necessary, the time is adjusted by the operator to promote the appropriate rheological properties of the dough, although dough temperature may also be important.

Some doughs, such as those for cream crackers, normally require a period of yeast fermentation, which may vary from 4 hr for a straight dough process to 24 hr for a "sponge and dough process." Intense mechanical work, equivalent to the CBP, has the same effect (66).

MANIPULATION OF THE DOUGH PIECE

Different machines form the required shape of dough piece for baking. A laminating machine is used for the hard doughs of the cream cracker and butter puff types, with the object of producing a sheet of dough with many alternate laminations of dough and a flour/fat mixture. The "puff" in such biscuits arises from the steam produced within each lamination during the early stages of baking. In the case of cream crackers (67), two fairly thick sheets of dough are extruded through sets of rollers, and a thin layer of a fat/flour mixture is spread over one sheet, while the second sheet is placed on top to produce a sandwich of one layer of fat/flour mixture between two sheets of dough. The thickness of this sandwich is then gradually reduced by passage through rollers. The thin dough sandwich is then built up into six or eight layers, usually by layering the continuous dough sheet backward and forward across another conveyor, moving much slower and at right angles to the first conveyor. This multilayer sheet of dough is then progressively reduced in thickness by rollers until after the final conveyor, which travels more slowly to allow some relaxation of the elastic tension built up during manipulation; dough pieces of the required size, shape, and design are cut out by a reciprocating head cutter or a cylinder cutter.

Hard sweet doughs are initially made into a relatively thick dough sheet, which is then reduced by passing through several pairs of heavy rollers. Dough of the required thickness is then relaxed before dough pieces are cut.

Short doughs may also be formed by reciprocating or cylinder cutters, but in the alternative rotary molding process, the pieces are formed by compresssing the dough into suitable dies, engraved on a phosphor-bronze roller, and extracting the resultant dough pieces by pressing with a specially treated web.

BAKING

Traveling-band, tunnel-type ovens are used exclusively in the biscuit industry; biscuit dough pieces are loaded at one end and the baked biscuit is unloaded at the other. Ovens range in length from 45 to 90 m, and baking time can vary from 3 to 8 min depending on the type and size of the biscuit dough piece. The actual baking time is controlled by adjusting the speed of the traveling band onto which the dough pieces are fed continually. The oven performs several functions, and the humidity and temperature throughout its length have to be carefully controlled (37). In the first oven sections, the dough structure is raised by action of steam or by carbon dioxide gas formed by yeast or chemical action. In the middle sections of the oven, the dough structure is set either by coagulation of the protein network or by the early stages of drying of the dough. In the final section the formed biscuit is dried out, and the crust color and flavor are developed. The hot biscuits leaving the oven are cooled thoroughly by transfer to a canvas conveyor, on which they travel for a time roughly equal to twice their baking time, before they are packed or processed further.

FURTHER PROCESSING

Some of the finishes applied to the biscuit base are creams, fudges, marshmallows, caramel, jellies, fruit, nuts, icing, coconut, and chocolate. These finishes can be used alone or in combination, and the same biscuit base may be used with different finishes to manufacture several types of product. Different automatic machines are used for applying these finishes; for example, an enrobing machine that covers the product with chocolate. The biscuits are passed along a horizontal wire mesh conveyor into an enrober and then through a thin vertical curtain of falling melted chocolate. After passing through the chocolate curtain, air blowers remove the excess chocolate from the upper surface of the biscuit, the speed of the blower determining the final thickness of the coating. Vibration of the conveyor shakes excess chocolate from the base. The chocolate-coated biscuits then pass along a refrigerated tunnel to emerge ready for packing, with the chocolate coating set.

PACKAGING

Essentially, all biscuits are now sold prepackaged to protect them from entry of foreign matter, odor contamination, mechanical damage, moisture changes, and oxidative rancidity. The design, pictorial display, and color of the packaging are also intended to increase customer eye appeal. There is a wide variety of shape and size of biscuit packets and of the appropriate packaging machinery; in one of the most common types, a column of biscuits is wrapped directly in a flexible, moisture-permeable material. After wrapping, the packaging film is heat-sealed.

D. Wafers

Wafers need to be treated separately because the formulation and method of manufacture differ greatly from those of biscuits (44–46). Because of the amorphous structure of the wafer, the protein level in the flour used in its manufacture is immaterial.

Wafers are made from a very fluid batter and typically consist of 1,000 parts flour, 1,450 parts water, 24 parts vegetable oil, 2.4 parts salt, 9.5 parts powdered lecithin, and sodium bicarbonate, 3.1 parts by weight. (The oil assists the release of the wafer from the baking plates, the lecithin acts as an emulsifier in the batter, the salt improves the flavor of the wafer, and the sodium bicarbonate helps raise the structure during baking.)

The ingredients of the wafer are machine-whisked to produce a smooth, homogenous batter that will flow freely through the pumping system to the baking plates. The special ovens used have between 12 and 30 plates, carried by wheels through the insulated main body of the oven. Each plate is in two halves, hinged on one side, and there is a pattern of grooves inside each half to form the surface design on the wafer sheet. Wafer batter is pumped from a storage tank to a depositing arm, which partially spreads the batter onto the base plate. As the top plate closes, the spreading is completed by pressure. Vent holes in the wafer plates and a slight lifting of the top plate during the early stages of baking allow escape of the very large volume of generated steam. After traveling through the oven, the plates open and the baked wafer sheet is ejected by a jet of compressed air. The sheets are then trimmed, before feeding into secondary processes.

A wide range of products is formed by sandwiching the wafer into various types of creams, fudges, caramel, etc., cutting into different shapes, and coating some varieties with chocolate.

PART TWO: SOFT WHEAT USES IN FRANCE

I. WHEAT

A. Production and Use

The wheat harvest in France over the past few years has varied from 12 to 16 million tonnes (te). The home market represents a consumption of 4.3 million te, to which must be added that of the European Community, approximately 3.5 million te. The remainder of the production is taken up by exports (both wheat and flour) and animal feedstuffs. In the home market, most wheat is used for French bread—more than 80%—and this consumption is no longer decreasing (the 1975 figures are 0.5% higher than those for 1974). The rest of the market, about 20%, represents various other sectors of the baking industry, i.e., the production of 1) *biscottes* (rusks), where the very rapid postwar increase has now stabilized, with the same holding true for *pain grillé*, introduced more recently; 2) *pain de mie* (soft, white bread), the comsumption of which has increased in the past few years; 3) biscuit making, which is following a development similar to that of *biscottes*, but where production has diversified, particularly in salted snacks; 4) cakes, where consumption is noticeably increasing; 5) ready-to-use. flour (ready mixes); and 6) sundry uses (flour prepared to use in the home, the manufacture of cooked meats, canning, etc.).

B. Usage and Grading

In France there are 2,700 cooperatives concerned with storage of wheat and their capacity is constantly increasing. Great improvements have been made in drying, and now whatever the condition of the crop, grain can go into storage at a moisture level compatible with good conservation.

A well-stored, healthy grain is not the only condition for its marketability. Its utilization value must also be determined. This is the aim of the classification developed by the National Interprofessional Office for Cereals (ONIC) in 1969, for French wheat. This classification is based upon the Zeleny Index, on the protein content of the grain, and (since 1975) on its amylase activity (Hagberg falling number).

The classification system shown in Table III has had the merit of making producers and elevator operators aware of the technological value of wheat, but it has not received the complete approval of users, because the Zeleny Index does not express the real value of French wheats. For this reason it has not been used commercially, most millers preferring to rely on the results of the alveograph,[5] of the baking test;[6] and more recently, on a method of identifying cultivars by electrophoresis of the gliadines.[7]

Wheat quality depends on many factors (soil, climate, fertilizer), but in France the cultivar is considered to be of paramount importance. The selection of wheat cultivars is made by the farmer for whom the idea of quality is primarily that of the yield per hectare. The same does not hold true for the miller, who looks for wheat that mills well and has technological characteristics suited to the different needs of the user. Consequently, on the current French market is a group of

TABLE III
Classification of French Wheat According to the ONIC[a]
(minimum test data for each class)

Class	Protein Content (N×5.7) (%)	Sedimentation Index (Zeleny Value)	Falling Number, Hagberg (sec)
I	13	38	180
II	12	28	180
III	11	18	180
IV	10	13	180

[a] ONIC = National Interprofessional Office for Cereals.

[5] The alveograph measures the deformation characteristics of a dough by inflating a thin piece into a bubble. The pressure/extension curve is recorded, and from its shape certain factors can be obtained, such as the "W-value" (in relation to the baking force) and the swelling-value, which gives information on extensibility.

[6] The bread-making test, defined by the National Co-ordinating Center for Studies and Research in Nutrition and Food, allows the different stages of French bread making to be evaluated using a 2-kg sample of flour. Results are evaluated by scoring the characteristics of the dough (on a 100-point scale) and of the exterior appearance and volume of the breads (on a 200-point scale).

[7] A method perfected by Dr. J. C. Autran, National Institute of Agronomic Research.

satisfactory bread-making wheats represented by cultivars such as Capitole, Hardi, and Remois; another group of cultivars valued by the farmer but which give very mediocre results when made into bread, such as Champlein and Joss (the latter being of possible interest for the manufacture of biscuits); and wheats that are totally unusable for bread making but have a high field yield, such as Maris Huntsman and Clement.

Cultivation of wheat of the last category has induced the formation of a commission of experts, within the framework of the EC, whose task has been to devise a baking test capable of eliminating samples of wheat that cannot make a satisfactory loaf. A current recommendation of the commission is a test based on the assessment of the "machinability of doughs."

The areas sown to these different wheats, as well as their climatic conditions, affect the average quality of a given harvest. Crop quality as designated by the alveograph W-value progressively increased from 1968 to 1972, primarily owing to the cultivation of wheat such as Capitole and Hardi. Quality decreased appreciably after 1972, undoubtedly because of the introduction of poor cultivars and unfavorable climatic conditions. In the past two years, however, the W-value of wheat received for milling has increased because of users' demands for wheat of better baking cultivars and the increase in the average protein level of the wheat. (The higher protein levels of the past two years result primarily from climatic conditions that greatly diminished the yield per hectare.)

In addition to the groups of cultivars, there are some strong types of wheat such as Rex, Florence/Aurore, and Magdalena, which are sought for certain types of manufacture (finest wheat flour, *biscottes*) but in limited production.

II. FLOUR

A. Definition

A code of practice connected with bread in France (28) gives the following definition:

"The description of *wheat flour*, or *flour* without any other designation, indicates exclusively the powdery, or essentially powdery substance obtained from marketable wheat of the species *Triticum aestivum* subspecies *vulgare* prepared, conditioned for milling and which is industrially pure."

B. Milling

Roller mills have been used in France for almost a century. The process involves the classic operations of cleaning the grain and adjusting its moisture content to about 17%; grinding, in the course of which the bran is separated from the kernel; converting the semolina and fines to flour; sifting; and winnowing, which, for certain kernel fragments (semolina), completes the sifting process by grading according to purity.

For 50 years, the French milling industry has been progressing, and the production capacity of the factories has thus increased considerably. This increase in production has occurred in the face of a reduction in milling facilities through the increase in the speed of the roller mills and improved stock handling, which have increased their efficiency, thus bringing about the closing of one roller mill of every two. Sifting is concentrated in fewer but more effective plansifters with larger capacity. In some mills, winnowers (purifiers)have been eliminated.

Streamlining of modern mills has also been aided by the pneumatic transport of products, elimination of lifts and spiral conveyors, and increase in the use of labor-saving equipment. These changes have been sufficiently effective for some plants to eliminate the traditional tall installations (with products transported by gravity) in favor of single-level, prefabricated units. The most marked innovations in the French milling industry have been the spread of bulk transport of flour and in the use of air-classification.

In France, a "continuous industrial selector" has been developed for air-classification that separates a flour into 20 fractions with protein levels from 25% to less than 5%. In practice, the flour is grouped into only three fractions, clearly distinguished by their protein contents and corresponding technological properties: a) Fine fraction. In bread making, this fraction brings about better development of the product and greater tolerance in the manufacture of leavened doughs in general as well as of certain cakes and puff pastry. b) Medium fraction. The medium fraction is very well suited for certain types of cakes with an egg base, of which *Genoese* is the most representative. In mixtures (in different percentages according to use), it is perfectly suited to all biscuit or cake products the quality of which necessitates a weak flour, which is obtained with some difficulty even with weak types of wheat. c) Coarse fraction. The analytical characteristics of this fraction resemble those of classic flour, but the size of the particles improve its fluidity.

Air-classification of flours therefore improves French wheats and makes new industrial uses possible.

C. Flour Classification

Official French regulations classify flours according to "type" or possibly their baking strength as in fine wheat flour (*Gruau*). Table IV shows that flour type is determined by the ash level. Although the ash level is influenced by the extraction rate, there is no precise relation that allows the extraction rate to be determined with any exactitude from the value of the ash content.

The protein content and protein quality (expressed in France by the W-value) permit the precise classification of flours within a given type. Thus, a flour of Type 45 and one of Type 55 may each have a W-value of 220 and swelling (G) value of 19 on the alveogram, but the former may have a protein content of 11.0% and the latter 11.5%. These values serve to exclude flour not in analytical compliance. From a practical point of view, the quality differences between the flours delivered by the French milling trade mainly reflect the choice of wheat mixture.

Flour in France is made almost totally of "straight run flours," excluding tailings, so that the extraction level is generally between 75 and 80%.

The following categories of flour may be found in the French market:

Quality flours for bread making usually come from country wheats, to which can be added varying percentages of corrective wheat (e.g., strong- or weak-swelling). Usage calls for bread-making flours to belong to Type 55, but on the technical level nothing prevents making bread from another type, provided that its characteristics allow a good-quality bread to be baked (especially the color of the interior of the loaf).

Special purpose flours, further subdivided as follows:

- Fine wheat flours (*Gruaux*) are made exclusively from strong wheats such as

TABLE IV
Classification of French Flour by Type and Ash Content

Type	Flour Ash Content (d.m.b.) (%)
45	< 0.50
55	0.50–0.60
65	0.60–0.75
80	0.75–0.90
110	1.00–1.20
150	> 1.40

French Rée or North American CWRS of Types 45 and 55. Regulations define the minimum baking strength. In practice, a proper fine wheat flour should have a protein level and a W-value above this norm. Fine wheat flours are usually used in rich formulas in cake making and Viennese baking.

- Strong flours are a mixture of domestic wheats and strong wheats, usually Types 45 and 55. These flours are indispensible for fine bread making and especially for *biscottes.*
- Biscuit flours can be made, totally or partially, from special wheat such as Etoile de Choisy, Joss, and others with a low baking value.
- Ready-to-use flour mixes have been made in France only during the past few years and are being increasingly used by bakers and cake makers. They generally contain all, or part, of the ingredients making up the recipe and are really a guarantee of the recipe and of success, without taking away the craftsman's creative initiative in the presentation or decoration of cakes.

III. BREAD PRODUCTS

A. Traditional French Bread

EVOLUTION AND CHARACTERISTICS

Bread baking in France is done almost entirely by bakery craftsmen, as shown in Table V.

Although there is a great selection of bread in France, the one traditionally eaten, the *baguette* (you crunch the crust on your way out of the bakery!) is a white bread with a distinctive appearance (Fig. 11). It is similar to other bread in that is a mixture of crust and internal crumb, but the proportions in a *baguette*

TABLE V
Distribution of Bakeries in France by Output in 1971 and 1976[a] (flour made into bread)

Output (cwt/year)	1971	1976
< 550	24,706	21,980
550–1,099	17,263	15,700
1,100–1,649	2,590	2,441
1,650–2,199	525	618
2,200–3,299	269	326
> 3,300	221	308
Total	45,574	41,373

[a]Data from La Meunerie Française (1).

are different. This loaf has up to 50% crust; in contrast, in many other loaves, the crumb forms the main part of the bread and the crust forms only a thin protective film. These different proportions of crust and crumb contribute to the bread's special character, called "chew." That of French bread owes more to the dry and crisp crust than the interior, and the crumb has a largely open structure

Figure 11. *Baguettes.*

and elastic consistency. The "blade cut" made by craftsmen, even in a partially mechanized production, assures that the *baguette* retains its original appearance. The "blade cut" results in the characteristic "crack,"[8] which is the signature of the baker and also the focal point of the contrasting colors and of the crispness of the loaf. These qualities derive mainly from the baking, which is performed directly on the base-plate of the oven in an atmosphere saturated with water vapor.

Considered a delicacy throughout the world, the *baguette* symbolizes French bread, although it is only one of the specialties of Parisian bread baking.

French bread making evolved slowly. In the 19th century, bakers' show windows already displayed a large selection of round, short, and long loaves with weights varying from half a pound to 12 pounds. As the "bread markets" disappeared and night work spread (apparently this dates back to the 18th century), fresh bread and then hot bread appeared as a new service for urban customers who gradually abandoned "ordinary bread" in favor of the so-called fantasy breads. This development was facilitated by the spreading of the baking process in which yeast is used directly and is well suited to crusty breads, since their quality depends more on a good balance between the crunchiness of the crust and the so-called nutty taste of the inside crumb than on the flavor yielded by sour dough. This method of using yeast, introduced in Paris at the end of the 19th century by Viennese workers and therefore known as "Viennese bread-making," gradually replaced the sourdough method and rather paradoxically is contributing to the worldwide reputation of French bread. In the 1930s, it also replaced the process of fermentation called "Polish," which was based on a liquid preparation of flour, water, and yeast from which loaves were kneaded after fermentation. Since then, the direct technique has become widespread in France, gradually replacing the sourdough and "Polish" techniques.

Development of new methods of bread making has continued. The new technique was followed by the mechanization of some of the steps in bread making, resulting in improved working conditions for bakers without appreciably altering the craftsman structure of the profession. Nevertheless, the quality of the bread has changed imperceptibly.

METHOD OF PRODUCTION

Originating from a bread-making process based on weak wheat flours and involving under-kneading, a slow fermentation, and manipulations (weighing, shaping) that had to be gentle, French bread has kept its traditional form, even though it is made differently. Indeed, the quality of the *baguette* depends as much on its formula as on the manual dexterity of the baker.

The formula is simple, with four basic materials: 100 parts flour; 58 to 63 parts water, depending on the quality of the flour; 1.8 to 2.2 parts salt; and approximately 2.0 parts yeast, with fermentation time of 4 hours. Optional additives are legal in France.

The first stage in the production involves preparation of the ingredients so that the optimal dough temperature of 25°C is obtained at the end of kneading.

"Intensified kneading" is the technique generally used in France for mixing the

[8]The "crack" is the bulge brought about on the surface of the bread by the expansion of gas at the place where the baker has made a cut called the blade cut at the moment of putting the bread into the oven. This blade cut creates a point of least resistance on the surface of the dough and allows a larger swelling at this point, hence better development of the loaf.

dough ingredients. It is based on energetic kneading of the dough for about 20 minutes, using a kneading trough with an arm turning at a minimum of 80 rpm, and supplied with water cold enough to compensate for the heating dough. This method of mixing incorporates a large quantity of air (hence oxygen) and alters the dough and the resultant bread. The glutinous mixture takes on stronger elasticity, which gives the dough better resistance to the expansion of gases and greater tolerance during the final fermentation. This physicochemical transformation of the dough is essential for good bread. The effects on bread quality are far superior to all procedures previously used in classic kneading at slow speed, whether in the field of baking techniques or of additives. The most spectacular effect concerns the volume of the bread; it is practically doubled. It is not surprising that consumers are attracted by this better looking bread, which also gives the impresssion of a large quantity for the same price.

This argument in favor of the intensified kneading technique is often countered by the complaint that the bread tastes insipid and has an artificially white crumb. Deterioration of flavor and color of the crumb is caused by an enzyme, lipoxygenase, which in the presence of oxygen acts on the free linoleic acid to form hexanal (26). Hexanal, which is not a usual component of bread, changes the flavor characteristics drastically, especially when there is an excess of it.

To compensate for this effect, more and more bakers are using a method where, although the speed of kneading is rapid (80 rpm), its duration is reduced to about 12 min. Under these conditions, dough development does not result in so much bleaching, and the loaves remain more palatable. Unfortunately, this method has the disadvantage that it does not give such voluminous loaves, and hence it is not popular because the idea of bread lightness and volume has been confused with that of quality.

Although open to criticism, it is nonetheless true that intensified kneading, which has much in common with the U.K. Chorleywood Bread Process, has brought new ways of manufacture to bakers.

Fermentation of the dough then takes place in three stages. Immediately after kneading, the entire batch is fermented in a vat; this is the first, or mass, fermentation. The batch is then divided into small pieces and the dough is "knocked back," which is indispensable so that the sections of dough can recover the suppleness necessary for a good shaping. The final fermentation stage is the fermentation that occurs between shaping the dough and putting it into the oven and occurs in baskets, on "sheets," or on "beds." During this final fermentation the cellular structure of the inside of the loaf is formed.

Manual weighing or "volumetric division" is made on the mass of dough that is more or less fermented. The weight of dough of a *baguette* is 350 g; after baking, 250 g. After "knocking back," these pieces of dough are shaped either by hand or by machine, into a roll approximately 80 cm long.

Traditional French bread is baked directly on the base-plate of an oven heated to approximately 250° C in an atmosphere saturated with water vapor (the temperature of the interior of the loaf never exceeds 100° C). Baking time varies according to the weight and shape of the bread, and in practice baking is complete when the crust is sufficiently formed and has taken on its characteristic color. When taken out of the oven, the loaf should have a solid but still crisp crust, and the inside should be supple, elastic, and not sticky. The inside of the loaf has about the same moisture as dough when taken out of the oven but loses 2 to 3% by evaporation while cooling. This period called "sweating" precedes the staling, which normally does not happen until much later. Keeping qualities do not exceed 24 hours for a *baguette*.

ADDITIVES

French bread can be made without any improver, but under French regulations four products can be added that allow the miller or baker to correct possible deficiencies in wheat flour.

- Horse-bean flour which contains lipoxygenase that enhances the whiteness sought in the intensified method, may be incorporated in flour at the mill at a maximum dose of 2%.
- Ascorbic acid, which has well-known effects on gluten, may be added to flour by the miller or at the kneading stage by the baker. The maximum dose allowed is 300 ppm, but in practice it does not exceed 100 ppm.
- Malt, permitted in milling and baking, is used to correct the amylase activity of wheat (or of flours). In milling, 0.3% is permitted. Malt is also used in certain Vienese baking formulas, where it is valued for the flavors it allows to develop in these products.
- Lecithin is only permitted for baking (0.3% of flour weight); in milling, only ready-to-use flour may contain it.

B. Regional Breads

In general, these are large-sized loaves arising from the time when the French had a reputation of being "big bread eaters" and when bread was expected to last for several days after purchase. More or less obscured by the success of the

Figure 12. *Courrones.*

baguette, these products are again becoming fashionable, especially in towns where, without a doubt, consumption of country bread is now the greatest.

The French "Code of Use" (28) gives the following definition: "Country bread is a product conjuring up the simplicity of breads made before the introduction of the direct method. Its manufacture involves the use of a bread-making flour, a whole-meal flour, or a mixture of these flours with or without rye flour. Kneading is done in such a way as to avoid "bleaching" of the dough. Fermentation is started by bringing together bakers' yeast, combined with a sour dough; it is carried out in such a way as to develop a slightly acid taste and achieve a longer shelf life."

The French craftsman makes a special shape from the dough piece that very often is characteristic of a region (Fig. 12). If country bread, whatever its shape, has become rather paradoxically almost exclusively an urban product, other breads are representative of a region; their dough is largely the same as that of the *baguettes*. They include "dumpy" bread from the north, or "sandwich bread" with a finely cellular inside texture; split bread; twisted bread from Limoges; and various "rings." These breads are frequently made of doughs in which leaven is combined with yeast.

C. Fine Bread Products

FINE WHEAT BREAD

Fine wheat bread, *Pain du Gruau* (Fig. 13), should be made with finest wheat flour and may contain milk powder. This bread has a very fine crust with a pleasant taste that results in part from the nature of the wheat used for the flour and its low extraction level.

Figure 13. *Pains fendu de gruau*.

The formulation for fine wheat bread is: fine flour, 100 parts; water, 65; yeast, 0.8; salt, 2.1; milk powder, 1.8; and malt, 0.9 parts. The dough temperature is 25° C. The duration for the first fermentation is 4 hours and that of the second, 1 hour. Baking is done in an oven at 250° C after dough has been shaped in "threads" (*ficelles*), rolls, or sometimes "split" bread weighing about 200 g.

VIENNESE BREAD

This product, often confused with fine wheat bread, is baked from a dough enriched with fats. The ingredients of Viennese bread (quantities based on flour weight) are: fine wheaten flour, 100 parts; water, 60; salt, 2; milk powder, 1.8; malt, 1.2; sugar, 2.5; fats, 6; yeast, 6; improver (ascorbic acid or lecithin), variable. The dough is fermented at 25 to 26° C for 30 minutes in a vat, kneaded at intervals for 1 hour, and given a final fermentation of 1 1/4 hours. Viennese breads (Fig. 14) are generally shaped in the form of *baguettes*, but the blade-cuts are made crosswise immediately after shaping (the so-called *coupe saucisson*). The baking is preferably done on sheets.

The inside of Viennese bread is like that of *pain de mie*, but it has a slightly sugary taste differing from that of traditional French bread.

BISCOTTES

The *biscotte* or rusk has been in existence for no more than 50 years and is comparable to the German Zwieback or the Dutch Toast. In its present form, it has become a real French specialty. It is a loaf cut into slices and toasted, but not of the standard bread to which one is accustomed in France. Bread made for

Figure 14. *Baguettes viennoises.*

Figure 15. *Croissants.*

Figure 16. *Brioches.*

biscottes is baked in tins and has fine, regular holes. To attain this regular cellular texture, fat and sugar equal in quantity to yeast are added to the dough. The amount of yeast used is relatively large, about three times greater than the amount used in ordinary bread.

A common formula for *biscottes* is (based on flour weight): fine flour, 100; sugar, 5; fats, 5; malt extract, 1; salt, 1.5; yeast (bakers'), 5; water, 50; ascorbic acid and lecithin, variable. Production is highly mechanized and takes place in factories where several tonnes of flour are transformed into bread every hour. In the process, fermentation is very short, there is no fermentation in vats, and less than 2 hours are required for the other stages.

CROISSANTS

Croissants (Fig. 15) were introduced into France from Austria at the beginning of the 20th century, but only the form has been preserved and its formulation has been completely modified. The French *croissant* is made with a raised flaky dough. Its texture is therefore the result of both leavening during fermentation and the flakiness brought about by the successive rolling of the dough with fat.

The following formulation is used for *croissants*: fine flour, 100 parts; salt, 2.5; sugar, 6; milk powder, 5; malt, 1; invert sugar, 1; yeast, 2.5 to 5.0; and water 60. After 4 to 5 hours of fermentation, 53 to 55 parts of butter (based on flour weight) is added, and the dough is rolled and folded (*tours*) three times. Dough pieces of 40 to 50 g each are cut, shaped, brushed with eggs, and baked in a warm oven.

BRIOCHES

Brioches are also made from a raised dough, fermented slowly with direct or leaven-yeast (sponge) fermentation. Its original formula is very rich because hydration is brought about only by egg liquid. The quantity of fat varies but usually is about 400 g of butter per kilogram of fine wheat flour. It is made into various forms, and garnishes and decorations can be added (Fig. 16).

The formulation for typical *brioches* is: fine flour, 100 parts; yeast, 6; salt, 2.5; sugar, 10; eggs, 60; shortening, 50; malt, 0.5. The dough is kneaded to a soft consistency and the shortening is added without excessive kneading. Dough is permitted to ferment for 4 hours, after which it is "knocked down" and cut into pieces. After a final one-hour fermentation the dough is baked in a warm oven.

D. The French and Their Bread

The view of the French people on the quality of their bread varies with where they live and with their habits, but there are criteria valid not only in France but wherever French bread is made.

French bread should be sufficiently light so that the crumb is never close-textured. The holes in the crumb should be large in Paris, but in many other regions of France such as the north, people prefer bread with many holes inside. Tastes differ on the appearance of the crust also. In the south of France, people want a pale crust. In some parts of the country, people like the top of the bread to be floury. In the Paris region, they like a golden crust that they require to be crunchy, for the bread is eaten fresh. A *baguette* can be eaten stale, but a fresh and crunchy one is preferable.

Whatever its shape or composition, in France bread remains closely linked with gastronomy. Bread has been said to be the oldest "food course" in the world; consequently, it is difficult for a Frenchman not to eat any, even if only in the form of a *biscotte.*

IV. CONFECTIONERY PRODUCTS

Confectionery products abound in France. Some are not of French origin, but those that follow are among the more famous.

Puff pastries such as *mille-feuilles* or *chausson* (apple slipper) are a type of confection (Fig. 17) with a dough up to 100 parts flour, 58 parts water, and 2 parts salt. Shortening equal to 50% of the dough weight is added, and the dough is given five *tours* (rolling and a three-way fold to laminate the shortening and dough). The dough is then broken into small pieces and baked at 230 to 240° C. *Chaussons* are garnished with apple jelly before baking, whereas *mille-feuilles* are filled with pastry cream and decorated with melted sugar or fondant after baking.

Choux pastry products are filled with custard cream, flavored with coffee, chocolate, vanilla, or various alcoholic beverages or liqueur. They are generally decorated with a fondant the color of which is given by the same constituents as the cream (coffee, chocolate, etc.) (Fig. 18).

The pastry dough is prepared from: water, 100 parts; salt, 2; shortening, 40; flour, 60; eggs, 90 to 100 parts. The flour is added (without heating) to the hot water, salt, and melted shortening in a pan and worked into a dough with a spatula. After the dough has cooled, the eggs are added a little at a time, with

Figure 17. *Feuilletage.*

mixing in between. The dough pieces are laid on a tray and baked at 230 to 240° C.

Of the articles made from sweetened dough, the most common are fruit tarts (Fig. 19). A sweet dough is prepared from flour, 100 parts; sugar, 30; salt, 1; baking powder, 0.5; shortening, 25; eggs, 40 parts. The eggs, sugar, and salt are beaten together. The shortening, sifted flour, and baking powder are then added, and the mixture is blended until smooth. The dough is cut into pieces and shaped, the fruits are added, and the dough is baked at 200 to 210° C.

Genoese is another product of a sweetened dough. The formulation consists of flour and sugar, 100 parts each; and eggs, 160 parts. The eggs and sugar are beaten together until they form a ribbon. Sifted flour is then added by hand. The batter is poured into a greased and floured mold and baked at 190 to 200° C.

The *savarins* or *babas* are made from a fermented dough similar to a *brioche* but with a lean formula to which rum syrup is added and which can eventually be topped with a pastry cream (Fig. 20). Ingredients and their quantities follow: flour, 100 parts; yeast, 1.5; salt, 2; sugar, 6; eggs, 30 to 40; shortening, 25. All of the ingredients except the shortening are mixed 2 to 3 minutes at low speed and then 7 to 8 minutes at high speed. The melted (but not hot) shortening is then blended in. After a primary fermentation of 30 minutes, the dough is given a *tour*, followed by a secondary fermentation of 30 minutes. The dough is then arranged in molds, given a final fermentation of 45 minutes, and baked at 200 to 210° *C*.

A regional specialty product is a cake made of equal weights of flour, sugar, shortening, and eggs. The composition gives the product the name "Four Quarters." The sugar and eggs are well beaten together, the shortening and flour added, and the batter is baked at 200 to 210° C.

Figure 18. *Pate à choux*.

Figure 19. *Tartes aux fruits.*

Figure 20. *Babas* or *savarins.*

V. BISCUIT PRODUCTS

Biscuits are produced industrially on automated production lines, from the initial stocking of raw materials to the packaging of products.

The French biscuit industry is one of variety and diversity and new and original goods are constantly being added to the industrialized recipes.

Among biscuits, several types can be distinguished:

- Dry-sugared or salted biscuits, made from hard or semihard dough, such as *petit beurre, sablé, casse-croûte,* which may be filled (*goûters fourrés*).
- Cake biscuits and assortments made from soft dough. In this category are dry products such as *boudoirs* and *champagne* biscuits, *langues de chat,* etc., and soft biscuits such as *cuillers, madeleines,* macaroons, *petit fours,* etc.
- Small wafers are prepared from a fluid batter; they may be consumed plain (*éventails, cigares, cigarettes*) or filled and/or enrobed.
- Spiced cakes comprising regional specialties (*Couques du Nord, nonettes de Dijon*) the composition of which includes rye and honey.
- In addition, expanded or extruded articles are increasingly important in the consumption of accessory foods, especially appetizers.

Few additives are used in biscuit making, but they may include baking powder (mixture of a salt and an organic acid), emulsifiers (lecithin, mono- and diglycerides), and antioxidants (ascorbic acid or BHT). The use of artificial color is now practically abandoned, not because of possible toxicity, but in response to campaigns conducted by consumers associations.

All biscuits are packaged with particular care. In most cases, packages are made of small quantities of products (mostly at 100 to 200 g) that permit sealing and have a shelf life of six months.

VI. THE FUTURE OF BISCUIT AND PASTRY PRODUCTS IN FRANCE

Although large supermarkets have developed in France during the past 20 years, the artisan's effort in baking or cake making has scarcely been affected. Paralleling this transformation, manufacture of bread and some types of cake has been slowly industrialized, but this only represents a small part of the market.

The evolutionary changes in the marketplace are intimately related to the way of life of the French people. Even though changing, they seem tied to their traditional foods, especially fresh and crusty bread. French researchers today are seeking improved quality in the diversity and freshness of products. In this way they expect to maintain the important part they have in the bread and cake industry. It will, however, be difficult to maintain a commercial structure that will of necessity be influenced by the economic and social conditions in the country.

PART THREE: SOFT WHEAT USES IN THE FEDERAL REPUBLIC OF GERMANY

I. WHEAT PRODUCTION AND WHEAT QUALITY

A. Production and Yield

From 1967 to 1980 (except 1975), the acreage devoted to wheat production in the Federal Republic of Germany has continually increased (10–13) (Table VI).

This increase in wheat production was undoubtedly caused by the policy of the European Community guaranteeing farmers a higher intervention price for wheat than for other cereals. In addition, there are fewer problems in the production of wheat than of rye, barley, or maize, and the yield per hectare is higher as well. Not only is the yield of rye less, but there is always the uncontrollable problem of sprouting, so farmers do not know during the growing season whether the crop can be sold as "bread-rye" for a higher price or as "feed-rye" at a lower price. In the case of maize, growing conditions do not assure low moisture contents, and therefore this crop may have to be dried immediately after harvest.

The recent large increase in wheat yield has been produced by the breeding of new cultivars and the optimization of conditions for fertilizer application. Until 1973, there was no production of high-yielding cultivars characterized by poor baking properties. In the future the average yield per hectare can be expected to increase further if more of this nonbaking wheat is grown (14).

The figures in Table VII (16,17) show that the overall quality depends on the climatic conditions of the crop year. Because only one-third of the West German crop is used for milling and baking, the miller can choose wheat according to his own specifications. As a result, the average quality of wheat for milling is higher than that of the average for the crop.

B. Quality Evaluation of Wheat Cultivars

The quality of the soft wheat crop for processing depends directly on the accomplishments of wheat breeders. During the last 10 years, breeders have been successful in increasing yield and also in improving quality in new cultivars, especially spring wheat.

In West Germany, all licensed wheat cultivars have been tested for milling and baking quality by the Federal Institute of Cereal and Potato Processing in

TABLE VI
Soft Wheat Production and Yield in West Germany, 1971-1977

	Area (ha)	Yield (te/ha)
1971	1,550,000	4.6
1973	1,610,000	4.5
1975	1,570,000	4.5
1977	1,590,000	4.5

TABLE VII
Average Baking Quality of Wheat in West Germany

	Protein Content (N × 5.7, d.m.b.) (%)	Sedimentation Value
1974	11.5	27
1975	12.4	28
1976	14.3	40
1977	12.1	31

Detmold. The methods used are described in Arbeitsgemeinschaft Getreideforschung (2). Originally, baking strength evaluation was based on gluten content, swelling number, and Pelshenke test number, but after World War II, a standard baking test was introduced. In 1964, however, the testing system was again modified because of the poor relationship between laboratory evaluation and the experience of the milling and baking industries (7–9).

In 1964 the protein content-sedimentation value system was introduced, supplemented by the so-called "Rapid-Mix-Test" (2,8,9). This baking test takes into account new developments in baking technology. The dough (5% yeast, 1.5% salt, 1% sugar, 1% fat, and 20 ppm ascorbic acid on flour basis) is developed by high-speed mixing; the fermentation time is short and uses high dough temperatures (32° C). (The enzyme activity is adjusted to a falling number of 250 sec by adding malt flour.) The dough is baked not in a tin but as small rolls. The final result is expressed as the volume in ml/ 100 g of flour, and in addititon, the elasticity of the crumb and the percentage of rolls with satisfactory appearance (Ausbund) are reported.

Using this baking test with German wheat cultivars, we found the following correlation values (10): protein content-loaf volume, $r = 0.85$ and sedimentation value-loaf volume, $r = 0.81$. The loaf volumes are in good agreement with the findings of the Canadian Remix-baking test.

All new cultivars with good baking quality are also tested with a special baking test to determine their so-called blending value. If a mixture of 25% of the tested cultivar with 75% of a standard wheat reaches a significantly higher loaf volume than that calculated from the percentages of the two components, this cultivar is evaluated as *Aufmischweizen* (wheat for blending purposes). Based on the results of the baking test, the wheat varieties are classified into grades A (bread volume > 107% of Grade B), B (bread volume of wheat cultivars Jubilar/ Caribo), and C (bread volume < 87% of Grade B).

Group A is divided into subgroups A_{Q+}, where Q+ means an excellent blending value and Q a medium blending value. If a new cultivar shows sticky dough after mixing, the variety is marked T– (for example, B_{T-}). After being licensed, each new variety is described by the Bundessortenamt, a federal institution responsible for the varietal register (20).

C. Production of Quality Wheat

In the last 15 years, the extent to which better wheat cultivars are grown in West Germany, has improved remarkably as shown in Table VIII (11–13).

Grade A wheats have increased 100%; grade C wheats have decreased dramatically. This was encouraged by a higher price for quality wheat being paid

TABLE VIII
Distribution of Grades (%) of Wheat Grown in West Germany, 1965-1977

Grade	1965	1970	1972	1975	1976	1977	1979	1980
A	20	26	30	28	30	36	4	4
B	52	68	66	67	64	56	52	54
C	18	6	4	5	6	8	7	5

on a voluntary basis by the milling industry.

This quality movement can also be demonstrated by the concurrent changes in the average composition of the milling grist (Table IX). The percentage of imported U.S. and Canadian quality wheat (basis 14% protein) has decreased; French wheat is an important component in the German milling grist, especially as a result of the French wheat price policy. Currently, there is a tendency to use a decreasing proportion of French wheat.

D. Grading of Wheat

The European Community (EC) has no official grading system for soft wheat. There are only minimum figures for wheat for intervention (Table X). All wheat parcels within these limiting figures and without any negative odor have to be accepted for intervention at the guaranteed intervention price.

Until recently, farmers and trade people were not interested in a classification system. Therefore, grades were established depending on the buying departments of different mills.

The unexpected production of so-called nonbaking wheat has changed the intervention regulations. This nonbaking or nonmachinable wheat manifests itself only after mixing to remarkably sticky doughs. Mechanical dough handling is therefore very difficult—almost impossible—because these doughs adhere to machines and transport systems. Stickiness, not only a varietal factor, also depends on the area of production (51). Therefore, it is not sufficient to name only the wheat cultivar with sticky doughs.

The EC has decided to take the wheats with sticky doughs into intervention. The differentiation between machinable and nonmachinable wheat is made by means of the so-called European Baking Test, developed as a result of very close cooperation among official EC grain institutes. In this test, wheat is classified as nonmachineable if a dough shows sticky properties after mixing and/or rounding.

At present, production of wheat with sticky doughs is not a major problem in West Germany. The production of "nonbaking" wheat was 2, 7, 9, 7, 6, and 5% of

TABLE IX
Average Composition of Wheat Blends Milled in West Germany[a]

	Wheat (%)		
	German	French	U.S. or Canadian
1965-1966	70	6	24
1970-1971	67	12	21
1974-1975	76	14	10
1976-1977	74	21	5
1978-1979	82	10	8

[a]Adjusted to give a loaf volume of 650 ml/100 g of flour (75% extraction; ash content 0.55% d.m.b.) using the standard Rapid mix test baking procedure.

total wheat production in 1973, 1974, 1975, 1978, 1979, and 1980, respectively. The proportion is still less than 10% (15). The impact of such wheat may become severe if it becomes necessary for all wheat to be used for baking purposes in the future. In 1979, only one third of the German wheat production was used in milling and baking, so a percentage of the crop could be nonbaking wheat. Since dough stickiness is dominant in wheat blend, this wheat must be kept separate from baking wheat. All blending of nonbaking wheat with baking wheat should be forbidden.

II. SOFT WHEAT MILLING

A. Milling Grist

There are two standard German milling grists: one for bakery purposes and the other for family flour. The standard milling grist for bakery flour consists of U.S. or Canadian wheat, French, and home-grown wheat. The quality level is adjusted to give a loaf volume of 630 to 650 ml/100 g of flour, and at the moment, about 10% U.S./Canadian wheat is included in the grist (Table IX). Only home-grown wheat is used for production of family flour, and the quality is adjusted to 580 to 600 ml/100 g of flour. Other milling grists with higher percentages of imported quality wheat may be prepared for special purposes (18) (Table XI). Another range of flour qualities is necessary for bakeries producing confectionery goods (69) (Table XII).

B. Principles of Milling

During the last 10 years soft wheat milling technology has not changed significantly. All technical developments have been aimed at reducing production costs and improving the quality of the flour, meal, and by-products.

In the screen room, instead of washing machines and stone separators, special equipment for intensive cleaning of the kernel surface is used. The development of milling techniques during the last 25 years is reflected by the following figures (50):

	1949	1974
Length of rolls:	45 to 40 mm/te/day	15 to 20 mm
Area of sifting:	2.5 to 3 m^2/te/day	0.5 to 1.0 m^2
Passages:	25 to 30 flour streams	15 to 20 flour streams
Space requirement:	350 m^2/te/day	150 m^2
Energy requirement:	70 kwh/te	60 kwh/te

C. Flour and Whole-Meal Regulations

Flour and whole-meal standards in the German system are based on ash content (d.m.b.), and only flour and whole meal with a certain ash content can be sold (Table XIII). If the ash content is between the flour types, the flour must be blended to fall within the range of the higher or the lower ash type (19,43). Flour

TABLE X
Characteristics for Wheat for Intervention (EC Regulation) Stock

hl (100 L) weight	min. 72kg
Moisture content	max. 16%
"Besatz" content	
Broken kernels	max. 5%
"Korn"-besatz	max. 5%
Sprouted kernels	max. 6%
"Schwarzbesatz"	max. 3%

[a] Data from Schäfer (48).

TABLE XI
Flour Characteristics for Roll and Bread Production

	Flour Type		
	550		1050
	(Rolls)	(Toast bread)	(Mixed Bread)
Protein content (%)	11.0–12.0	12.5–13.5	12.0–13.5
Maltose (%)	1.8–2.2	1.8–2.2	2.5–3.2
Falling number (sec)	230–300	250–300	230–300
Water absorption (%)	53.0–55.0	54.5–56.5	56.0–58.0
Bread volume (ml)	620–650	650–700	500–550

TABLE XII
Flour Characteristics for Confectionery Goods

	Flour Type			
	405	550	550	550
			Yeast Doughs	
	Cake	Biscuit	Light (10–20% fat)	Heavy (20–50% fat)
Protein content (%)	6.0–9.0	9.5–11.0	12.0–13.0	13.0–14.0
Maltose (%)	0.8–1.5	1.5–2.0	1.5–2.0	1.5–1.8
Falling number (sec)	> 300	200–300	250–300	250–300
Water absorption (%)	48.0–50.0	51.0–53.0	53.0–55.0	55.0–57.0

TABLE XIII
Official German Flour and Whole-Meal Types

Type	Average Ash Content (%)	Production 1978-1979 (%)	Ash Content Lowest (%)	Ash Content Highest (%)
405	0.405	15	0.38	0.44
550	0.55	44	0.49	0.58
630	0.63	4	0.60	0.70
812	0.815	6	0.75	0.87
1050	1.05	12	1.00	1.150
1600	1.60	2	1.55	1.75
1700 (whole meal)	1.70	1	1.60	1.90
Export flour	0.52	16	. . .	0.52

type 405 is used for confectionery goods, especially cakes, whereas type 550 and sometimes type 630 is family flour or all-purpose baking flour for rolls, white bread, and toast bread. Type 812 and especially types 1050 and 1600 are mixed in the bakery with different percentages of rye flour to bake the typical German mixed bread.

Only a very small percentage of milled wheat is used as whole meal to produce a whole-meal bread or Graham bread. Whole-meal flour contains all parts of the kernel including the wheat germ. The baked product is called whole-meal bread.

The remainder of production (16%) is used for export purposes.

D. Flour Treatment

The Federal Republic of Germany has very strict and restrictive food laws, and the use of additives in a basic foodstuff like bread is especially limited.

Currently, enzymatic treatment with amylases and proteases is permitted. If necessary, the main flour types are treated in the mill with malt flour (mostly barley). The added quantity is adjusted to a falling number of 250 sec. For oxidative treatment, only ascorbic acid is allowed, between 1 and 4 g/100 kg of flour (10 to 40 ppm). For the treatment of whole meal, up to 10 g/100 kg meal may be added. Addition of potassium bromate or ammonium persulfate is strictly prohibited. Besides the cost, there is a great advantage in using ascorbic acid instead of potassium bromate; overtreatment in connection with long flour storage time is avoided and the flour develops optimal baking strength 2 to 3 days after milling.

Bleaching and chlorination of flour are also strictly forbidden. This restriction causes some trouble, especially with the production of high-ratio cake flour, but this problem also can be solved by selecting the right wheat and mixing special flour streams in the mill.

It is our opinion that we should allow as few additives as possible. Very often consumers speak about chemicals in basic foods but do not understand the technological necessity. The restricted use of additives for flour treatment in West Germany has demonstrated that consumer reaction can be accepted by millers and bakers.

E. Flour and Whole-Meal Quality

Flours for rolls, white bread, and toast bread are fairly standardized, so bakeries can use flours from different mills without great difficulty. Also, generally the quality of flour is a little higher in northern than in southern Germany.

For baked products other than bread, flour quality must be adjusted for the equipment. Therefore, there is no standard flour for biscuits or rusks, but the quality level must fit the machines used for the particular baked goods.

With whole meal, the particle size distribution is one of the most important quality factors. A finer whole meal has a higher water absorption and a lower loaf volume and vice versa.

If a mill produces wheat flours the quality of which is acceptable to the customers, the mill should maintain uniform quality levels. When a change in flour quality is necessary (for example, a new wheat crop), the mill should inform its customers about the change and advise them how to modify their production procedure.

F. Production of Ready Mixes

Soon after World War II, some large German milling companies started—influenced by the U.S. market—to produce ready mixes for housewives and bakers. Although there are no official figures, an estimated 5% of the total flour production is sold as ready mixes.

The principal ready mixes are made for rolls, toast bread, yeast doughs (5 to 10% fat), sponge cake, cakes, and yeast-raised donuts. Ready mixes can be produced only by large milling companies, for there should be a well-equipped laboratory to monitor the raw materials for the mixes and a test bakery to check the production daily. Ready mixes can be sold if backed up by a good service program, and the producer needs skilled bakers to introduce and demonstrate the products to the baker.

Depending on the mix, water and yeast or eggs must be added. The overall quality of the mixes is good, but bakers should follow the production procedure given by the producer. It was feared that the mixes would cause uniformity of baked products, but 20 years' experience has shown that skilled bakers are able to produce products with individual form and taste from the same ready-mix.

III. BAKED PRODUCTS

According to the German Food Law, baked products are divided into two types, i.e., the bread recipe allows a maximum of 10 parts sugar and/or fat to 90

Figure 21. Wheat roll.

parts flour; products with more than 10 parts sugar and/or fat to 10 parts flour in the recipe are called *Feine Backwaren*.

A. Bread

Rolls with a recipe similar to that of bread belong in the bread category.

WHEAT ROLLS (*Weizenbrötchen*)

Rolls are eaten for breakfast every day by one of every three German families. Because the shelf life is only several hours, many bakers produce rolls (Fig. 21) twice a day, or the rolls are stored at home in the deep-freezer. A typical formulation is: wheat flour, 0.980 kg; additive mixture, 0.020 kg; sugar, 0.005 kg; fat, 0.010 kg; yeast, 0.050 kg; salt, 0.020 kg; and water, about 550 ml. The dough is mixed for 10 minutes in a normal mixer (dough temperature; 26 to 27° C; dough yield: 155%). After a fermentation time of 10 to 15 minutes, the dough (1.8 kg) is divided into 30 dough pieces and rounded by a special machine. After 3 minutes, the dough pieces are passed through a sheeting roll and rolled up. For proofing, the dough pieces are placed on a wooden board covered with cloth (right side to the bottom) for 35 to 45 minutes at a room temperature of 30 to 35° C (relative humidity 75 to 80%). The baking time is about 20 minutes at 250° C, and at the start plenty of steam has to be added (42).

HUNTER'S ROLLS (*Jägerbrötchen*)

Hunter's rolls (Fig. 22) are produced from a blend of wheat and rye flour: wheat flour (type 550), 0.700 kg; rye flour (type 997), 0.150 kg; sourdough, 0.285

Figure 22. Hunter's roll.

kg; baking improver, 0.015 kg; fat, 0.030 kg; caraway (ground), 0.004 kg; yeast, 0.060 kg; salt, 0.020 kg; water, 550 ml. The sourdough consists of 150 g of rye flour (type 997), 15 g of sour concentrate (Anstellgut), and 120 ml of water. The maturing time is 18 hours. The flour mixture therefore consists of 70% wheat flour and 30% rye flour.

The dough is mixed for 10 minutes at a temperature of 27° C (dough yield: 167%). After fermentation of 10 minutes, the dough (1,150 g) is divided and rounded to form 30 dough pieces. During the proofing (35 to 45 minutes), the dough pieces are cut twice. The baking time is about 18 minutes at 240° C with a medium amount of steam at the beginning of baking.

WHEAT BREAD

Wheat bread (Fig. 23) is prepared according to the following formulation: wheat flour (type 550), 1 kg; baking improver, 0.030 kg; yeast, 0.040 kg; salt, 0.020 kg; and water, about 550 ml. The dough is mixed for 2 minutes with a high-speed mixer (dough temperature: 28 to 30° C) or for 10 minutes with a normal mixer (dough temperature: 25 to 27° C). After fermentation of 10 to 15 min, the dough is scaled to the appropriate weight (580 g for a 500-g bread, or 1,150 g for a 1,000-g bread) and molded. After proofing of about 50 minutes at a room temperature of 35° C and 70 to 80% relative humidity, the dough is baked at a decreasing temperature of 240 to 220° C for 30 to 35 minutes (500-g bread) or 40 to 45 minutes (1,000-g bread). At the beginning of the baking process, a medium amount of steam must be added.

TOAST BREAD

Toast bread (Fig. 24) formulation includes the following ingredients: wheat flour (type 550), 1 kg; baking improver, 0.020 kg; fat, 0.050 kg; milk powder, 0.020 kg; sugar, 0.020 kg; Ca-propionate, 0.003 kg; yeast, 0.060 kg; salt, 0.018 kg; and water, about 540 ml. The dough is mixed for 2 minutes with a high-speed mixer (dough temperature: 30° C) or for 10 minutes with a normal mixer (dough temperature: 26 to 28° C). After fermentation of 8 to 10 minutes, the dough is scaled to the appropriate weight and rounded. After a second fermentation of 10

Figure 23. Wheat Bread.

minutes, the dough is molded, twisted, and placed in a tin. The tins are 26 to 28 cm long, 9 to 10 cm wide, and 9 cm high. After a proofing of 50 to 60 minutes at 35° C and 80% relative humidity, the tins are covered. The baking time is 35 to 40 minutes at a decreasing temperature of 240 to 220° C and a medium amount of steam at the beginning. After cooling the bread is sliced and wrapped (56,57).

WHEAT MIXED BREAD (*Weizenmischbrot*)

Wheat mixed bread (Fig. 25) is the German standard bread. About 50% of the total bread production is of this type. If the proportion of rye flour in the recipe exceeds 50%, the bread is called rye mixed bread.

Wheat mixed bread consists of the following ingredients: wheat flour (types 550 and 1050), 0.700 kg; rye flour (type 997), 0.100 kg sourdough, 0.380 kg,

Figure 24. Toast bread.

Figure 25. Wheat mixed bread.

yeast, 0.020 kg; salt, 0.020 kg; and water, 400 ml. The sourdough consists of: rye flour, 0.200 kg; water, 0.180 kg; and sour concentrate, 0.040 kg. This bread thus consists of 70% wheat flour and 30% rye flour.

The dough is mixed for 10 minutes with a normal mixer (dough temperature: 27° C). After a 20-minute fermentation, the dough is scaled to the appropriate weight and rounded. Proofing takes 50 minutes at 30 to 35° C and 65 to 70% relative humidity. The baking time is 50 minutes at temperatures of 250 to 220° C.

B. Other Baked Goods (*Feine Backwaren*)

RAISIN BREAD (*Rosinenstuten*)

Raisin bread (Fig. 26) is produced mostly for consumption on weekends and has a shelf life of several days. It is prepared using a formulation of wheat flour (type 550), 1 kg; sugar, 0.060 kg; glucose, 0.030 kg; margarine, 0.120 kg; yeast, 0.070 kg; salt, 0.016 kg; water, about 460 ml; and raisins, 1 kg. The dough is mixed for 10 minutes in a normal mixer at a dough temperature of 24 to 26° C. The washed raisins are then carefully mixed in. After a fermentation of 10 minutes, the dough is scaled to 850 g and rounded. Five minutes later the dough is molded, put into a tin, and proofed for 50 minutes at 28 to 30° C. Raisin bread is baked for 35 minutes at an oven temperature of 210° C (52,54).

STOLLEN

This product is traditionally baked for Christmas and is also called Christmas-*Stollen*. *Stollen* (Fig. 27) is now available throughout the year.

Stollen is usually baked by a sponge dough method. The sponge consists of wheat flour (type 550), 1 kg; yeast, 0.150 kg; water, about 500 ml; and sugar, 0.012 kg. The dough is mixed (dough temperature: 27 to 28°) and allowed to rest 20 minutes.

A premixture is prepared from wheat flour (type 550), 1 kg; margarine, 0.750 kg; sugar, 0.200 kg; marzipan, 0.075 kg; citrus fruit peel, 0.010 kg; salt, 0.025 kg; water, about 75 ml. The ingredients are mixed together slowly for 5 minutes (dough temperature: 26 to 28° C) and allowed to rest for 20 minutes. The

Figure 26. Raisin bread.

following mixture is then added and mixed in carefully: raisins, 1.250 kg; candied orange peel, 0.125 kg; candied lemon peel, 0.125 kg; almonds, 0.250 kg; rum 25 ml. The fermented sponge dough is added to this mixed dough, mixed further, scaled to 75-g pieces, rounded, and fermented for another 10 minutes. The dough is placed on a sheet for a final 15-minute proof at 32° C and 70 to 80% relative humidity. The baking time is 35 to 40 minutes in a decreasing temperature of 230 to 190° C. Steam is added at the beginning of the baking. After cooling for an hour, the product is spread with liquid margarine and covered later with a special sugar.

SPONGE CAKE

The typical German cake is *Sandkuchen*, shown in Figure 28. This product has at least 20% of egg and 20% butter or margarine, calculated on batter weight.

Figure 27. *Stollen.*

Figure 28. *Sandkuchen.*

Equal quantities of flour and wheat starch are included in the formulation as follows: wheat flour (type 550), 0.150 kg; wheat starch, 0.150 kg; margarine, 0.180 kg; sugar, 0.240 kg; whole egg, 0.195 kg; water, 45 ml; dry stabilizer (with 30% monodiglycerides), 0.009 kg; baking powder, 0.005 kg; salt, 0.002 kg; and flavor according to taste.

All ingredients are combined, and the batter is produced according to the "all-in" procedure. First, the ingredients are mixed at low speed for 1 min, to achieve homogeneity. Then, the mixing is continued until a batter weight of 0.730 to 0.760 g is reached. The batter is scaled into pans and baked for 45 minutes at between 190 and 210° C.

TARTS

Tarts are shown in Figure 29. A so-called biscuit tart has at least 66% egg in the recipe, calculated on flour plus starch basis.

Typically, tarts are prepared from a formulation of wheat flour (type 550), 0.150 kg; wheat starch, 0.150 kg; sugar, 0.225 kg; whole egg, 0.280 kg; dry stabilizer, 0.009 kg; baking powder, 0.014 kg; water, 95 ml; and flavor according to taste. The production of the batter corresponds to that of *Sandkuchen*. The mixing is continued until a liter weight of 0.380 to 0.440 g is reached. The baking time is about 12 to 15 minutes at a temperature between 200 and 230° C (40,53).

BISCUITS (*Keks*)

Each type of biscuit is characterized by compositional limitations. Thus, butter biscuits (Fig. 30) have a minimum of 10% butter in the dough, milk biscuits contain at least 200 ml of milk per kilogram of flour, and egg biscuits must include at least four eggs per kilogram of flour.

The basic formulation for biscuits includes wheat flour (type 550), 0.300 kg;

Figure 29. Tarts.

margarine or butter, 0.030 kg; sugar, 0.075 kg; water, about 70 ml; baking powder, 0.004 kg; salt, 0.002 kg. The dough is mixed for 25 to 30 minutes in a mixer at slow speed (40 to 60 rpm). The dough temperature reaches 38 to 42° C. The dough is rolled several times and laminated. After repeated rolling, it is cut into appropriate pieces. The baking conditions (with little steam) are 8 to 9 minutes at 200 to 220° C (55).

LEBKUCHEN

Lebkuchen (Fig. 31) is another Christmas product; its production is similar to the sponge dough method. The "sponge" is a mixture of wheat flour (types 550 and 1050), 0.200 kg and invert syrup, 0.130 kg. This dough mixture may be stored as long as 5 months at 5 to 10° C.

For *Lebkuchen*, the following formulation is used: "sponge," 0.750 kg; invert syrup (80% d.m.b.), 0.150 kg; wheat flour, 0.600 kg; powdered sugar, 0.160 kg; water, 90 ml; ground biscuit, 0.300 kg; and baking powder, variable. After mixing for 10 minutes in a mixer at slow speed (as for biscuits), the dough is rolled and cut into appropriate pieces. The baking conditions are 15 minutes at 200° C (49).

RUSK

The rusk shown in Figure 32 is a typical German product. The Dutch type is round. Rusk is prepared using the recipe: wheat flour (type 550), 1.000 kg; sugar, 0.150 kg; fat, 0.080 kg; egg, 0.005 kg; water, about 440 ml; lecithin, 0.005 kg; and salt, 0.010 kg. Rusk may be produced by the sponge dough or straight dough method. After mixing, fermenting, and final proofing, the dough is baked in long

Figure 30. Biscuits.

tins at 210 and 220° C for 30 minutes. After 24 hours, the baked product is cut into 8 to 9 mm slices and roasted.

C. Consumption

There are not official figures in West Germany for the total consumption of bread and other baked goods. Official data are based on the consumption of the

Figure 31. *Lebkuchen*, plain (above) and glazed (below).

Figure 32. Rusk.

so-called four-person family and do not take into consideration the bread consumption outside the family in, for example, restaurants, canteens, and hospitals.

Based on flour production, the consumption per capita per year is estimated to be: wheat and rye mixed bread, 41 kg; rolls, 7 kg; wheat bread, 7 kg; toast bread, 7 kg; rye and wheat meal bread, 4 kg; rye bread, 2 kg; and other baked goods 7 kg, for a total of 75 kg.

LITERATURE CITED

1. Anonymous. Classification des boulangers-Année 1967 à 1976. 1976. La Meunerie Française. 23:30 May.
2. Arbeitsgemeinschaft Getreideforschung. 1971. Standard-Methoden für Getreide, Mehl und Brot., 5th ed. Verlag Moritz Schäfer, Detmold.
3. Axford, D. W. E., Chamberlain, N., Collins, T. H., and Elton, G. A. H. 1963. The Chorleywood bread process. Cereal Sci. Today 8:265-266,268,270.
4. Axford, D. W. E., Colwell, K. H., Cornford, S. J., and Elton, G. A. H. 1968. Effect of loaf specific volume on the rate and extent of staling in bread. J. Sci. Food Agric. 19:95-101.
5. Bell, B. M., Daniels, D. G. H., and Fisher, N. 1977. Physical aspects of the improvement of dough by fat. Food Chem. 2:57-70.
6. Bell, B. M., and Fisher, N. 1977. The binding of model shortenings during mixing of mechanically developed bread dough from fresh and stored flours. J. Am. Oil Chem. Soc. 54:479-483.
7. Bölling, H. 1964. Zur Bewertung analytischer Untersuchungsergebnisse bei deutschen Aufmischweizen. Mühle 101:879-882.
8. Bölling, H. 1969. Bestimmung des Aufmischwertes mit dem Rapid-Mix-Test. Mühle 106:2-4.
9. Bölling, H. 1973. Über die Substituierbarkeit ausländischer Weizenprovenienzen durch deutsche Qualitätsweizensorten unter besonderer Berücksichtigung ökonomischer Aspekte. Getreide, Mehl, Brot 27:22-28.
10. Bölling, H. 1974. Wandel und Fortschritt im Qualitätsweizenanbau in der Bundesrepublik Deutschland. Getreide, Mehl, Brot 27:197-206.

11. Bölling, H., and Gerstenkorn, P. 1975. Die Qualität der deutschen Weizenernte 1975. 1. Mitteilung: Quantitatives und qualitatives Ergebnis in Bund und Ländern. Mühle und Mischfuttertechnik 112:533-540.
12. Bölling, H., and Gerstenkorn, P. 1976. Die Qualität der Weizenernte 1976. 1. Mitteilung: Quantitatives und qualitatives Ergebnis in Bund und Ländern. Mühle und Mischfuttertechnik 113:585-595.
13. Bölling, H., and Gerstenkorn, P. 1977. Die Qualität der deutschen Weizenernte 1977. 1. Mitteilung: Quantitatives und qualitatives Ergebnis in Bund und Ländern. Mühle und Mischfuttertechnik 114:603-611.
14. Bölling, H., and Meyer, D. 1975. Zur Verarbeitungsqualität neuer Weizensorten - ein Beitrag zu Rohstoffproblemen der Müllerei. Getreide, Mehl, Brot 29:135-141.
15. Bölling, H., and Zwingelberg, H. 1975. Die Qualität der deutschen Weizenernte 1975. 2. Mitteilung: Mahl- und Backqualität. Mühle und Mischfuttertechnik 112:591-594.
16. Bölling, H., and Zwingelberg, H. 1976. Die Qualität der Weizenernte 1976. 2. Mitteilung: Mahl- und Backqualität. Mühle und Misch futtertechnik 113:601-604.
17. Bölling, H., and Zwingelberg, H. 1977. Die Qualität der deutschen Weizenernte 1977. 3. Mitteilung: Mahl- und Backqualität. Mühle und Mischfuttertechnik 114:649-652.
18. Bracht, TH. J. 1976. Weizenmehl für Brot und Kleingebäck. Getreide, Mehl, Brot. 30:80-82.
19. Bundesanzeiger. 1961. 17. DVO zum Getreidegesetz vom 21. Juli 1961. Bundesanzeiger Nr. 244/1961.
20. Bundessortenamt. 1975. Beschreibende Sortenliste 1975. Alfred Strothe Verlag: Hannover.
21. Chamberlain, N., and Greenwood, C. T. 1975. Yeast and bread flavour in Great Britain. Proc. 6th Meeting Int. Prob. of Modern Cereal Processing & Chem. Institut für Getreideverarbeitung, Bergholz-Rehbrücke, D.D.R. pp. 317-323.
22. Colwell, K. H., Axford, D. W. E., Chamberlain, N., and Elton, G. A. H. 1969. Effect of storage temperature on the aging of concentrated wheat starch gels. J. Sci. Food Agric. 20:550-555.
23. Cornford, S. J. 1970. Importance of cooling in loaf quality control. Baking Ind. J. 3(6):26, 29-30.
24. Daniels, D. G. H., and Fisher, N. 1976. Release of carbon dioxide from dough during baking. J. Sci. Food Agric. 27:351-357.
25. Daniels, N. W., Frazier, P. J., and Wood, P. S. 1971. Flour lipids and dough development. Bakers Dig. 46(4):20-26.
26. Drapron, R., and Beaux, Y. 1969. Sur l'oxydation des acides gras essentiels par la lipoxygénase (E.C. 1.13.1.13) en panification et la formation de composés volatils intervenant dans l'arôme du pain. Compt. Rend. Acad. Sci. Paris 268:2598-2601.
27. Fisher, N., Bell, B. M., and Rawlings, C. E. B. 1973. Lipid binding in flour, dough, and bread. J. Sci. Food Agric. 24:147-155.
28. French National Center for Coordination of Studies and Research on Nutrition and Food. 1977. Recueil des usages concernant les pains en France. Colloque 'Le Pain' et journées d'information, 14–15 Nov.
29. Gough, B. M., Whitehouse, M. E., and Greenwood, C. T. 1978. The role and function of chlorine in the preparation of high-ratio cake flour. Crit. Rev. Food Sci. Nut. 10:91-113.
30. Greenwood, C. T. 1976. Rôle de l'amidon endommagé en meunerie. Bull. Anciens Elèves d'Ecole de Meunerie. ENSMIC 271:13-21.
31. Greenwood, C. T., and Ewart, J. A. D. 1975. Hypothesis for the structure of glutenin in relation to rheological properties of gluten and dough. Cereal Chem. 52:146r-153r.
32. Greenwood, C. T., and Stewart, B. A. 1976. Wheat grading in the United Kingdom. Cereal Foods World 21:477-478.
33. Hodge, D. G. 1977. A fresh look at cake staling. Baking Ind. J. 9:17, 19.
34. Hodge, D. G., Woodward, R. J., and Wade, P. 1972. Studying the mixing process for sponge cakes. Baking Ind. J. April pp. 28-32, 197.
35. Hodge, D. G., Woodward, R. J., and Wade, P. 1972. Control of batter specific gravity. Baking Ind. J. May pp. 26-28, 30-32.
36. Hodge, D. G., Woodward, R. J., and Wade, P. 1972. Establishment of a small scale test procedure and the effects of changes in conditions of continuous mixing. Food Trade Rev. 42(10):9-17.
37. Lawson, R. 1972. Mathematical modelling of commercial (biscuit) ovens. Dechema-Monographien 70:367-387.
38. Lawson, R. 1974. Evaluation of bread oven performance. Bakers Dig. 48(1):52-56.
39. Lawson, R., and Maris, P. I. 1971. Focus

on automation: The baking industry. J. Inst. Electrical & Electronic Tech. Eng. July pp. 18-26.

40. Ludewig, H.-G., and Bretschneider, F. 1974. Standardisierung eines Backversuchs für Sandkuchen. Getreide, Mehl, Brot 28:124-128.
41. McDermott, E. E. 1974. Measurement of the stickiness and other physical properties of bread crumb. J. Food Technol. 9:185-193.
42. Morgenstern, G. 1970. Massnahmen zur Erzielung und Erhaltung einer röschen Kruste bei Brötchen. Bäcker-Ztg. 24(14):8-10, 9-11.
43. Mühlenstelle. 1973. Übersichten über die Vermahlung von Brot-getreide im Getreidewirtschaftsjahr 1972/73. Mühlenstelle, Bonn.
44. Pritchard, P. E., and Stevens, D. J. 1973. The influence of processing variables on the properties of wafer sheets. Food Trade Rev. 43(5):11-16.
45. Pritchard, P. E., and Stevens, D. J. 1974. The influence of ingredients on the properties of wafer sheets. Food Trade Rev. 44(6):12-18.
46. Pritchard, P. E., and Wade, P. 1972. Development of a test baking procedure for wafer sheets. Food Trade Rev. 42(8):9-15.
47. Robb, J., and Cornford, S. J. 1974. The use of freezing and refrigeration by the baking industry. Br. Baker March 1. pp. 1-8.
48. Schäfer, W. 1974. Die EG-Gesetzgebung für Mehl, Brot und feine Backwaren—Situation und Probleme. Getreide, Mehl, Brot 28:29-32.
49. Seibel, W. 1971. Mehlqualitäten für Lebkuchen. Zucker- und Süsswarenwirtschaft 24:107-109.
50. Seibel, W. 1974. Entwicklungen in der deutschen Müllerei wänrend der letzten 25 Jahre. Technica Molitoria 25:46-51.
51. Seibel, W., and Bölling, H. 1976. 'Nichtbackfähinge' Weizen—ein Rohstoffproblem für das Müllerei- und Backgewerbe. Lecture, ICC Cong., Wien, 1976.
52. Seibel, W., and Bretschneider, F. 1971. Standardbackversuch für Hefefeingebäcke in Kastenform. Brot und Gebäck 25:122.
53. Seibel, W., and Bretschneider, F. 1973. Standardisierung der Biskuitmassenherstellung. Getreide, Mehl, Brot 27:237-242.
54. Seibel, W., Ludewig, H.-G., and Bretschneider, F. 1975. Standard isierung des Backversuchs für Hefeteige. Getreide, Mehl, Brot 29:127-130.
55. Seibel, W., Ludewig, H.-G., and Bretschneider, F. 1976. Standard isierung eines Backversuchs für Hartkeks. Getreide, Mehl, Brot 30:101-105.
56. Seibel, W., and Stephan, H. 1970A. Bäckereitechnologische Probleme bei Toastbrot-Herstellung. Brot und Gebäck 24:116-117.
57. Seibel, W., and Stephan, H. 1970B. Lebensmittelrechtliche Fragen bei Buttertoastbrot und bäckereitechnologische Probleme bei der Toastbrot-Herstellung. Brot und Gebäck 24:21-28.
58. Seiler, D. A. L. 1971. How the microbiologist can help the baker. Br. Baker, Aug. 13:21-26, 32.
59. Seiler, D. A. L. 1971. How the microbiologists can increase shelf life. Br. Baker, Sept. 3. pp. 18, 20, 24.
60. Seiler, D. A. L. 1971. Emphasis on hygiene in handling cream. Baking Ind. J., Oct. pp. 18-20.
61. Thewlis, B. H. 1971. Fate of ascorbic acid in the Chorleywood bread process. J. Sci. Food Agric. 22:16-19.
62. Thewlis, B. H. 1974. The fate of potassium bromate when used as a breadmaking improver. J. Sci. Food Agric. 25:1471–1475.
63. Wade, P. 1970. Technological aspects of the use of sodium metabisulphite in the manufacture of semi-sweet biscuits. Food Trade Rev. 40(5):34-39.
64. Wade, P. 1971. Technology of biscuit manufacture: Investigation of the process for making semi-sweet biscuits. Chem. Ind. pp. 1284-1293.
65. Wade, P. 1972. Flour properties and the manufacture of semi-sweet biscuits. J. Sci. Food Agric. 23:737-744.
66. Wade P. 1972. Technology of biscuit manufacture: Investigation of the rôle of fermentation in the manufacture of cream crackers. J. Sci. Food Agric. 23:1021-1034.
67. Wade, P. 1972. Technology of biscuit manufacture: Investigation of the process for making cream crackers. J. Sci. Food Agric. 23:1213-1219.
68. Wade, P. 1972. Flour properties and the manufacture of cream crackers. J. Sci. Food Agric. 23:1221-1228.
69. Weiland, P. J. 1976. Weizenmehl für Feine Backwaren. Getreide, Mehl, und Brot 30:3, 82.
70. Willhoft, E. M. A. 1971. Bread staling—Experimental study. J. sci. Food Agric. 22:176-180.
71. Willhoft, E. M. A. 1971. Bread staling—Theoretical study. J. Sci. Food Agric. 22:180-183.

CHAPTER 7

SOFT WHEAT USES IN THE ORIENT

SEIICHI NAGAO
Nisshin Flour Milling Co., Ltd.
Tokyo, Japan

I. INTRODUCTION

In most nations of the Orient, rice rather than wheat is the staple food. Wheat is, however, the primary caloric source of the people of India, Afghanistan, and Pakistan, and wheat use is gradually increasing in almost all other nations. Displays of soft wheat flour confections may have been unusual in rice-consuming nations in the past, but they are becoming more common each year.

Some soft wheat is produced in nearly all Oriental countries, but most is imported, principally from the United States and Australia. U.S. western white, Australian standard white from Western Australia and Victoria, and Australian soft comprise the principal classes of imported soft wheats. Some of this wheat is milled to short extraction, high-quality flours, and these are processed into special products. Others are ground to all-purpose flours or even to whole-wheat meals. Occasionally, a blend of hard and soft wheats is milled.

An increasing diversity of confectionery products will offer a larger selection of more attractive and appetizing foods to the people in the Orient, thus enriching their culinary experience.

II. SOFT WHEAT CONSUMPTION

A. Japan

Total wheat imports by Japan increased from 4,685,000 metric tons (te) in 1970-1971 to 5,926,000 te in 1979-1980. More than 99.6% of the wheat imported in the latter period was distributed to flour mills by the Food Agency of the Japanese government. Of this, U.S. western white, Australian standard white from Western Australia and Victoria, and Australian soft wheats comprised the principal classes of soft wheat. Trial imports had been made of U.S. soft red winter and French soft wheats. However, factors of price, quality, and

TABLE I
Sources of Soft Wheat Used in Japan (in 1,000 te)

	1975	1976	1977	1978	1979
Domestic production	241	223	236	367	541
Imports					
U.S. western white	964	1,063	1,219	1,094	1,068
Australian standard white (Western Australia)	702	623	628	668	645
Australian standard white (Victoria)	17	83	0	0	0
Australian soft	34	0	0	0	0

TABLE II
Estimated Consumption of Flour in Japan by Wheat Class and Product Type

From soft wheat (34% of all flour)	
Confectionery products	14%
Japanese noodles	16%
Family flour	4%
From hard wheat (65% of all flour)	
Bread	36%
Chinese noodles	8%
Instant Chinese-type noodles	7%
Macaroni and spaghetti	2%
Other food uses	8%
Industrial uses	4%
From durum wheat (1% of all flour)	
Macaroni and spaghetti	1%

availability promoted the use of Australian and western U.S. white wheats as the principal imported soft wheats. Domestic Japanese wheat production, almost all soft red winter, totaled only 541,000 te in 1979, grown in Kyushu, Kanto, and Hokkaido areas. Domestic production and imports from 1975 to 1979 are given in Table I.

The estimated percentages of wheat flour use by class and product in Japan are shown in Table II. Japanese noodles, confections, and most family flour are produced from soft wheats, which comprise 34% of total flour consumption. The Chinese-type noodle and its "instant" counterpart are not produced from soft wheat flour, because some gluten strength is required for a quality product.

It would not be an exaggeration to say that any soft wheat confectionery product indigenous to any part of the world can be found today on a commercial scale in Japan. During Japan's long history, many soft wheat products were introduced from foreign countries and became integrated into the diet of the Japanese. Some of them were modified to become uniquely Japanese to accommodate the subtleties of flavor and texture; others have retained their original appearance.

The general classification of confections made from soft wheat flours in Japan are shown in Table III, which was prepared by rearranging the Japanese

TABLE III
General Classification of Soft Wheat Flour Confections in Japan

Main Group	Subgroup	Item[a]
Baked goods	Biscuits	Hard biscuit
		Soft biscuit
		Cookie
		Cracker
	Baked small goods	*Saga*[b]*-bolo*
		Soba-bolo
	Japanese cracknels	Tile-shaped cracknel (*Kawara-senbei*)
		Nanbu cracknel (*Nanbu*[b]*-senbei*)
	Wafers	Wafer
Fried goods	Donuts	Cake donuts
		Yeast donuts
	Fried dough cookies	*Karintou*
Fresh Japanese-style goods	Lightly baked buns	Chestnut bun with bean jam filling (*Kuri-manju*)
		Cathay bun with bean jam filling (*Kara*[b]*-manju*)
		Bean jam pancake (*Dora-yaki*)
		Japanese muffin with bean jam filling (*Imagawa-yaki, tai-yaki*)
	Steamed buns	Leavened bun (*Saka-manju*)
		Chemically leavened bun (*Mushi-manju*)
Fresh Western-style goods	*Castillas*	Nagasaki[b]*-castilla*
	Cakes	Sponge cake
		Butter cake
		Waffle
		Hot cake
		Chou à la crème
		Baumkuchen
Fresh Chinese-style goods	Steamed buns	Chinese bun with jam filling (*An-manju*)
		Chinese bun with meat filling (*Niku-manju*)

[a] Representative confections. Japanese names in parentheses.
[b] Names of locations (city, area, region, etc.).

standard classification of commodities to clarify the somewhat complicated soft wheat usage (8,37).

Several types of noodles are found in Japan, but they are customarily classified into four main categories: 1) Japanese, 2) Chinese, including the "instant" products, 3) buckwheat noodles, and 4) macaroni and spaghetti. Macaroni and spaghetti are produced from durum semolina and hard wheat flour, whereas buckwheat noodles are made from a blend of buckwheat and hard wheat flour. Thus, only Japanese noodles, a product of soft wheat flour, will be discussed in this chapter.

Family flour in the Japanese market, used mainly for cooking *tempura* (to be described later), is milled from soft wheat. In recent years, cake baking has become popular among housewives and soft wheat family flour is used for this purpose also. Hard wheat flour for home bread baking can be purchased at retail outlets.

A great many kinds of flour for a variety of Japanese noodles and confections are produced by the 209 flour mills located throughout Japan. The four largest milling concerns produce more than 65% of the flour used in Japan. Although most flour millers have a good command of their milling techniques to meet the complicated requirements of their customers, the most important and influential parameter in milling remains the quality of the wheat that is milled.

Soft wheat flour milled in Japan may be divided into two general categories: low-protein flours of various extractions for confections, and the type that may be termed Japanese noodle or all-purpose flour. Figure 1 shows the yields of various types or grades of flour resulting from these two milling processes, and Table IV presents typical ash and protein content ranges for the flours. Slightly different extractions may be obtained by different flour mills processing various wheat blends. The extraction rate of flour normally amounts to 72 to 80% of the wheat kernel. Usually, first-grade, second-grade, and low-grade flours are obtained. For the production of certain confectionery flours, however, a special or top-grade cake flour is often extracted from high-quality soft wheat, of which the most important is the western white wheat from the United States. A low protein content, desirable softness or mellowness of gluten quality, and high amylograph viscosity are the primary requirements in a flour of this type. Most confectionery flours are presently milled from 100% U.S. western white wheat. As defined in the Official Grain Standards of the United States (50), the subclass western white is a mixture of soft white and white club wheats. The greater the content of white club in the mix, the more favorable it is for the production of confectionery flours (31). A small quantity of Australian soft wheat, softer than

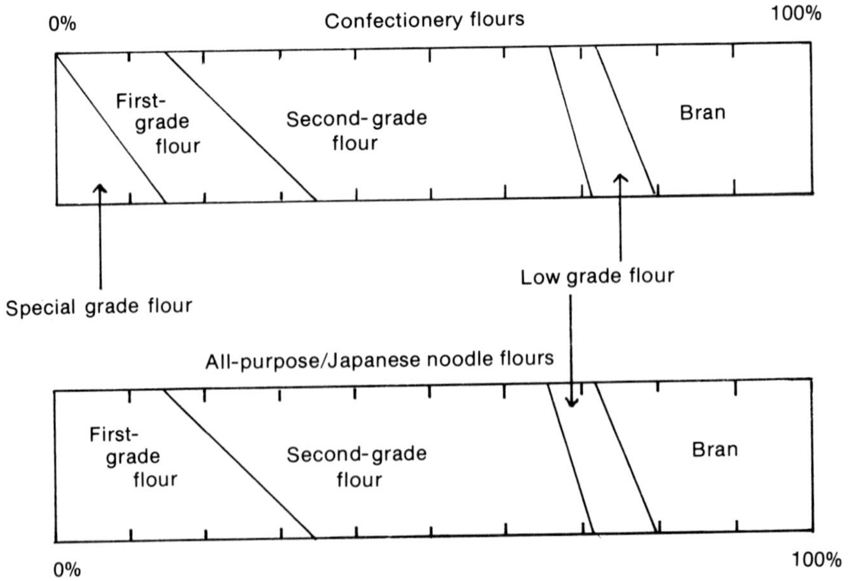

Figure 1. Approximate yields of various grades or types of flours milled from soft wheat.

other Australian wheats, can be blended with the western white wheat. The suitability of flour for confectionery in Japan may be estimated by test-baking sponge cakes and cookies. The sponge cake test is especially valuable in evaluating the confectionery processing quality of soft wheats (30).

Several extraction patterns are followed in the milling of noodle or all-purpose flour. The first- and second-grade flours are combined and used either as an all-purpose or as Japanese noodle flour, or the first-grade flour only may be used for Japanese noodle production and the second-grade flour finds its way into all-purpose flour. In all of these millings, the low-grade flour is used for industrial purposes or as feed. Since all-purpose flours are used principally for confections, soft gluten quality and moderately low protein content constitute primary requirements.

Japanese noodles were developed over the years using domestic wheat, and such products have long been accepted by the people. Thus, Japanese wheat flour is considered to have all the attributes essential for high quality in noodles, but the short supply of domestic wheat has forced flour millers and noodle manufacturers to find substitute wheats with the required protein, starch, and other compositional features (31). A Japanese noodle test is most valuable in evaluating the noodle processing quality of soft wheats (30). For the somewhat complicated needs indicated here, Australian standard white wheat from Western Australia and/or U.S. western white wheat are blended with Japanese wheat for Japanese noodle and/or all-purpose flours. In the wheat-producing areas of Japan, however, the indigenously grown wheat is still the main source of noodle flour. Australian standard white wheat from Victoria may be blended with domestic wheat for noodle flour, the ratios varying widely according to availability.

With the above milling and flour use considerations as background, the quality requirements in soft wheats may be expressed in a few words. First, the grain should be healthy and have favorable milling characteristics. The needs are common to mills throughout the world. Second, the flour should have quality characteristics favorable for processing into both confections and Japanese noodles. Wheat suitable for only one or the other purpose can be used to advantage, but the quantity required would be limited. Soft wheat with characteristics suitable for processing into confections is particularly desired.

TABLE IV
Average Protein and Ash Contents of Soft Wheat Flours in Japan

	Grade			
	Special	**First**	**Second**	**Low**
Protein (%)				
Confectionery flours	6.5–7.5	7.0–8.0	8.0– 9.0	
All-purpose flours		7.5–8.5	9.0–10.0	
Japanese noodle flours		8.0–9.0	9.5–10.5	
Ash (%)				
Confectionery flours	0.33–0.37	0.36–0.40	about 0.5	
All-purpose flours		0.36–0.40	about 0.5	1.5–3
Japanese noodle flours		0.36–0.40	about 0.5	

Western white wheat from the United States owes its dominance over other soft wheats imported into Japan to both its availability and its possession of quality attributes desired in Japan.

The domestic wheats have been grown principally in the Kanto area of Honshu, the main island, and in Kyushu. They are mostly of the soft red winter type and average 10 and 9% protein content, respectively. The cultivar Norin 61 comprised 39.5% of the 1979 acreage (36). Recently wheats grown in Hokkaido to the north have increased. These are mostly red winters, with some red springs, and are intermediate in kernel texture between hard and soft. Some quality improvement is expected.

B. South Korea

Total wheat milled and consumed in South Korea approximated 1.9 million te in 1976, 2.1 million in 1977, and 1.7 million in 1978. Of these quantities, soft wheat comprised about 60%. Domestic wheat production is small, only 36,000 te in 1978, and is mostly of the soft red winter type averaging 10 to 11% in protein. This wheat normally does not enter commercial marketing channels but is consumed locally.

South Korea imports most of its wheat from the United States, principally western white from the Pacific Northwest. Some of this imported soft wheat is blended with U.S. hard red winter wheat at a ratio of 60% soft and 40% hard and milled to an all-purpose flour that is used for almost all baked goods. Noodles account for 40 to 50% of flour use, bread for 20 to 30%, and confectionery products for 5 to 10%, with 10 to 30% taken up by home and industrial uses. Larger bakeries often blend flour with wheat or corn starch for production of quality confectionery products.

Only insignificant quantities of 100% soft wheat flour are produced, but its consumption is increasing yearly. A first-grade flour of 0.40 to 0.45% ash content and a second-grade flour of 0.60 to 0.65% ash are often extracted. The flour protein content varies from 8.5 to 9.5%. Most of the flour is bleached.

In spite of the use of wheat, the principal source of caloric intake in Korea remains rice.

C. Hong Kong

Hong Kong's three flour milling companies milled about 147,000 te of wheat in 1977, 125,000 in 1978 and 140,000 in 1979. Wheat is not produced in Hong Kong, and the United States, Australia, and Canada are the principal suppliers.

Both U.S. western white wheat and Australian standard white wheat are used in the production of soft wheat flours. Australian F.A.Q. wheat had traditionally dominated the soft wheat flour market, but the share of U.S. western white wheat has progressively increased in recent years so that the latter is now the predominant source of soft wheat flour. Soft wheat flour originating in Japan and other countries is used by the industry.

Soft wheat flour in Hong Kong is used in confections (cakes, pastries, etc.), in biscuit manufacture, for certain types of noodles and in various forms of Chinese food, including steamed Chinese buns. The population of Hong Kong is slightly more than 4.44 million of whom 98% are Chinese. The colony hosts many tourists, most of whom are highly wheat-food oriented. These facts explain the wide variation in soft wheat uses.

A wide range of ash contents can be found in locally milled and in imported flours, from that of a typical high-quality "short patent" to that of the low-grade flour. Protein content also varies considerably, ranging from about 7.5 to 11%. A manufacturer usually obtains flour with the protein content best suited to his product needs and production technique; production procedures range from simple hand operation to complete automation. Moreover, the user might well be doing his own blending of several flours from different sources; consequently the exact nature of the flour may not be readily determined.

The manufacture of biscuits, cakes, and Western-style pastries are reasonably standardized worldwide, but many variations exist in the manufacture of the wide range of Chinese noodles. For the innumerable Chinese dishes, the variations are almost endless; moreover, most cooks have their own "secrets." There is very little flour sold for household purposes.

D. Taiwan

Total wheat consumption in Taiwan (all imported) was about 0.54 million te in 1977, 0.59 million in 1978, and 0.65 million in 1979. Historically, the United States, Australia, and Canada have been the suppliers.

It is estimated that of total wheat flour consumption, 51% goes into various Chinese-style foods, 33% to noodles, 8% to cakes, biscuits, and cookies, 8% to bread, and 0.1% to others. Soft wheat comprises about 25 to 35% of the wheat milled in Taiwan. U.S. western white wheat and Australian standard white wheat (Western Australia and Victoria) are most often used.

Soft wheat flours of various grades account for 15 to 20% of total flour production. They are used for making Western-style cakes, cookies, and crackers; Chinese-style cakes, cookies, and crackers; Cantonese-style steamed dumplings; and fresh oiled noodles (Taiwanese noodles). Some flour is blended with rice flour in the manufacture of rice noodles.

E. The People's Republic of China

China is a large wheat-producing country. Average production in recent years amounted to 40 to 45 million te per year, all of which is grown in the northern half of the country. The southern part of this wheat area produces considerable soft red winter wheat. Approximately 2.7 million te of wheat, about 60% of it soft wheat, was imported from Australia in 1978-1979.

Although rice is still the major staple in China as a whole, more than one-third of the Chinese population consumes wheat as its main staple food (15). Most of the soft wheat is milled into all-purpose flour and consumed as steamed Chinese buns with or without filling, dumplings with meat, fish, and vegetable, deep-fried dough biscuits, noodles, and various forms of Chinese food.

F. The Philippines

Total wheat imports were 765,000 te in 1977, 721,000 in 1978, and 732,000 for 1979. The nation continues experimental work on domestic wheat production, but yield, quality, and production cost results have not been promising. The United States and Canada are the major wheat suppliers.

A large portion of the wheat consumed in the Philippines is in the form of hard wheat, with only about 20% of the total imports being of the soft type, mostly U.S. western white. Soft wheat flours are also imported from Italy, France, Belgium, and Japan, but the quantity has progressively decreased in recent years, following the increase in capacity of flour mills in the Philippines.

It is estimated that of total wheat flour consumption, about 8% is used as *miki* (Philippine noodle), 12% as Western-style cookies and crackers, and the balance as *pan de sal* (a typical local bread), cakes, home use, etc. Items made with soft wheat flour include noodles, cookies, crackers, cakes, and pastries.

Two types of soft wheat flour are produced. One is an unbleached regular grind used for cookie and cracker production and for blending purposes by some bakeries and noodle manufacturers. A bleached, finely reduced cake flour is also produced, but along with the recent decrease in consumption of pastry items, its production has been going down. A very small quantity of cake mixes is also produced, for home use. Generally, however, use of flour in the home, whether as cake mixes or all-purpose, is small, because not many homes have ovens.

G. Indonesia

Indonesian wheat imports were 940,000 te in 1976,1,070,000 in 1977, and 1,154,000 in 1978. Indonesia produces no wheat and attempts to develop a wheat program are still in the research stage.

Indonesia imports wheat from the United States, Australia, and Canada. The ratio of hard to soft wheat varies every year, but the soft wheat consumed in Indonesia is U.S. western white wheat and Australian standard white wheat from Western Australia.

Per capita consumption was 8.5 kg of wheat equivalent in 1978 and is increasing in all major wheat products including bread, biscuits, and noodles. About 85% of the flour milled by the three flour mills of the nation are all-purpose, made from a 50-50 blend of hard and soft wheats. About 5% of production consists of soft wheat flour; 10% is 100% hard wheat flour. The demand for bread flour and soft wheat flour is increasing. Most bakeries are not large and market within their immediate area. Cookies and crackers are also produced extensively. New and modern biscuit plants making quality products have come into operation in recent years.

H. Singapore

Singapore imported a total of 248,000 te of wheat in 1978. Singapore grows no wheat.

About two thirds of the imported wheat in 1978 came from Australia and the balance from the United States, Canada, and Mozambique. About half of the wheat consumption is in soft wheat. The three flour mills of Singapore supply soft wheat flour of 72 to 75% extraction to the manufacturers of confections (cakes, pastries, etc.), noodles, and various forms of Chinese food. Although Singapore has a few modern bakeries, much of the wheat flour products continue to be manufactured in small units. However, Singapore is a quality-conscious nation, with flour use patterns similar to those in Japan and the United States.

I. Malaysia

Malaysia's wheat imports for 1978 were 491,000 te, up 29,000 te from the 462,000 te imported in 1977. Malaysia does not produce wheat in commercial quantities.

Historically, Australia has been the main wheat supplier to Malaysia, and Australian standard white wheat from Western Australia, a type of soft wheat, dominates the wheat flour market. Today, about 60–70% of the imported wheat is soft.

Four flour mills supply wheat flours to the domestic market. Soft wheat is used for the production of pastries and certain steamed Chinese buns. General-purpose flours are usually milled from a 50-50 blend of hard and soft wheats. These flours are used by commercial bakeries and noodle manufacturers. Very few modern bakeries exist in Malaysia, and most wheat flour foods are produced in small family shops.

J. Thailand

Wheat imports into Thailand increased from 93,000 te in 1973 to 145,000 te in 1978. Thailand does not produce wheat in commercial quantities.

Australia and the United States are the principal suppliers. A fairly large portion of the imported wheat is soft, consisting of Australian standard white wheat (Western Australia) and U.S. western white wheat.

Breads, pastries, noodles, and steamed Chinese buns are popular wheat foods. Three flour mills produce pastry and all-purpose flours of 72 to 75% extraction from the imported soft wheats, but most of the consumption is still limited to urban areas. Although many new bakeries have come into existence in recent years, most of the manufacturing is still done by rather traditional methods. Pastries, noodles, and steamed Chinese buns are popular soft wheat products. As in other nations, hard wheat flour is used for bread baking. Generally speaking, Thailand is a quality-conscious nation.

K. India

Wheat production and consumption in India are as old or older than the wheat culture in Egypt (21). At present, about 90% of the wheat is consumed in the form of *chapati* and its variations. Punjab, Haryana, Madhya Pradesh, Rajasthan, Uttar, and Himachal Pradesh states in northern India are the most important *chapati*-consuming regions, although the product is becoming increasingly more popular in other states as well. A whole-wheat meal known as *atta* is used in making *chapaties* (22). *Atta* is produced in small flour mills (*chakki*), as well as in larger mills. Wheats vary markedly in their *chapati*-making quality, which has been extensively studied recently (11).

In southern India, basically a rice-producing and -consuming area, flour milling is not as extensive as in the north. White flour is produced for Western-style bread and *nan*, which is more popular than *atta* in the region.

At present, more than 200 roller flour mills, excluding *chakki*, operate in India, and flour millers are becoming increasingly quality-conscious. They prefer soft wheats for cookies, crackers, and confectionery products.

L. Sri Lanka

Although Sri Lanka is historically a rice-producing and -consuming nation, products from imported wheat flours have become a staple in the diet of the people. Sri Lanka has been a major wheat flour importer, with only one moderate-sized flour mill in the nation, but a second mill, with the largest capacity in Southeast Asia, will be completed in 1980.

Soft wheat flours are imported because they are usually less costly than hard wheat flours. France, the United States, and Australia supply the needed wheat and flour, the main use of which is Western-type bread and noodles. Because the bread is baked from imported soft wheat flour, its quality differs somewhat from that in Western countries.

M. Pakistan

As in India, the principal food use of wheat in Pakistan is in the form of *chapati* or *nan*. About half of the mills grind *atta* in the form of whole meal; the remainder produce *atta* of about 91% extraction, bran, and *maida* (fines). Although Pakistan grows considerable wheat, much is imported. Generally speaking, the wheat supplied to the mills is uncategorized by class, grade, or point of origin. However, soft wheat is preferred, because the *chapati* from it is thought to be more tender than that made from hard wheat *atta*.

N. Bangladesh

Wheat in Bangladesh is primarily used for *chapaties*. Flour mills grind the wheat (which is mostly imported) to produce *atta* of whole meal or high extraction ratio. Soft wheats are preferred for making *chapaties*.

O. Afghanistan

The staple food of Afghanistan is wheat. Much of the indigenously grown wheat is soft; hence, the consumption pattern of wheat products is based on the use of this class of wheat. As in India, Pakistan, and Bangladesh, the primary wheat foods in Afghanistan are *chapati* or *nan*.

Production of the large flour mills in the nation are 3% white flour and 97% whole-wheat meal *atta*. The latter is milled in small country mills as well.

III. BAKED PRODUCTS FROM SOFT WHEAT FLOUR

A. *Castilla*

Castilla, a form of Spanish cake named for a kingdom in Spain (2) was introduced into Japan by the Portuguese in the 16th century. Originally *Castilla* was a variety of sponge cake of Western countries, but modifications in formula, baking process, and flavor have made it a true Japanese cake. The fact that *Castilla*, also called "Nagasaki-Castilla" for the port where the Portuguese first landed, could be produced from domestic (Japanese) soft wheat flour seems to have aided its development into one of the most delicious and highest quality confectionery goods in Japan.

Ingredients for *Castilla* include: flour, 100 parts; eggs, 200 to 210; sugar, 180 to 210; starch syrup, 15 to 30; honey, 15 to 30; small quantities of sweet sake (*Mirin*); and water (2,34,35,38,44,48). The whole eggs are beaten (35), sugar added, and the mixture blended at low speed. The remaining ingredients are then added. Mixing is critical. The dough is poured into a wood frame placed on sheets of kraft paper and a steel plate. The frame is then covered with a second heat-intercepting plate. During the initial baking for 25 minutes at 180° C, the upper plate is removed. A second wood frame then is placed on the first, and the temperature is reduced to 150° C. Total baking time is 55 to 60 minutes. The cake is often eaten as a dessert or snack. It is also served during tea ceremonies. Sliced *Castilla* is pictured in Figure 2.

Short patent flour milled from U.S. western white wheat or low-protein wheat grown in the Kyushu area has been favored for *Castilla* production. A typical *Castilla* flour ranges from 7.0 to 7.8% in protein and 0.33 to 0.38% in ash content and has a desirably mellow gluten property. High amylase activity in the flour is considered undesirable, as it causes the product to collapse during cooling.

Although U.S. soft red winter wheat has not been used extensively, it is presumed to be one of the best sources of *Castilla* flour. *Castilla* and sponge cake

Figure 2. Sliced *Castillas*.

formulations differ in the higher relative ratios of sugar, eggs, honey, and/or starch syrup in the former. The sugar and honey and/or starch syrup are thought to aid in maintaining tenderness and moistness; the egg yolk enhances the attractiveness of the cake through its contribution to cake crumb color. The desirably soft and moist masticatory sensation is a most important characteristic of *Castilla* cake.

B. Sponge Cakes

Sponge cake formulations may be applied in the baking of layer cake, shortcake, and fancy cake. Moreover, the baking technique for sponge cake has been applied to the production of roll cake. Today, sponge cake is one of the most popular confections in Japan, and its delicate flavor, adapted to the Japanese taste, makes the cake a favorite among all age groups.

Other nations in the Orient also have sponge cakes. Basic baking techniques for them are similar to those of the West, although quality may differ owing to variations in formulation.

Two sponge cake formulas are typical (1,34,35,43,48). Wide variation in ingredient proportions is to be expected from plant to plant, depending on the flour, type of product, and other factors. Certain minor additives may also be used. Emulsifying and foaming agents containing sucrose fatty esters are often added. The first formula consists of flour, 100 parts; eggs, 100; sugar, 100; and milk and water, 30 to 35. The second has flour, 100 parts; eggs, 170 to 200; sugar, 120 to 160; and butter and milk, 30 to 50.

Sugar is added to the beaten eggs and mixed thoroughly. In the first formula, water is added during mixing. The specific gravity of the batter at this stage should be 0.25 ± 0.01 and its temperature 30° C. Sifted flour is added and the batter is gently mixed for a short time. Milk and/or butter of the second formula

Figure 3. Shortcakes.

are added during the final mixing stage. Batters are baked at 180° C for 30 minutes. Shortcake, an example of sponge cake, is pictured in Figure 3.

Short patent flour from U.S. western white wheat has been the principal source of sponge cake flour. Analytical and gluten tenderness properties characteristic of *Castilla* flours are considered desirable for sponge cake baking.

According to the Official Grain Standards of the United States (50), the sub-class western white contains more than 10.0% wheat of white club varieties and more than 10.0% other white wheat. Nagao et al (31) showed the sponge cake baking quality of white club wheat to be superior to that of soft white, even if the protein content of white club is higher than that of soft white. Thus, the more white club is blended into western white wheat, the more favorable it is for sponge cake production (27,29,31).

Australian soft wheat does not appear to be the equal of U.S. western white for sponge cake flour production, but a blend of the two appears to be satisfactory. Domestic wheat of southern Japan may also be blended with western white. The quality fluctuation of U.S. soft red winter wheat seems to be fairly large, and the better ones would seem to be superior to western white wheat for sponge cake flour purposes.

To bake a good sponge cake, finely granulated flour is required. Some flours are therefore subjected to additional treatments such as impact grinding and air-classification (45). High amylase activity (low amylograph viscosity) in flour has resulted in poor cakes (28).

A sponge cake of high quality will have cells in the crumb that are uniformly small and have thin walls. The cake should be moist, tender, and soft and have good volume and desired flavor characteristics.

C. Baked Japanese Buns

Manju is the generic term for Japanese buns with or without filling. First known in the 13th century, it has become a representative Japanese confectionery product and a favored snack food. As indicated in Table III, *manju* may be classed either as lightly baked or steamed buns.

Kuri-manju (chestnut bun with bean jam filling, Figure 4) is the most popular of the baked *manjus*. The name comes from the similarity in shape of the buns to

Figure 4. *Kuri-manjues* (chestnut buns with bean jam filling).

a chestnut, as well as to the chestnuts that are included in the filling. *Castilla-manju* (*Castilla* bun), *Kara-manju* (*Cathay* bun), and *dora-yaki* (bean jam pancakes, Fig. 5) are also popular baked buns. *Imagawa-yaki* or *kaiten-yaki* (Japanese muffin containing bean jam) and *tai-yaki* (sea bream-shaped Japanese muffin containing bean jam) differ from the other *manju* products in that they are served hot at the small bakeries where they are made.

The production of *kuri-manju* is described as an example of baked *manjus*. Many variations exist in ingredient proportions for dough preparation. A typical formula is (20,35,38): flour, 100 parts; eggs, 35 to 55; sugar, 50 to 60; starch syrup and honey, 5 to 15 (optional); and baking powder (sodium bicarbonate, ammonium chloride, cream of tartar, and baking alum), 0.5 to 1.0. A formula for bean jam filling containing chopped chestnuts is: white bean jam, 100 parts; sugar, 10 to 15; starch syrup, 5 to 10; and chestnuts, soaked in honey and finely chopped, 10 to 20. The bean jam is prepared by cooking white kidney beans, sugar, and salt. The beans are boiled, the skin removed, the boiling continued until the beans soften, and the beans are filtered through a coarse screen. Finally, 100 parts of filtered beans, 60 to 70 parts of sugar, and 0.05 parts of salt are cooked together.

The dough is formed by beating the eggs slightly; adding sugar, starch syrup, honey, and baking powder; and mixing. Sifted flour is added and the dough mixed at low speed. The jam is prepared by heating bean jam with sugar and starch syrup. Chestnuts each chopped in eight pieces are added as the final step.

In most large bakeries and confectioneries, encrusting machines (Fig. 6) are used to fill the bean jam into the dough sheet and to mold this combined material into the desired shape. The continuous supply of dough and filling to the machine makes possible the production of a quality product, lowers unit cost, and meets sanitation requirements. The operation is performed manually in small

Figure 5. *Dora-yakis* (bean jam pancakes).

confectioneries. Egg yolk mixed with sweet *sake* (*Mirin*) is brushed on the *manju* surface as it exits from the machine or after hand molding, in order to add a glaze to the product, which is baked at 180° C.

Short patent flour from domestic Japanese wheat formerly comprised the main material for *kuri-manju*. Today, short patent flour from a blend of U.S. western white and Australian standard white (Western Australia) or domestic Japanese wheat is commonly milled. Flour ash content ranges from 0.33 to 0.40% and protein from 7.5 to 9.0%. Some bakeries prefer to blend 70% low-protein soft and 30% hard wheat flours. Flour requirement for making *Castilla-manju*, *Kara-manju*, and *dora-yaki* are similar to those above. However, second-grade all-purpose flour is usually employed for making *Imagawa-yaki*, *kaiten-yaki*, or *tai-yaki*.

A soft and tender texture in the baked dough is highly desirable. However, overmixing the dough usually results in an objectionable tough product. The external appearance is a very important quality factor, as is the flavor of the chestnut containing bean jam filling.

Figure 6. Encrusting machine. (Courtesy of Rheon Automatic Machinery Co., Ltd.)

D. Baked Chinese Buns

Yit bien (moon cake) is a representative of baked buns with filling popular in China. It is well-known, being served at the Festival of the Moon of the Chinese calendar (about August 15) (14,48). It is also popular in other nations with significant Chinese population. The crust of the bun is decorated with embossed figures of rabbits, frogs, pagodas, etc. (Fig. 7).

A typical *yit bien* formula, subject to wide variations in ingredient proportions, follows (14,38,48): flour, 100 parts; sugar, 50 to 55; starch syrup, 25; lard oil, 35 to 40; sodium bicarbonate, 1; alum, 0.04; and water, 15 to 25.

The filling for *yit bien* consists of: chopped almond, cashew, walnut, and watermelon seeds, 100 parts; sugar, 100; white sesame seeds, 20; steamed flour (prepared by steaming flour for about 30 minutes), 80; powdered rice, 50; pork fat, 40; and Chinese watermelon cooked with sugar, 30. All ingredients except flour and rice are combined and heated; the two are added during the final cooking stage.

The dough is prepared by mixing sugar, starch syrup, alum, and water; adding liquefied lard oil; and mixing. Soda dissolved in water is then added. Finally, sifted flour is added and the ingredients are mixed at low speed or manually to a fairly firm dough. The dough is allowed to rest for about an hour and is divided into small balls that are flattened, the filling is placed in the cavity, which is sealed by folding the edges. It is then molded in a wooden form. Egg yolk mixed with sweet wine or caramel solution is brushed on the dough, and the dough is baked at 190°C for about 15 minutes.

Moderate protein content and gluten mellowness are required in the flour. Short patent all-purpose flour milled mostly from soft wheat is suitable for *yit bien*. A mixture of 80% confectionery flour and 20% hard wheat flour is also

Figure 7. *Yit biens* (moon cakes).

employed. Product quality is determined by the clarity of the pattern embossed on the surface. Shrunken, cracked, or blistered surface detracts from product attractiveness and is considered undesirable. The baked dough should have a smooth, fine texture that tends to melt in the mouth.

E. Biscuits and Cookies

Biscuits are baked in most countries in the Orient, although the production scale varies.

Biscuits were first introduced into Japan during the 16th century (19). In 1871, they were imported to Japan on a commercial scale for the first time, and within a year, domestic biscuit production was initiated. Growth was slow, but between 1914 and 1924, the biscuit industry expanded rapidly, aided by modifications in formulations and processing techniques to blend with the flavor and textural requirements of the public. With the return to industry of control of wheat flour sales in 1952 and the subsequent importation of modern equipment and techniques (51), a large number of products, some unique to Japan, are being manufactured in modern plants. Annual biscuit consumption totals 220,000 to 240,000 te (42).

Biscuits are classified in Japan into hard and soft types. Only a portion of soft biscuits made by a richer than usual formula are termed cookies. Most are produced mechanically by the use of rotary machines, wire cutters, and depositers. Numerous varieties of cookies introduced from the West and modified to appeal further to domestic taste are on the market. Examples of hard and soft biscuits, and cookies manufactured in Japan are shown in Figures 8, 9, and 10.

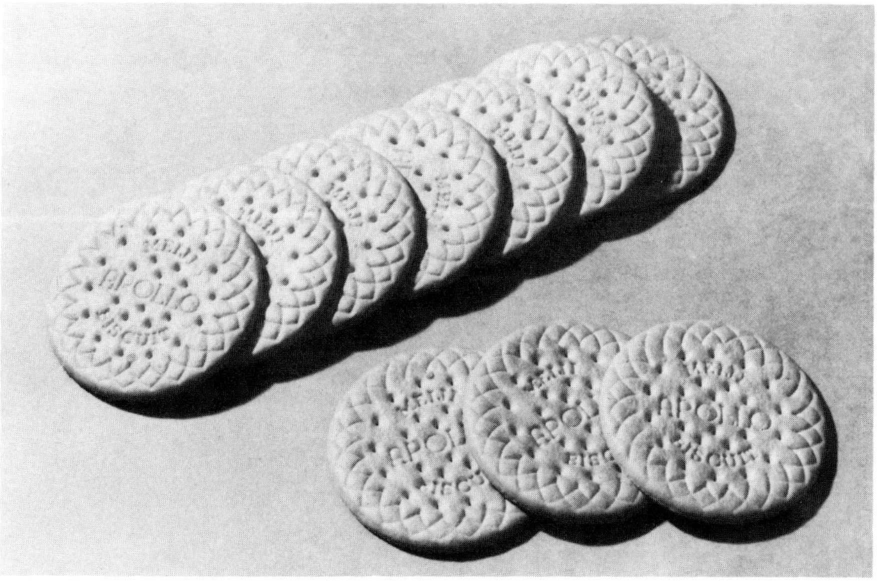

Figure 8. Hard biscuits.

Typical formulations for hard and soft biscuits and for cookies follow.

Hard biscuits: flour, 100 parts; sugar, 20 to 25; shortening, 10 to 15; condensed milk, 0.5 to 1.0; eggs, 0 to 5; salt, 0.5 to 1.0; leavening (ammonium carbonate, sodium bicarbonate, and cream of tartar), 1.5; flavoring; and water.

Figure 9. Soft biscuits.

Figure 10. Cookies.

Soft biscuits: flour, 100 parts; sugar, 30 to 40; shortening, 25 to 35; condensed milk, 3 to 5; eggs, 0 to 10; salt, 0.5 to 1.0; leavening, 0.5 to 1.5; flavoring; and water.

Cookies: flour, 100 parts; sugar, 40 to 60; shortening, 25 to 50; milk, 5 to 10; eggs, 10 to 15 for rotary or wire-cut, 25 to 30 for deposited cookies; salt, 0.5 to 1.0; leavening, 0 to 1.0; and flavoring.

Manufacturing processes for cookies and hard and soft biscuits are quite similar to those employed in Europe and the United States (12,18,46,47,52).

First-grade confectionery flour of low protein content, fine granulation, and mellow gluten properties is suitable for soft biscuits. Cookie flours are similar to those for soft biscuits, but occasionally first-grade all-purpose flour may be employed for special products. For hard biscuits, first or second grade all-purpose flour of moderate protein content (8 to 9%) is satisfactory. Since most biscuit manufacturing plants are highly automated, flour property uniformity at a high quality level is a most important requirement.

F. Chinese Cookies

Chinese cookies are very popular with the Chinese worldwide. Many variations of the product may be found; Figure 11 pictures some of them.

Hsing jen ping (Chinese almond cookie) is a typical item. A characteristic formula for the almond cookie is: flour, 100 parts; sugar, 50 to 60; eggs, 10 to 30; lard, 60 to 80; salt, 0 to 0.5; sodium bicarbonate, 0.02 to 0.5; almonds, split in halves, 20 to 40; and essence, to taste. The lard and sugar are creamed. This mixing is important to prevent the lard from separating in the finished product. The eggs, essence, and sodium bicarbonate are then added and mixed. The sifted flour is added at low speed or by hand to a fairly firm dough. The dough is formed

Figure 11. Chinese cookies.

into small balls and placed on a baking pan. A cavity is formed in the center with a dowel and an almond half placed therein. Baking is for 10 minutes at 180 to 190°C. If desired, the dough pieces may be brushed with egg yolk film to form a glaze.

An all-purpose flour from soft wheat is usually used for Chinese cookies. Flour quality requirements are similar to those for Western cookies.

G. Crackers

Crackers produced in Japan more nearly resemble the American type than the English cream crackers. Until the mid 1950s, cracker production was negligible, but with the introduction of band ovens in 1955, consumption has progressively increased. Each year 25,000 to 31,000 te of crackers are consumed (42).

For the sponge fermentation, the following formula is applicable: flour, 70 parts; shortening, 5; yeast, 1.0 to 1.5; yeast food, 0.05 to 0.1; and water, 25 to 30. The composition for the doughing process follows: flour, 30 parts; shortening, 3 to 5; sugar, 1.0 to 1.5; salt, 2.0 to 2.2; and sodium bicarbonate, 0.5 to 0.8. Processing in cracker manufacture in Japan is almost identical to that in the United States (12,19,46,47,52).

A flour corresponding to first-grade confectionery or all-purpose milled product is suitable for crackers. It typically has an ash content ranging from 0.36 to 0.45% and a protein content of 7.5 to 9.0%. If two flours are employed, the sponge stage fermentation is carried out with the flour of higher protein content.

H. Japanese Crackers

Senbei is the generic name for Japanese crackers, which may be classified into two categories: wheat flour cracknel and rice cracker. *Senbei* was originally made by parching wheat flour in sesame oil (26). Until the 16th century, *senbei* was principally of the rice cracker type; but a century later wheat flour cracknel had become popular (35), and even now it is eaten as a snack food.

A great many variations of wheat flour cracknel are known. *Kawara-senbei* (tile-shaped cracknel), the most popular form, is shown in Figure 12. A typical formula for *kawara-senbei* includes flour, 100 parts; sugar, 110 to 130; eggs, 10 to 20; carbonic acid, 0.03 to 0.1; and water, 60 to 80. The sugar and eggs are whipped, and the carbonic acid (dissolved in a little water) is added with the flour and mixed well, with the remaining water added to a uniform batter. The batter is poured into baking dies that are then heated over a fire, the dies being inverted carefully midway in the heating. Usually 10 dies are heated simultaneously.

A wide range of soft wheat flour may be used for making *senbei*. It appears that flour with 0.35 to 0.55% ash and 7 to 10% protein content processes with little difficulty. However, high-quality *senbei* requires a flour of moderate protein content and soft gluten quality. For such premium *senbei*, a soft wheat flour ranging in ash from 0.35 to 0.40%, with protein content between 7 and 9%, is recommended.

A crispness peculiar to the product is a primary quality requirement. At the same time, the *senbei* should readily soften in the mouth without becoming gummy.

I. *Chapati* and *Nan*

Chapati and culinary variations such as *tandoori roti, nan, paratha,* and *poori* are the staple food in India, Pakistan, Afghanistan, and some areas of neighboring nations. Indians probably have eaten wheat in its present form for at least 3,000 years, and consumption of *chapati* is as extensive as ever.

Chapati is a type of unleavened pan-baked cake made from a whole wheat meal called *atta.* When the *chapati* is deep fat fried, it is known as *poori.* If vegetable shortening is rolled in the dough ball and the ball fried on a hot iron plate, the product is known as *paratha. Nan* is leavened *chapati* made from white flour, water, and a special leavening agent.

Chapati is prepared simply and consists of *atta* (whole-wheat meal), 100 parts, and lukewarm water, 50 to 70 parts. For cooking *paratha, ghee* is used. The vegetable oil for *poori* is prepared by cooking it over low heat for a long period to clarify it and enrich its flavor (5).

For making *nan*, flour, 100 parts, and leavening agent plus water to total 50 to 80 parts are used. The leavening agent may be curd or buttermilk and baking soda, or eggs, baking powder, bakers yeast, or boiled potatoes and wild yeast. The quantity used varies widely. In the preparation of *chapati* and *poori, atta* and water are mixed and kneaded thoroughly, and enough water is added to form a dough of moderately stiff consistency. *Paratha* dough is formed by rubbing *atta* and *ghee* to a uniform mixture. About one-half of the water is added and the mixture is stirred vigorously with a spoon or by hand. Enough of the remaining water is added to form a dough that can be gathered into a firm, compact ball.

Figure 12. *Kawara-senbeis* (tile-shaped cracknels).

The dough is made smooth and elastic by folding and pressing a number of times. It is then covered with a damp cloth and allowed to rest for about 30 minutes, after which it is divided into small balls that are kneaded for a short time, rolled, and flattened. Cooking is done on a hot iron plate or frying pan for a minute or two until the surface begins to darken. The dough is turned so that both sides are cooked.

Examples of *chapati, poori,* and *paratha* are shown in Figure 13. They are eaten with curry or gravy made with vegetables, meat, or lentils.

Nan dough is prepared in almost the same way as *chapati* dough. However, *nan* is baked in a clay oven known as *Tandoor,* which is usually heated by firewood. Many *nans* can be baked simultaneously, depending on oven size (21).

Extensive studies were recently conducted to evaluate wheat varieties for *chapati* quality and to determine factors contributing to *chapati* quality (11). It is now generally known that the *chapati*-making properties of wheats vary markedly.

Soft amber or white wheat such as the U.S. western white with 8 to 10% protein of mellow nature and with high diastatic value (21) are preferred by consumers. *Chapati* produced from such wheats are light in color, remain soft longer, and are slightly sweet. The popular Indian wheats of Sharbati grade make relatively well-puffed, soft, silky, and pliable *chapaties*, and these favorable characteristics are retained for some time (11). Those made from imported hard red American and soft white Australian wheats do not generally have these pleasing attributes of appearance and texture. Moreover, *chapaties* made from American wheat become stiff and brittle after standing for an hour or so.

Chapaties from modern, improved high-yielding varieties are thought not to

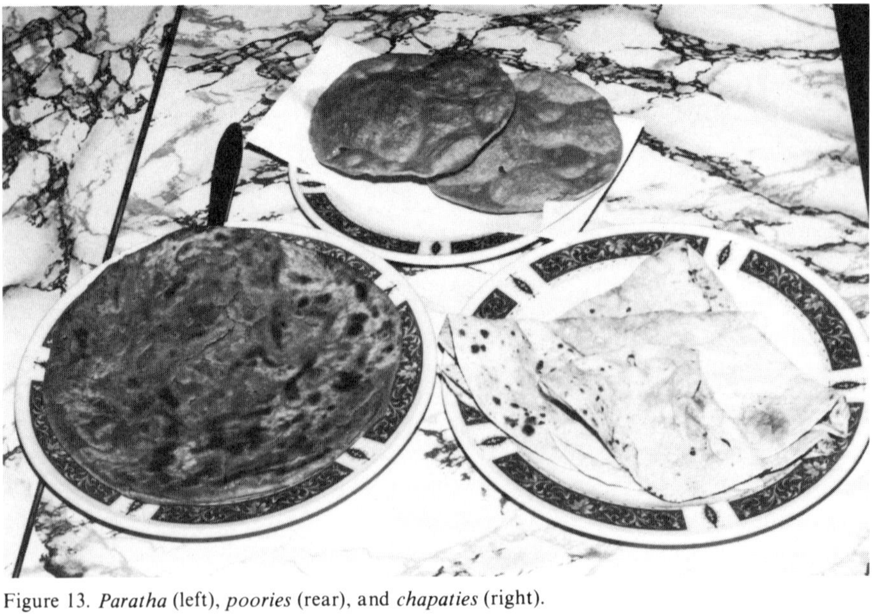

Figure 13. *Paratha* (left), *poories* (rear), and *chapaties* (right).

have desirable characteristics to the same extent as *chapaties* from indigenous wheats, especially Punjab varieties. However, results of an opinion poll on *chapati*-making properties showed that improved domestic varieties such as Sharbati, Sonora, and Sonalika are comparable or even superior to the Punjab variety C. 306 in this respect.

Austin and Ram (11) indicated that almost all wheat varieties that made good *chapaties* from the standpoint of texture and keeping quality had water absorption values of 68 to 70%. Additionally, varieties with Pelshenke values between 115 and 150 made the best *chapaties*. In all probability, the most important factors in flavor and palatability are the sugar content and diastatic activity of the *atta*.

In Afghanistan and Pakistan, soft wheat is considered to make desirably tender *chapaties* and products from hard wheat are often tough.

In India, normally milled white flour is used for making *nan*. Wheats that produce *atta* of good *chapati*-making quality are also considered to be a satisfactory source of flour for *nan*.

Complete and rapid puffing is an important *chapati* quality. The surface of *chapati* should be untorn and have a silky appearance. A soft, silky, smooth, and pliable product consistency is desirable. Moreover, the texture should remain soft and pliable for several hours. The most desirable color from the standpoint of consumer acceptance is a creamy hue. In addition, the *chapati* should have a hint of sweetness and a pleasing, subtle flavor.

J. Other Baked Products

WAFERS

Wafers, another introduction from the West, are favored especially by the very young. They are frequently served with ice cream.

Two parts sugar, 8 parts egg yolk, 0.5 sodium bicarbonate, 0.8 ammonium carbonate, 180 parts water, and a little salt and flavoring per 100 parts flour may be considered a typical formula. Another is 2.5 parts nonfat milk solids, 0.75 parts salt, 0.5 sodium bicarbonate, 150 water, and a little flavoring based on 100 parts flour. Batter mixing and baking processes are very similar to those employed in the United States.

First-grade flour or a blend of first and second grade from U.S. western white wheat is preferred for wafer production. The flour should mix to a batter that flows freely and gives a high yield of quality finished products. The wafer should be tender and dissolve readily when eaten.

ICE CREAM CONES

Cones, or containers for ice cream, are classed as sugar cone and ordinary cone. They are usually baked in three shapes: cup, bar, and plate.

An example of ordinary cone formulation is 10 parts starch; 3 to 5 parts sugar; 2 to 3 vegetable oil; 0.3 to 0.5 salt; 0.06 to 0.08 calcium carbonate; 0.08 to 0.12 sodium bicarbonate; 0.08 to 0.12 baking alum, coloring, and flavoring; and 180 to 200 parts water based on 100 parts flour. All ingredients other than flour and starch are blended. Starch and sifted flour are added and mixed briefly at low speed. The resulting batter is poured into baking dies through a feeding tank and baked at about 200° C.

Flour quality requirements and product quality factors are similar to those for wafers. In addition, product transportability is a quality requirement.

BUTTER CAKES

In Japan, butter cakes comprise one of the major products of Western-style confections. Pound cakes, cup cake, madeleines, and lemon cakes are included in this category.

A typical formula for butter cake (16) follows: 100 parts, sugar; 100, egg; 100, shortening; 2 to 5, rum or whiskey; 1.0 to 1.5, baking powder; and a little flavoring based on 100 parts flour. Mixing and baking procedures are similar to those in the United States and Europe and may follow the flour-batter, sugar-batter, separate whipping, or all-in-mix method.

Flour quality requirements are almost identical to those of sponge cakes.

HOT CAKES

Most hot cakes are made in the home, but some shops have opened that serve several varieties of hot cakes with topping, and some coffee shops are serving frozen hot cakes that are reheated.

Hot cake mixes are available in Japan for both home and commercial use.

A formulation for hot cakes follows: sugar, 30 to 40 parts; eggs, 70 to 80; milk, 40 to 50; butter, 0 to 5; baking powder, 3 to 4 parts based on 100 parts flour. First-grade confectionery flour is most suitable for hot cakes.

WAFFLES

Soon after the introduction of waffles into Japan, semi-Western style waffles with jam or cream filling were developed. Today, they are favored by many as a quality snack food.

A typical formula consists of 100 to 110 parts sugar; 100 to 120, eggs; 10 to 20, starch syrup; 1 to 2, baking powder; 30 to 40, water; and flavoring based on 100 parts flour. Mixing and baking processes are similar to those used in the West. Jam or cream is spread on the baked waffle, which is then folded. Short patent flour of low protein content and soft gluten quality such as that from U.S. western white wheat is most suitable for waffle preparation.

BAUMKUCHEN

Baumkuchen (tree cake) was introduced into Japan from Germany. This quality confection is eaten in slices or may serve as a base for decorating. A typical *baumkuchen* formula consists of 100 parts eggs; 40 to 60, sugar; 35 to 60, butter or margarine; 10 to 30, corn starch; 5 to 10, rum; and flavoring based on 100 parts flour. Mixing and baking processes are the same as those employed in Germany. A special grade of confectionery flour is used.

CHOU À LA CRÈME

Chou à la crème, a cabbage-shaped confection with cream filling, was introduced into Japan from France (13). The delicious flavor derived from the combination of the puffy *chou* and smooth custard is favored by many.

A formula for the puff includes 70 to 80 parts shortening; 200 to 300, eggs; 0.5, salt; and 100 to 180 parts water based on 100 parts flour. After the shortening and salt are heated, flour and water are added and the eggs gradually blended into the mixture. The paste is pressed from a bag onto a baking pan and baked at 210 to 220° C for about 18 minutes. The puff is then filled with a specially prepared custard cream.

Both hard and soft wheat flours are used in the production of *chou à la crème*. For soft puffs, first-grade confectionery or all-purpose flour is suitable. The use of short patent or semilong patent flour from hard wheat is recommended for large volume (and tougher) puffs.

PIES

Both American and French type pies are produced in Japan. A formula for an American type pie crust consists of 60 to 80 parts butter or margarine; 1.5 to 2.0, salt; 0 to 3, sugar; and 55 to 60 parts water based on 100 parts flour. On the other hand, crust of a French pie contains 80 to 100 parts butter or margarine; 1 to 2, salt; 0 to 3, sugar; and 60 to 65 parts water to 100 parts flour. Hard and soft wheat flours are blended for making pie crusts; the ratio depends on the masticatory feel required for the finished product. First-grade flours are used.

BOLO

Bolo was introduced into Japan by the Portuguese in the 16th century. Today, *saga-bolo* is a well-known confection in the Kyushu area. *Soba-bolo*, made from a blend of wheat and buckwheat flours, is a variety of *bolo* noted in Kyoto. *Saga-bolo* consists of 80 parts sugar; 50 to 60, eggs; 15 to 25, starch syrup and/or honey; and 1 to 2 parts sodium bicarbonate based on 100 parts flour. Production processes consist of dough mixing, proofing, sheeting, cutting, oil brushing, and baking. A second-grade all-purpose flour is usually employed in *saga-bolo* production.

IV. NONBAKED USES OF SOFT WHEAT FLOUR

A. *Tempura*

Prepared by dipping fish, shellfish, or vegetables in a batter and deep frying, *tempura* is a light, delightful, and appetizing Japanese food enjoyed all over the world. The term "*tempura*" is said to have come from the Portuguese "*tempero*" or the Spanish "*templo*," after the introduction of fried food into Japan in the 16th century (3). However, blessed with abundant and diverse marine life, the Japanese developed a unique fried food that now completely differs from the original. *Tempura* is served in restaurants that feature individual variations in preparation and presentation and is also often prepared in the home. Because *tempura* is one of the principal outlets for family-type soft wheat flour in Japan, the flour needs quality characteristics suitable for it as well as for the usual soft wheat flour products.

Tempura batter formulas are quite simple. One example includes 100 parts flour, 50 parts eggs, and 150 parts water; another example, 100 parts flour, 25 parts eggs, and 135 parts water. The eggs are lightly beaten, the water added, and the light beating continued. Finally, the flour is sprinkled in with light mixing. Over mixing causes an undesirably sticky batter to form, so the mixing should be done very carefully. The water should be cool (especially in summer) in order to maintain a fresh batter for as long as possible. The batter is normally made after all other food preparatory steps have been completed for cooking of *tempura*. Aeration of flour by sifting just before batter preparation appears to improve batter quality.

Almost any food is suitable for *tempura*, although fresh fish, shellfish, and vegetables (including sweet potatoes) are usual. Items that do not have excessively strong flavor or that are not high in moisture are preferred. For frying, high-quality vegetable oil is heated and maintained at 160 to 190°C, at which temperature range frying takes on the combined characteristics of boiling and baking. The food item is dipped in the batter and the coated food placed gently in the hot oil. Cooking is complete when bubbles generated by the frying become small and few. The item is withdrawn from the oil and shaken vigorously a few times to remove excess oil. Cooked *tempura* is shown in Figure 14.

Tempura is served hot and is eaten after dipping it in a specially prepared soysauce soup (*ten-tsuyu*) to which grated radish has been added. In a modified version, *tempura* is served on steamed rice (*ten-don*), a quite popular dish in Japan. Short patent flour from U.S. western white wheat is most extensively used for batter preparation. Its low protein content and soft gluten quality are

Figure 14. *Tempura*.

particularly advantageous for a crisp, tender fried coating. A typical *tempura* flour ranges from 0.33 to 0.43% ash and 7.0 to 8.5% protein. High diastatic activity (low amylograph viscosity) in the flour can make it unfit for *tempura* use. Because of the large quantity of flour used for the purpose, prepared *tempura* mixes are available for both home and commercial applications.

The batter coating plays an important role in protecting the food from the high oil temperature and in preventing excessive moisture evaporation. It thus aids in maintaining food flavor and even enhances palatability through absorption of some of the oil.

Regional variations are apparent in *tempuras*. In western Japan the food item to be coated is cut or sliced into smaller portions, the batter coating is a bit thinner, and the cooking temperature is lower than in Tokyo.

B. Japanese Noodles

Noodles were introduced into Japan from China more than 1,000 years ago (35). In early days, noodles were highly valued as an offering to the Kami or Buddha. Variations of noodles were introduced over the years, but most of the noodle forms seen today were available by the 16th century. The development of a noodle making machine by Masaki in 1884 revolutionized the manufacture of noodles (49).

Japanese noodles are classified into *so-men* (very thin noodles), *hiya-mugi* (thin noodles), *udon* (standard noodles), and *himo-kawa* (flat noodles), according to the width of the noodle strands (Table V). *Udon* and *himo-kawa* are sold wet or dry, depending on the requirements of the market. *Kan-men* (dry noodles) is the generic term for dry *so-men, hiya-mugi, udon,* and *himo-kawa* and is the storable form produced by drying wet noodles. *Yude-men* (boiled noodles) is the generic name for boiled *udon* and *himo-kawa*, which must be consumed soon after manufacture.

Much of Japanese noodle manufacture is mechanized, but handmade or semihandmade noodles are available in all categories. These are referred to as *tenobe so-men, tenobe hiya-mugi, teuchi udon,* and *teuchi himo-kawa.* Handmade noodles are highly valued for their eating quality, presumably owing to the mode of gluten formation.

About 16% of wheat flour produced in Japan is consumed in the form of

TABLE V
Classification of Japanese Noodles

Class	Strand Width (mm)	Cutting Rolls (no. 30/mm)	Dry or Boiled	Manufacturing Method
So-men (very thin)	0.50-1.11	60-27	Dry	Hand-made (*Tenobe*); Semihand-made
Hiya-mugi (thin)	1.36-1.67	22-18	Dry	Semihand-made (*Tenobe*); Machine-made
Udon (standard)	2.31-3.00	13-10	Dry or boiled	Semihand-made (*Teuchi*); Machine-made
Himo-kawa or *Kishi-men* (flat)	4.29	7-3	Dry	Machine-made
			Boiled	Semihand-made (*Teuchi*); Machine-made

noodles. Although noodle consumption has not increased recently, it is still one of the major products of soft wheat flour. Their bland taste makes them a favored staple food and snack food. A large variety of Japanese noodles may be seen in the market.

The formulation for Japanese noodles is simple: flour, 100 parts; salt, 2 to 3 parts; and water, 28 to 40 parts. The quantity of salt is adjusted according to noodle type, market requirement, and climate; dry noodles require more salt (2 to 3%) than boiled noodles (2%). In the winter, less salt is required. The salt should be dissolved in the water before contact with flour. The volume of water should be adjusted to flour quality, type of mixer, and climate.

Flour and salted water are doughed in a mixer for 15 to 20 minutes. The stiff and crumbly dough is sheeted through rolls spaced 2 to 3 mm apart. Two such sheets are laid together and passed through rolls of larger diameter, which further develop the dough. This step is followed by passage through a sequence of rolls, each a little smaller in diameter and set a little closer than the previous one. The passage of the dough through the series of rolls completes gluten development and also adjusts the sheet to the desired thickness. Excessive sheeting may lead to sticky dough and disintegration during cooking. The sheets are then cut into strips of the desired width (Table V).

Some of the uncooked product at this stage of manufacture is sold to retail shops and restaurants under the generic name of *nama-men* (wet uncooked noodles). However, most is further processed to boiled or dry noodles.

Boiled noodles are prepared by cooking in boiling water of pH 5 to 6 (6,10) for 10 to 25 minutes, with the timing adjusted precisely to give optimal masticatory texture to the product. Under optimal conditions, the noodles should weigh 250

Figure 15. Boiled noodles ready to be cooked with soy sauce soup at a restaurant.

to 400% of flour weight, retain their shape after cooking, and have a smooth surface. The boiled noodles are washed and cooled in running water. Figure 15 shows boiled noodles ready to be cooked with soy sauce soup.

For the manufacture of dry noodles, wet noodles from the cutting machine are hung in special chambers equipped to remove humid air and to introduce fresh air. The rate of drying must be closely controlled. In the first drying stage, the moisture content of the noodles is reduced from about 40 to 27%. In the second stage, corresponding to sweating in macaroni drying, internal noodle moisture is diffused to the surface. Salt added to the dough plays an important part in moisture diffusion. Because of its hygroscopicity, concentrated salt on the dried surface extracts moisture from the interior of the dough strands. However, the role of salt is not only to diffuse moisture properly but also to enhance flavor, to strengthen the gluten, to inactivate enzymes in the dough, and to inhibit dough fermentation during processing. Too rapid drying is undesirable at this stage, but air heaters may be used as it is considered advisable to accelerate internal moisture diffusion and surface evaporation.

The third drying stage is the cooling of noodles. Cool air lowers the humidity around the goods, making the drying effect more complete. Typical dry *so-men* is pictured in Figure 16.

In 1975, Nisshin Flour Milling Co., Ltd. developed a continuous noodle processing system (9). The introduction of bulk flour handling systems, the improvement in processing through the introduction of a new mixer, the high efficiency of mechanized production, and the reduction in labor requirements are advantages (40). The newly developed speed mixer has a rotary shaft with special

Figure 16. Dry *so-men* (very thin noodles).

blades that contribute to uniform dough mixing and to some gluten development through the beating action of the blades.

All noodles must be cooked before being eaten. The cooking time for *so-men* is 6 to 8 minutes, for *hiya-mugi* 11 to 13 minutes, and for dry *udon* or *himo-kawa*, 19 to 21 minutes. Boiled noodles are also recooked for 2 to 3 minutes.

So-men and *hiya-mugi* are served cool with a cup of special soy sauce soup. The

Figure 17. *Udon* served in hot soy sauce soup with one or two eggs, *tempura*, fried bean curds, boiled fish pastes, vegetables, etc.

noodles are dipped in the soup and eaten. *Udon* and *himo-kawa* are usually served in hot soy sauce soup with boiled eggs, *tempura*, fried bean curds, boiled fish paste, and/or vegetables (Figure 17). In Osaka, *udon* is occasionally served with *sukiyaki* (*udon-suki*).

Japanese noodles owe their favorable "mouth feel" to the combination of wheat flour and processing techniques. And, because of the simple formula, the influence of flour quality is greater than is the case with other wheat flour products. However, because of wide variations in noodles according to regional and individual taste, quality requirements for noodle flour are not simple. Protein content and mellow gluten quality are very important. Although hard wheat flours of 10 to 12% protein are sometimes used in the northern districts of Japan, soft wheat flours of 8 to 10% protein are commonly used throughout the country (33).

Since Japanese noodles had been developed with domestic (soft red winter) wheat, it is a favorable flour source, and domestic wheat is still used. However, the short supply of Japanese wheat has forced flour millers and noodle manufacturers to look elsewhere for flour. At present, Australian standard white wheat from Western Australia and/or U.S. western white wheat are often blended with Japanese wheat for noodle flour.

There appears little doubt that mellow gluten quality is required for satisfactory noodles, but starch quality has recently become the focus of attention. The similarity in noodle processing characteristics of Australian standard white wheat (Western Australia) flour to that of flour made from Japanese wheats appears to be related to its gluten properties and starch gelatinization characteristics. Amylograph gelatinization temperature for Australian standard white (Western Australia) flour is lower than that from similar wheat from Victoria, which in turn is lower than that of U.S. western white wheat. This relatively low gelatinization temperature is presumed to be a factor in soft and pliable noodles (32). Hard wheat flour, even at low protein content, is not suitable for noodles. For the manufacture of strong noodles, however, hard and soft wheat flours are blended.

Flour with 0.33 to 0.55% ash content has been used for noodle manufacture. However, most noodles are made from flour of 0.36 to 0.40% ash, accompanied by bright color. A dull white hue is not desirable, but a bright yellowish tone is acceptable.

Minimal cooking loss (solids dispersed in cooking water) and high noodle volume increase through water uptake are important noodle quality factors associated with flour. These attributes are also related to the masticatory property of the product. Extremely low protein content, abnormally weak gluten quality, low amylograph viscosity, and similar flour shortcomings cause problems in noodle manufacture and in the flavor and acceptability of the finished products. Fine flour granularity is also a desirable trait in noodle flour.

Noodles have long been a staple food in Japan and are produced in large quantity, but improvements in noodle quality have been pursued by all segments of the industry. Handmade noodles were and are very flavorful, but the flavor elements associated with older production methods have been more or less dissipated since the introduction of mechanical procedures. The merits of handmade products are now being reviewed, and efforts are being made to develop processing methods or machines that will incorporate the advantages.

Differences between manually and mechanically prepared *so-men* noodles were studied by Ogawa (39), who attributed the quality superiority of the handmade products to the following: 1) Cottonseed oil added as an ingredient to handmade noodles coats the surface and contributes to flavor. 2) The greater salt and water requirements in their production aids in obtaining greater uniformity in noodle texture. 3) Although handmade noodles are somewhat coarse, the retention of starch granules in the gluten network makes for a more uniform dough, thus leading to quicker cooking and smaller cooking loss. 4) Handmade products remain firm during cooking, and their deterioration is thus minimized. The aging of the handmade product is believed to result in improved eating quality (7). The mechanism of aging has not been studied extensively.

The rapid growth of supermarkets has resulted in a decrease in the number of small retail noodle shops. This has led to accelerated interest and progress in preservative technology. Frozen or low-temperature distribution systems will probably receive more attention in the future.

C. Soft Wheat Noodles in Other Countries

NOODLES IN KOREA

Typical Korean noodles are made from 100 parts all-purpose flour, 2 to 3 parts salt, 0.1 part carboxymethyl cellulose, and water. The manufacturing process is almost identical to that of Japanese noodles. The flour strength required is comparable to that in Japan.

NOODLES IN TAIWAN

A 60 to 70% extraction flour from U.S. western white wheat is commonly used for noodle manufacture. A typical noodle formula includes flour, 100 parts; potassium carbonate, 0.2; sodium carbonate, 0.3; salt, 2 to 3; and water, 35. The manufacturing process again is almost the same as that employed for the production of Japanese noodles. After the noodles are cooked and cooled, they are mixed with salad oil to prevent sticking and to impart a degree of brightness to the product.

RICE NOODLES IN TAIWAN

A blend of 75 parts rice flour and 25 parts soft wheat flour forms the principal ingredient for rice noodles. The wheat flour is normally a special short patent from U.S. western white wheat. The whiteness of the flour is especially desirable.

NOODLES IN THE PEOPLE'S REPUBLIC OF CHINA

In the manufacture of noodles in the People's Republic of China, hard wheat flour containing a generous proportion of gluten is desired. However, all-purpose flour is also suitable, if its protein content is sufficiently high. For quality noodles, about 30 to 50 parts egg based on 100 parts flour is added. The manufacturing process is quite similar to that for Japanese noodles.

Noodles are eaten in China in the form of a crispy product with mixed meats and prawns, as fried soft noodles, egg noodles cooked with chicken, ginger noodles, fried noodles with vegetables, soft noodles with hot vegetables and dressing, transparent noodles with meatballs, chicken liver, meat, or vegetables, and in other forms and combinations with other ingredients (14).

D. Steamed Japanese Buns

Steamed Japanese buns are of two types: *Saka-manju*, prepared by steaming a fermented dough containing a bean jam filling, and *mushi-manju*, a similar product but with chemically leavened dough. The leavening agent for *saka-manju* dough may be a fungus from a sweet fermented rice drink, from the lees after *sake* fermentation, or fresh yeast.

Mushi-manju (Fig. 18) is prepared by dissolving 65 to 70 parts sugar (based on 100 parts flour) in 30 to 35 parts water, adding 1 part ammonium carbonate and 1 part sodium bicarbonate dissolved in a little water, and mixing in sifted flour (100 parts) to a homogeneous dough. The dough is shaped into small balls that are flattened; the bean jam filling is placed in the center and enclosed by sealing the edges of the dough. Usually 7 parts of bean jam to 3 parts of dough are used. The buns are placed on racks and steamed for 10 to 12 minutes (23).

Long patent flour from domestic wheat was formerly used for preparing *mushi-manju* buns, but it has been replaced by the first-grade of either confectionery or all-purpose flour. Bright white color and soft gluten quality are desired. Protein content requirements vary according to the type of product, but soft wheat flour of 7.0 to 8.5% protein is suitable for *mushi-manju*. On the other

Figure 18. *Mushi-manju* (chemically leavened buns). Brown buns (left), often called *cha-manju*, are made from dough containing muscovado.

hand, first-grade all-purpose flour is preferred for *saka-manju* manufacture; whiteness combined with moderate protein content (8 to 9%) is required.

Product quality is distinguished by a soft and tender crust and a shiny white surface. Overmixing the dough results in a product with a tough coating and is therefore to be avoided.

E. Steamed Chinese Buns

Pao (steamed Chinese bun) is a representative Chinese food made from wheat flour. Today, steamed Chinese buns can be found in most countries in the Orient where they are eaten with a meal or as a snack.

Basically, there are three types of buns: those without any filling, with bean-jam filling, and with meat filling (Fig. 19). Formulations for the bun dough have been given previously (4,20,25,41,48). Typically, a dough is prepared with flour, 100 parts; fresh yeast, 2 to 4; sugar, 5 to 10; salt, 0 to 0.5; lard oil, 3 to 5; and milk and warm water, 50 to 60.

Active yeast with sugar in lukewarm water is added to sifted flour and mixed to a stiff homogeneous dough. Salt and lard oil are added during mixing. The dough is permitted to double in bulk and remixed. It is then shaped into a cylinder about 5 cm in diameter and sliced into pieces 2 to 3 cm wide. Each piece is shaped, filled, and sealed. The rounded dough, on a square of greaseless paper, is placed on a rack, permitted to rise for 7 to 10 minutes, then steamed for about 10 minutes.

Chinese bean jam filling consists of raw bean jam, 100 parts; lard oil, 6; black sesame seed, 2 to 4; sugar, 70 to 80; and water as required. It is prepared as described under Japanese baked buns.

Typically, a meat filling may consist of finely sliced pork, 100 parts; sugar, 0.4

Figure 19. *Paos* (steamed Chinese buns) with meat filling (left) and bean jam filling (right).

to 0.7; soy sauce, 3 to 9; vegetable oil, 2 to 5; and garlic and cooked vegetables as required. The ingredients are simmered for about an hour. Cornstarch may be added as thickener.

Since Chinese buns are produced in many countries of the Orient other than China, the quality of the flours used varies widely. Soft wheat flour is desirable to give softness and tenderness to the dough and to give it a fine, silky texture. If the flour is excessively weak or low in protein content, however, blending of hard wheat flour (usually 1 part to 2 parts soft wheat flour) is suggested. The buns' surface must remain white and glossy without turning yellow after steaming. Thus, a very high-quality, finely dressed flour is required.

A high-quality bun will have a fine-textured and shiny surface that enhances its attractiveness. The dough should be light and not gummy or sticky as it is being eaten. There should be minimal air space between the dough and the filling.

F. Donuts

The consumption of donuts has been increasing in most Oriental countries in recent years. In Japan, the use of prepared mixes has become popular since their introduction in the mid-1960s. Today, many varieties of donut mixes, for both yeast-raised and chemically leavened donuts, are available to bakeries and confectioneries.

Formulas, processing techniques, and equipment employed are similar to those in the United States (17).

Short or semilong patent flour from soft wheat is usually best suited for cake-donut production. Since moderate protein content and soft gluten quality are required, 10 to 30% hard wheat flour is often blended into the soft wheat flour. For yeast-raised donut production, the ratio of soft and hard wheat is reversed. Donut production lines are highly automatic in most bakeries, hence uniform quality is a most important consideration in flour specifications.

Figure 20. Hard *karintou.*

G. Other Nonbaked Products

KARINTOU

Karintou (fried dough cookie) is a confection originally imported from China. During the 18th century, the product was modified to differ considerably from the original.

Karintous are soft or hard, according to the difference in brittleness. Soft *karintou* is produced by yeast fermentation of a hard wheat flour dough that results in a light and airy finished item. Hard *karintou*, on the other hand, is made from a chemically leavened soft wheat flour dough, which processes into a coarser and slightly harder product (Fig. 20).

A typical formulation for hard *karintou* consists of 0 to 3 parts sugar; 0.3, skim milk; 0 to 1, salt; 4 to 5, sodium bicarbonate/ammonium carbonate; and 45 to 60 parts water based on 100 parts flour. The ingredients are mixed, sheeted, cut, fried, coated with sugar syrup, and cooled.

CURRY *ROUX*

Curry, introduced from India and characterized by its hot taste, has been modified in Japan by combining it with the flavor of French curry. Instant curry *roux*, or thickener, ready to be cooked with meats, vegetables, and water and eaten with boiled rice, is produced by large manufacturers. The instant product consists of frizzled flour, vegetable or animal oil, and spices, with a 1:1 or 2:1 flour/oil ratio. Second-grade soft wheat flour is considered most suitable for instant curry *roux* (24).

V. NONFOOD USES OF SOFT WHEAT

A. Industrial Uses

About a fifth of the low-grade flour produced in Japan is used as an ingredient of adhesive agents by the plywood industry. The flour is mixed with urea resins, etc., to make an adhesive paste. The blending proportion of flour to resins varies from 20 to 300 parts by weight based on 100 parts resin, according to the type of plywood to be made. Flour serves as a filler, as a means of fluidity control, and for cushioning the hardening caused by adhesive action. Flour for the industry is usually 1.5 to 2.5% in ash content, and its particle size is such that it passes through a 100-mesh screen. Coarse bran particles are contaminants causing poor adhesive action because they clog roller machine surfaces. The low-grade flour from hard wheat milling differs from that of soft wheat in its effect on adhesive viscosity. Whether plywood manufacturers prefer soft or hard flour appears to be a matter of individual manufacturing condition. In the Philippines, some plywood manufacturers have been using soft wheat long patent flour in adhesive paste.

In Japan, the use of wheat flour in the spinning industry has been consistently decreasing. Wheat flour serves as a substitute for starch in the cloth size process; thus the first grade of soft wheat flour is generally considered suitable (35).

B. Feed Uses

In most countries, flours that are not suitable for human consumption are sent to the feed market. In Japan, about four fifths of low-grade flour is used as an ingredient in formula feeds. The ash content usually exceeds 2%; particle size is not a factor. Low-grade flours from soft and hard wheat milling are equally satisfactory for feed use and may even be mixed.

ACKNOWLEDGMENTS

The author is very grateful to R. E. Maas, Y. Sone (Tokyo), B. Bratland (Seoul), F. Schneiter (Taiwan), S. D. Kambli (New Delhi), T. Templeton (Manila), A. Hunt (Singapore), all members of Wheat Associates, U.S.A.; G. A. Hollands, Y. Takahashi (Australian Wheat Board, Tokyo), H. Y. Kim (Samsung Moolsan, Korea), Y. K. Choi (Cheil Sugar, Busan, Korea), K. Tsao (Hong Kong Flour Mills, Hong Kong), and J. C. Yang (President Enterprises, Taiwan) for information on soft wheat uses in Asian countries other than Japan; and to H. Shoda, Y. Ishii, and H. Nohagi (Nisshin Flour Milling Co., Tokyo), for their approval of this contribution to the monograph. Thanks also go to C. Ooba for secretarial assistance and to H. Otsubo and his staff for aid in preparation of data.

LITERATURE CITED

1. Akutsu, S., Freundlief, H., and Matsumoto, H. 1963. Bakery Handbook. (In Japanese) Fujisawa Mfg. Co., Ltd., Osaka, Japan.
2. Anonymous. 1969. Encyclopedia of Bread Baking and Confectionery Making. (In Japanese) Shokuryo Times, Ltd., Tokyo.
3. Anonymous. 1969. Tempura. (In Japanese) Shibata Shoten, Ltd., Tokyo.
4. Anonymous. 1970. Recipes: The Cooking of China. Time-Life International (Nederland) N. V., The Netherlands.
5. Anonymous. 1970. Recipes: The Cooking of India. Time-Life International (Nederland) N. V., The Netherlands.
6. Anonymous. 1971. Noodle Technology. (In Japanese) Shokuhin Shizai Kenkyu Kai, Ltd., Tokyo.
7. Anonymous. 1972. Tasty and Marketable Noodles—Noodle Manufacture and Marketing. (In Japanese) Shokuhin to Kagaku Sha, Osaka, Japan.
8. Anonymous. 1974. Guide to the Laws and Regulations of Confections. (In Japanese) Zen Ka Ren Koho Center, Ltd., Tokyo.
9. Anonymous. 1975. Continuous noodle processing system—Outline of "super-turbo" and "micro-bulk" processes. (In Japanese) Shokuhin Kogyo 18(19):89.
10. Asano, R. 1971. Basics of Noodle Processing. (In Japanese) Shokuhin to Kagaku Sha, Osaka, Japan.
11. Austin, A., and Ram, A. 1971. Studies on chapati-making quality of wheat. Indian Council of Agric. Res. Tech. Bull. (Agric.) No. 31, New Delhi.
12. Bohn, R. M. 1963. Biscuit and cracker manufacture. (In Japanese) Seminar at Zenkoku Biscuit Kyokai and Nihon Shoku-seikatsu Kaizen Kyokai, Tokyo.
13. Bonte, C. 1975. Pâtisserie parisienne. (In Japanese) Nihon Yo-gashi Kyokai, Tokyo.
14. Burke, H. 1967. Chinese Cooking for Pleasure. Paul Hamlyn, Ltd., London.
15. Canadian International Grains Institute. 1975. Grain and Oilseeds—Handling, Marketing, Processing, 2nd ed. Can. Int. Grain Inst., Winnipeg, Manitoba.
16. Echigo, K. 1972. Manufacturing methods for 150 selected cakes. (In Japanese) Bakers Times Sha, Tokyo.
17. Echigo, K. 1973. Modern donut manufacture. (In Japanese) Bakers Times Sha, Tokyo.
18. Fry, W. 1972. Cakes and cookies. (In Japanese) Pan News Sha, Tokyo.
19. Iino, K. 1965. Guide to biscuit and cracker manufacture in Japan. (In Japanese) Zenkoku Biscuit Kyokai, Tokyo.
20. Kajitani, K. 1968. Mass-productive confectionery by machine. (In Japanese) Food Machinery Trade Center, Nagoya, Japan.
21. Kandhari, D. R. 1975. A review of Indian

bread. The Penang Seminar, Penang, India.
22. Kent-Jones, D. W., and Amos, A. J. 1967. Modern Cereal Chemistry. 6th ed. Food Trade Press, Ltd., London.
23. Koganei, Y. 1974. Manufacturing methods for fresh Japanese style confectioneries. (In Japanese) Pan News Sha, Tokyo.
24. Kuromatsu, Y. 1961. Instant curry. (In Japanese) 8th annual meeting, Nosan Kako Gijyutsu Kenkyu Kai. Tokyo. April.
25. Lin, H. J., and Lin, T. 1969. Chinese Gastronomy. Thos. Nelson & Sons, Ltd., London.
26. Morisue, Y., and Kikuchi, Y. 1953. The History of Food. (In Japanese) Dai-ichi Shuppan, Ltd., Tokyo.
27. Nagao, S. 1975. Cake and wheat flour. (In Japanese) Patisserie Confiserie Glaccric 65:51 (Tokyo).
28. Nagao, S. 1975. Why does low amylograph viscosity cause problems? (In Japanese) Grain Importers Assoc. J. 323:35.
29. Nagao, S. 1975. Why is Western white wheat so important as a source of confectionery flour? (In Japanese) Grain Importers Assoc. J. 321:33.
30. Nagao, S., Imai, S., Sato, T., Kaneko, Y., and Otsubo, H. 1976. Quality characteristics of soft wheats and their utilization in Japan. I. Methods of assessing wheat suitability for Japanese products. Cereal Chem. 53:988-997.
31. Nagao, S., Ishibashi, S., Imai, S., Sato, T., Kanbe, T., Kaneko, Y., and Otsubo, H. 1977. Quality characteristics of soft wheats and their utilization in Japan. II. Evaluation of wheats from the United States, Australia, France, and Japan. Cereal Chem. 54:198-204.
32. Nagao, S., Ishibashi, S., Imai, S., Sato, T., Kanbe, T., Kaneko, Y., and Otsubo, H. 1977. Quality characteristics of soft wheats and their utilization in Japan. III. Effects of crop year and protein content on product quality. Cereal Chem. 54:300-306.
33. Naka, S. 1974. Udon. (In Japanese) Shoku no Kagaku 2:42.
34. Nakae, K. 1975. Bakers index. (In Japanese) Shoku-ken Center, Ltd., Tokyo.
35. Nihon Mugirui Kenkyu Kai. 1964. Wheat Flour—Its Source and Processed Goods. (In Japanese) Nihon Mugirui Kenkyu Kai, Tokyo.
36. Nihon Mugirui Kenkyu Kai. 1975. Variety Distribution (in Japan). (In Japanese) Nihon Mugirui Kenkyu Kai, Tokyo.
37. Nihon Mugirui Kenkyu Kai. 1976. Wheat Flour—Its Source and Processed Goods (rev.) (In Japanese) Nihon Mugirui Kenkyu Kai, Tokyo.
38. Niimi, K. 1965. Encyclopedia of Manufacturing Methods for Japanese Confections. (In Japanese) Numata Shoten, Ltd., Tokyo.
39. Ogawa, G. 1974. Hand-made somen. (In Japanese) Shoku no Kagaku 20:50.
40. Ogawa, G. 1975. Continuous noodle making technology. (In Japanese) Shokuhin Kogyo 18(24):36.
41. Okamoto, S. 1969. Chinese Dishes. (In Japanese) Shibata Shoten, Ltd., Tokyo.
42. Ono, T. 1974. Biscuit and cookie industry. (In Japanese) Food Sci. (Special No., Production and marketing of bread and confectioneries) p. 85 (Osaka, Japan).
43. Pan Sangyo Gijyutsu Kaigi. 1975. Bread and Cake Baking Manual. (In Japanese) Bakers Times, Ltd., Tokyo.
44. Sakurai, Y. 1959. Encyclopedia of Foods. (In Japanese) Dobun Shoin, Tokyo.
45. Sakurai, Y. 1969. Basics of Cake Manufacture. (In Japanese) Korin Shoin, Ltd., Tokyo.
46. Smith, W. H. 1972. Biscuits, Crackers, and Cookies. Vol. 1. Technology, Production, and Management. Appl. Sci. Publ., Ltd., London.
47. Smith, W. H. 1972. Biscuits, Crackers, and Cookies. Vol. 2. Recipes and Formulations. Appl. Sci. Publ., Ltd., London.
48. Tokyo Koto Seika Gakko. 1971. Textbook of Japanese, European, and Chinese Style Confections. (In Japanese) Numata Shoten, Ltd., Tokyo.
49. Uehara, M. 1975. History of Japanese noodles. (In Japanese) Gekkan Shokudo (Special No., Soba to Udon) p. 94 (Tokyo).
50. United States Department of Agriculture. 1970. Official Grain Standards of the United States. U.S. Dept. Agric., Washington, DC.
51. Zenkoku Biscuit Kyokai. 1963. The Biscuit Industry Situation in Japan and the U.S. (In Japanese) Zenkoku Biscuit Kyokai, Tokyo.
52. Zenkoku Biscuit Kyokai. 1973. Report: U.S. Biscuit and Cracker Seminar. (In Japanese) Zenkoku Biscuit Kyokai, Tokyo.

Index

Adhesives from Flour, 203, 302
Aegilops
umbellulata ('Transfer'), 46
speltoides, 99
Aeration, mechanical dough, 223
Agropyron elongatum, 44
Air classification, 159, 172, 235
Alkaline water retention capacity test, 47
Alpha-amylase activity, 116
Alveograph test, 160, 233

Baguette, 161, 236
Baking test,
European, 251
United Kingdom, 116
Barley yellow dwarf, 54
Biscuit
butter puff, 230
cake, 248
dry-sugared French, 248
hard dough, 228, 283
hard sweet dough, 229
in France, 248
in Germany, 261
in Japan, 283
in United Kingdom, 227
short dough, 229
soft dough, 283
Bran products, 198
Bread
country, 241
crusty white, 219
fine wheat, 241
French (*Baguette*), 160, 236
gluten, 220
high protein, 219
in Germany (*Feine Backwaren*), 255
in Germany, 255
in United Kingdom, 209
malt, 220
milk, 220
raisin, 259
rye, 220
Scotch batch, 217
soft white, 219
sourdough, 238
starch reduced, 220
staling, 219
toast, 257
Viennese, 242
wheat, 257
wheat mixed, 257
wheat germ, 219
wheat meal, 219
white, 219
whole meal, 219
Breading, 201
Bread process, activated dough development, 215
bulk fermentation, 211
Chorleywood, 211
Polish, 238
Viennese, 238
Breakfast cereals, 198
Breeder's seed, 118
Breeding
α-amylase activity and, 116
artificial mutation in, 108
backcrossing in, 49, 70, 109
bulk selection technique in, 110
combining ability analysis in, 107
detailed pedigree selection in, 110
diallel crossing in, 107
F_2 progeny method in, 110
ideotype in, 105
irradiation in, 49
land races in, 102
milling quality in, 115
multivariate analyses in, 108
multiline cultivars in, 79
mutants in, 102
progeny trial system in, 110
protein content and quality and, 116
Brioches, 244
Bulgur, 202
Buns
chestnut, 279
Chinese, 282
in Japan, 279
steamed Chinese, 300
steamed Japanese, 299
Bunt (*Tilletia caries, T. foetida*), 52, 69, 75, 114
dwarf (*Tilletia contraversa*), 82
breeding for resistance to, 83
on semidwarfs, 82
wheat export and, 83

Cake
Baumkuchen, 290
butter, 290
Castilla, 275
foam, 172
Lebkuchen, 262
moon (Chinese), 282
Sandkuchen, 260
Savarins (*babas*), 246
shortening-based, 172
spiced, 248
sponge
(Japan), 275
(German), 260
test, 47
Cakes
formula balance in, 172
in Japan, 290

in United States, 171
Castilla (Japan), 275
Centro Internacional Mejoramiento Maiz y Trigo (CIMMYT), 107
Cereal leaf beetle *Oulema Melanopus,* 55
Cereals, ready-to-eat, 198
Chapati, 287
Chausson, 245
Chlorination, 171
Chorleywood bread process, 116, 211
Color, flour, 135
Cones, ice cream, 185, 289
Cookie
Chinese almond, *Hseng jen peng,* 285
fried dough, *Karintou* (Japan), 284
"home-made," 183
test, 47
Cookies
in Japan, 283
in United States, 180
Crackers
cream, 230
in United States, 177
Japanese, 286
Cracknel, 286
Croissants, 244
Curry *roux,* 302

Donuts
cake, 196
in Japan, 301
Dormancy, 35
Double cropping, 34, 58
Dough
ascorbic acid in, 240
blade-cut in, 238
development process, activated, 215
malt in, 240

Eggs, 193
Einkorn (*Triticum monococcum*), 99
Electrophoresis test, 143
Emmer (*Triticum dicoccum*), 99
Erosion control, 85
Evaluation
comparative, 46
early generation screening, 46
quality erosion in, 47
Eyespot (*Cercosporella herpotricchoides*), 11, 112

Fallow, summer, 75, 76, 101
Federal Institute of Cereal and Potato Processing (W. Germany), 249
Flavor, bread, 218
Flour
breading, 201
family, 169
fine wheat (*gruaux*), 235
horse-bean, 240
purity in France, 160
quality in European Economic Community, 135
self-rising, 190
strong, 236
soft wheat in Japan, 267
types
in France, 160, 235
in West Germany, 162, 252
Fly free date, 35
Foot rot (*Cercosporella*), 82, 83
Fusarium root rot, 115
Four quarters, 246

Genoese, 246
Gluten manufacture, 123
Goat grass (*Aegilops squarrosa*), 99

Hardness, grain, 121
Hessian fly (*Mayetiola destructor*), 35, 44, 55
Hexachlorobenzene, 75
Hot cakes in Japan, 290
Hybrid vigor, 51, 87

Irrigation, 58, 76

Kernel texture, 3, 131
inheritance of, 121
Kneading, intensified dough, 238

Leaf stripe (*Pyrenophora sativa*), 114
Leavening, 194

Mildew, powdery (*Erysiphe graminis*), 52, 112, 114
Milk, 194
Mille-feuilles, 245
Milling, 144
break system in, 150
flake disruption in, 132
flour treatment in, 155
garlic in, 36
germ in, 154
in Japan, 270
purifiers in, 154
quality test, United Kingdom, 115
reduction system in, 154
roller mill process in, 131
short surface system in, 156
split, 47
wheat, 49
Mills, impact, 158
Mixes
cake, 191
in West Germany, 255
Mixograph test, 41, 47
Mosaic
wheat, soilborne, 53
wheat, spindle streak, 54
Multiplication, breeding line, 117

Nan, 287
Nitropyrin, 55
Noodles
boiled, 294
Chinese, 298
dry, 295
fried, 202
Japanese, 293

Oven
setter, 217
traveling tunnel, 217

Packaging, 226, 231
Partha, 287
Particle size
and cake quality, 45
index test, 121
Pastry
Bolo, 291
Chou à la crème, 290
Choux, 245
Danish, 63, 184
puff, in France, 245
toaster, 192
Pearling resistance, 122
Pies, 175, 291
Plant Breeding Institute,
University of Cambridge, 103, 115
University of Sydney, 124
Poori, 287
Pretzels, 186
Protein quality tests, in United Kingdom, 116

Rapid Mix test, 116, 250

Relative humidity,
equilibrium, 227
Roll
Hunter's (*Jäger brotchen*), 256
wheat (*Weizenbrotchen*), 256
Rusk
in France (*Biscotte*), 242
in West Germany, 262
Rust
leaf (*Puccinia recondita*), 43, 51, 112
stem (*Puccinia graminis*), 52, 112
stripe (*Puccinia striiformis*), 79

Selection technique
F_2-progeny method in, 110
progeny trial system in, 110
single seed descent system in, 112
Septoria glume blotch
(*Septoria nodroum*), 53, 112, 115
(*Septoria tritici*), 35, 53, 112, 115
Shelf life, 226
Shortening, 192
Smut
flag (*Urocystis agropyri*), 82
loose (*Ustilago tritici*), 40, 44, 52, 112
Soft Wheat Millers Group (Australia), 120
Soft Wheat Quality Laboratory (U.S.), 40
Soup thickeners, 200
Sprout in wheat, 11, 90, 200, 204
Staling, bread, 219
Starch
damage, 132, 133
manufacture, 123, 124
lenticular, A-type, 130
spherical, B-type, 130
Steam
bread baking and, 217
as leavening, 223
Stickiness in flour, 251
Stickiness test, flour 141
Stollen, 259
Sweeteners, 192
"Sweet goods" in U.S.A., 184

Take-all (*Gauemannomyces graminis*), 54, 112, 115
Tart, fruit, 246
Tarts in West Germany, 261
Tempura, 291
Treaty of Rome, 139
Tri-State Soft Wheat Improvement Association, 40
Triticum timopheevi, 44, 49, 50, 52, 87

Viscosity test, acid, 41, 88

Wafer, ice cream, 182, 185
Wafer
in France, 248
in Japan, 289
in United Kingdom, 231
Waffles, 290
Western Wheat Quality Laboratory, 73
Wheat
blending value of, 250
club, 1, 67, 73, 78
conditioning, 148
durum (*Triticum durum*), 1
facultative, 74
hybrid, 50, 87
male sterility in, 50
ideotype, 106
import levy, 140
intervention price, 141
marketing, silo system, 139
semidwarf, 76, 80, 106
reference price, 139
shredded, 198
"soft" of Australia, 119, 121
soft, 33, 57, 61, 71
red winter, 41, 48, 64, 72
white spring, 66, 72, 78
white winter, 45, 66, 72, 77
spelt (*Triticum spelta*), 99
sprouting in, 90
target price of, 140
test weight in, 47
threshold price of, 139
"weak" of Australia, 119
Wild garlic (*Allium oleraceum*), 33

Yellow berry, 3